The New Government of London

The New Commercial London

The New Government of London

The First Five Years

Edited by
GERALD RHODES

with a Preface by
W. A. Robson

International Arts and Sciences Press, Inc
White Plains, New York

London School of Economics
and Political Science

Published in Great Britain by
Weidenfeld and Nicolson

First U.S. edition published in 1973 by
International Arts and Sciences Press, Inc.
901 North Broadway, White Plains, N.Y. 10603

Library of Congress Catalog Card Number:
73–79596

International Standard Book Number:
0–87332–042–5

Printed in Great Britain

Contents

Maps and Diagrams

Preface

An earlier volume by the Greater London Group, entitled *The Government of London : The Struggle for Reform*,[1] described the events leading to the establishment of an entirely new system of government in Greater London. That work was essentially an exercise in contemporary history. The present volume attempts the much more difficult task of trying to evaluate the reforms. It was obviously necessary to define the period to which the investigation is related, and for this purpose the first five years of the new system were chosen: namely from 1 April 1965 to 31 March 1970.

The research covers a wide ground. It includes a study of the changes which have taken place in the organization of the political parties, and describes the electoral system and the elections. It analyses in depth the way in which major functions have been affected, giving particular attention to planning, education, housing, the personal health and welfare services, the children's services, highways, traffic and transport. The financial consequences of the changes are traced as far as it is possible to do so. Lastly, a close look is taken at the internal organization of the Greater London Council in regard to its committee structure, departments and staff policy. A more generalized study is made of the London boroughs.

All this has constituted a very exacting programme of work which has occupied the time and energy of the Greater London Group for more than two years. There are, however, many topics we have been unable to study in the time and with the resources available. We have said almost nothing about the environmental health services or the multitude of regulatory functions which play an important part in maintaining standards of health, safety, honesty and even morals in urban communities. We have not attempted a systematic comparison of the performance of individual boroughs, and have generally had to be content with pointing out broad differences, such as those between the inner London and the outer London boroughs. Nor have we examined in detail the way in which the boroughs have carried out their functions in spheres such as planning and highways, where the main responsibility is centred on the GLC. This particular omission will be remedied in part by a detailed study of Camden

Borough Council which Mrs Enid Wistrich is making under the auspices of the Greater London Group.[2]

Again, we have not included a detailed investigation of the political process at the level of either the GLC or the boroughs, nor the effect of the reforms on the working of local interest groups. This is an omission we hope to be able to repair in a supplementary study.

The previous volume was written throughout by Mr Gerald Rhodes, who submitted each chapter to the Group for discussion, criticism, revision and approval. The present volume could not be produced by this method because the task of appraising the performance of the main functions called for a higher degree of specialized knowledge than any one person can possess. In consequence, individual members of the Greater London Group agreed to write the chapters dealing with the personal health and welfare services, the children's services, education, housing and planning. Professor Self wrote the chapter on planning. Dr Regan collaborated with Mr Sam Hastings, a member of the research staff, in writing the chapter on education. Mrs Betty Tate took charge of the chapter on the personal health and welfare services. Miss Della Nevitt and Mr Rhodes were jointly responsible for the chapter on housing. Professor Roy Parker, a member of the Greater London Group until he left LSE to become Professor of Social Administration in Bristol University, kindly agreed to write the chapter on the children's services despite the very exacting demands on his time and energy of his present department. Mr Ken Young, a former member of our research staff rejoined us to produce the chapter on political party organization, and (with Mr Rhodes) the chapter on the electoral system and elections. Mr David Southron, one of our research assistants, wrote the chapter on Finance. Each chapter, no matter by whom it was drafted, was submitted for discussion, criticism and comment to one or more of the Group's meetings; and was thereafter revised in the light of the comments before being finally approved.

Although much of the work was dispersed among members of the Group in the way described above, the main burden both of co-ordinating the research and of writing, alone or jointly, six of the thirteen chapters, has fallen on Mr Gerald Rhodes. I wish, on behalf of the Group as a whole, to express our warmest thanks to him and to all those who have contributed to the completion of the book.

Despite the active help and participation of leading experts in several spheres of municipal activity, the problem of evaluating the reforms by strictly objective criteria remains highly elusive. The appraisal had to take account of two separate aspects: the efficiency of the services and the health of local government. Each of these presents intractable problems. How does one evaluate a quite new municipal function such as the Greater London Council's power to regulate traffic throughout the area? How can one compare health and welfare functions carried out in London boroughs with those previously performed by the London or Middlesex County Councils? What basis of comparison exists between the disposal of house refuse by the former metropolitan boroughs, which sent the garbage by barge or road to the vast stinking, smoking rubbish tips in Essex and elsewhere, with the highly efficient but extremely costly incinerators now being constructed by the GLC? In this sphere, as in others, the standard of services provided has risen but so, in several instances, has the cost of providing them, often steeply. What criteria are there for judging how the quality of councillors in the new system compared with that in the old?

The reader must, therefore, not expect to find a clear line of distinction in terms of achievement drawn between the old regime and the new. Some things are better done in some respects but not in others. Functions transferred from the LCC and other county councils to London boroughs perhaps a twelfth as large have gained through being closer to the citizens; but they may have suffered certain disadvantages in having smaller resources of highly specialized staff available in, say, education. There is, indeed, much criticism in the present volume in regard to the results so far achieved, particularly in regard to planning, housing and further education. Mr Rhodes, in his concluding chapter, correctly remarks that it is not possible to draw up a comprehensive balance-sheet of comparisons between the old and the new systems covering the whole range of functions. In a later passage he writes that 'on balance there is little to choose between the two systems' in the performance of the personal health and welfare services and the children's services. While this may be strictly accurate at the present moment, it is worthwhile looking at the matter in a longer-term perspective. Mrs Tate concludes at the end of her detailed inquiry that the London boroughs are capable of carrying out the health and welfare functions of local authorities, but adds that neither the best hopes nor the worst fears of the protagonists in the discussion which preceded the reforms appear to have been justified. She points to a number of strong and weak points in the present position.

Professor Parker begins his chapter by pointing out that the London borough departments concerned with the children's services have existed for little more than five years during which time they have been exposed to continuous and radical change resulting from new legislation. He therefore considers that a 'before and after' comparison cannot satisfactorily be made; and that even if it were possible it would in any event be of dubious value at a time of further general upheaval.[3] What the reader should appreciate from these and the other chapters dealing with services is that the medley of local authorities in Greater London before 1965 had been in existence for seventy-one years in the case of the counties and county districts, and sixty-six years in the case of the metropolitan boroughs. They had long ago reached their zenith; and many of them were in a state of senile decay. The new local authorities in Greater London are still at the very beginning of their existence; and they should be appraised not only in terms of their present performance but also in the light of their potential achievements. The potential of the new system is immeasurably greater than that of the former regime, and I cannot believe that any responsible councillor, official or informed observer would wish to turn the clock back.

Nevertheless, it is now clear that the London Government Act of 1963 contained several major defects which have seriously impeded the benefits which the reorganization could yield. These relate, in most instances, to a failure to confer sufficient powers on the Greater London Council and an excessive insistence on the independence of the London boroughs. This applies above all to the planning function, where the arrangements laid down in the 1963 Act have proved extremely disappointing. The whole principle of a two-tier system of planning authorities, each virtually independent of the other, and both enjoying an equal status *vis-à-vis* the Minister is misconceived. It differs substantially from the recommendations of the Herbert Commission, which urged that the GLC should be responsible for the over-all development plan for Greater London. Development control was to be entrusted to the boroughs, but this was to be subject to the Minister's power to designate classes of application for development in the central area which were to be referred by the boroughs to the GLC.

The London boroughs might be given the power to work out the detailed structure and action plans for their areas, but in doing so and also in exercising their planning control functions, they should be subordinate to the GLC. One advantage which results from the exercise of detailed planning powers by the boroughs is that it enables them to co-ordinate all the services for which they are responsible in their respective areas.

The vital 'heart of Greater London' – the small central area containing almost all the monuments and institutions which give the capital its national and international importance – should obviously be placed under the planning control of the GLC rather than be divided among nine London boroughs; but Mr R. H. S. Crossman, the then Minister of Housing and Local Government, refused to use his powers to achieve this rational result, and so the fragmented authority over development continues.[4]

Another conspicuous defect is the failure to bring the planning of Greater London into an organic relation with the housing policies of the local authorities. A still more glaring mistake, to which the Greater London Group drew attention both in giving evidence to the Herbert Commission and later, is the inadequacy of the boundaries of Greater London. The need to plan the built-up area of the metropolis in conjunction with the outer metropolitan area, if not as part of an even wider region, was manifest long before the London Government Bill was published or the Herbert Commission appointed.

A lack of foresight is also to be found in the sphere of education. The dogmatic insistence of the Ministry of Education (as it then was) on the divine right of primary, secondary and further education to be regarded, like the Republic of France, as one and indivisible, and to be administered in all circumstances by a single local education authority may have been a reasonable proposition in Birmingham or Liverpool or Manchester. It was a far less reasonable proposition when twenty of the outer London boroughs became local education authorities for the first time. The disparity of resources in the field of further education formed a powerful argument for giving this service to the GLC. The unique range and quality of further education provided by the LCC should have at least have induced the Ministry of Education to question its own dogma.

An unfortunate consequence of the compromise reached in regard to education is the possibility of political conflict between the GLC and the ILEA. As a result of the municipal elections in 1970 the Inner London Education Authority has a Labour majority while the GLC is dominated by the Conservative Party. As the ILEA is authorized to determine the amount for which the GLC is required to precept on the inner boroughs for education purposes, and also the sums needed for capital expenditure, a conflict of political outlook on the two bodies could have very serious consequences. No attempt was made in the legislation to reconcile, overcome or avoid such a conflict, except that the position of the ILEA was to be reviewed within five years. This opportunity for reconsidering the matter has now been abandoned.

If, therefore, we cast our eyes forward rather than backward to the first five years of the new system, we can see a clear need to amend the 1963 Act in several respects. As Mr Rhodes remarks in his final chapter, the provisional conclusion from the evidence of the earlier chapters is that 'amendment of the 1963 Act by strengthening the GLC's housing and planning powers would go a long way towards eliminating weaknesses in the new system' – to which we may add the transfer of further education.

On rational grounds the Metropolitan Water Board should also be transferred to the Greater London Council. This was the intention of the Government when the London Government Bill was being drafted. Dr Charles (later Lord) Hill wrote to the Metropolitan Water Board and to the 90 local authorities represented on it informing them that he had come to the provisional conclusion that the Board should be abolished and its functions transferred to a committee of the proposed Greater London Council. There was nothing new about this proposal and many others had advocated a transfer of this kind for years.

Sir Keith Joseph, who succeeded Dr Hill as Minister of Housing and Local Government, expressed his intention to provide in the London Government Bill for the transfer of the powers of the MWB to the GLC, but he was prevented from doing so by the fact that it would have transformed the bill into a hybrid bill.

However, in 1970, the whole question of the organization of water supply was referred by the Minister to the Central Advisory Water Committee and at the end of 1971 the Government announced that water supply would be organized on a regional basis. Under this arrangement, the Metropolitan Water Board will be absorbed in a new regional authority covering most of the South East, with its headquarters at Reading. This new body will also take over the main drainage services of the Greater London Council which includes all the sewers for which the GLC is responsible, its sewage treatment works, pumping stations and the sludge disposal fleet. It will also take over other powers exercised by the GLC concerning control of pollution, etc.

This represents a remarkable change of policy. For what conceivable purpose was the Greater London Council set up as a regional authority if it is not to be made responsible for such fundamental municipal services as water supply, main drainage, pollution control, etc.? This latest move, which is in complete opposition to the intentions of the Government which brought about the reforms of 1963, is also quite contrary to the avowed intentions of the present Government to 'restore power to the people'. The new proposals will deprive Londoners of direct control of their water supply through their elected representatives.

Mr Rhodes also asks whether it is 'really feasible to divide local government powers so that there are distinct powers at each level and so that one level of authority does not become subordinate to the other? Both the Herbert Commission and the Government claimed that it was feasible, but . . . neither took into account the realities of their proposed systems'. He rightly observes that it is time to look at these and other questions which bear on the capacity of the reforms to meet the need of local government in London.

But it is not only London which now has a deep interest in such questions. The London Government reform has had a powerful influence on the reorganization of the local government structure in the rest of Britain. The mere fact that reform had taken place in the metropolis undermined the stupefying complacency which had paralysed all attempts at local government reform during the past forty years.[5] It also drew attention in a new way to the special problems of what had for long been known in the US as metropolitan areas and in this country by the less elegant name of conurbations.

The Royal Commission on Local Government in England was appointed in 1966 under the chairmanship of Lord Redcliffe-Maud to inquire into the organization of local government outside Greater London. The Commission identified three metropolitan areas – Merseyside, 'Selnec' and the West Midlands. For each of these they proposed a metropolitan area authority and a series of metropolitan district authorities, twenty in all. Two major characteristics of their proposals for these areas were:

(i) that the boundaries of the areas should be much more widely drawn than those of the conurbations; the Merseyside area, for example, including the conurbation extended from beyond Southport in the north to southern Cheshire (including Chester) in the south;

(ii) that the distribution of powers, between the metropolitan area authority and the districts, although basically similar to that between the GLC and the London boroughs, gave considerably greater powers to the metropolitan area authority, especially in planning and housing, than the GLC possessed. All planning and transportation powers were to be vested in the metropolitan area authority which was also to be responsible for metropolitan housing policy (including rent policy and policy for selecting tenants) and for building in the interests of the metropolitan area as a whole as well as to ensure the fulfilment of planning policies.[6]

In both these characteristics the Commission's proposals differed from the system in operation in Greater London. So far as the boundaries of their metropolitan areas were concerned they were not restricted, as the Herbert Commission had been, by their terms of reference to considering only the narrowly defined areas of continuous development of the conurbations. Apart from this, however, there are marked differences between Greater London and the provincial conurbations. Each of the three metropolitan areas identified by the Redcliffe-Maud Commission has a large core-city – Birmingham, Liverpool, Manchester – together with smaller but still substantial urban centres such as Wolverhampton or Stockport. London by contrast has tended to grow as a single city rather than a conurbation; the areas which were swallowed up by its expansion were on the whole relatively much smaller than those in the provincial conurbations, although there are a few exceptions (e.g. Croydon). A further obvious difference is the sheer size of London; taking the broader definition of a metropolitan area adopted by the Redcliffe-Maud Commission, the largest of their three areas, 'Selnec', had an area of just over 1,000 square miles and a population of $3\frac{1}{4}$ million against the 4,500 square miles and 13 million population which a London metropolitan area would have.

There is a more direct connection between the London reforms and the Redcliffe-Maud proposals in the distribution of powers between area and district authorities. The Commission deliberately set out to correct what they regarded as faults in the London Government system, and it is noticeable that they rejected the idea, so prominent in the Herbert Commission's plan and the White Paper of 1961, of making either the area authority or the districts 'primary'. At the same time, it is a striking fact that in spite of their strong predilection for a unitary system with a single local authority responsible for all local government services in each area of the country, the Redcliffe-Maud Commission accepted that as in London the provincial metropolitan areas required a two-level system of government, with the personal functions at district level and the large-scale planning and operational functions at area level. It is therefore highly significant that where they diverged from the London system was in strengthening the metropolitan area authority at the expense of the district authorities, in spite of the fact that the latter were to be generally larger and in some cases very much larger than the London boroughs.[7]

The Labour Government in February 1970 accepted the Redcliffe-Maud Commission's proposals as the basis for reforming local government in England. They made two important changes so far as the metropolitan areas were concerned; they proposed to add two further areas in which this two-level structure should apply – West Yorkshire and South Hampshire; and they proposed that education should be a metropolitan area and not a district function. The grounds given for this latter proposal were that the larger resources, especially of specialized manpower, which the area authorities could command would make for a more efficient education service; further and higher education were particularly important in influencing the decision.[8] Moreover, the Government appear to have seriously considered giving all housing powers to the metropolitan area authorities, but finally agreed with the Commission's recommendations, although stressing that the house-building powers of the metropolitan area authorities should be widely drawn 'and should include the power to supplement or reinforce the house-building programmes of the district councils'.[9]

Before these proposals could be implemented, the general election of 1970 brought the Conservatives to power pledged to implement a two-tier system of local government reorganization throughout the

country. The Redcliffe-Maud Commission's proposals for metropolitan areas differed significantly from the London system, and the
Labour Government's proposals moved even further away from the
system; but the Conservative Government's plan, which is likely to
come into operation in 1974, brings us back almost to the 1963 Act
again. The boundaries of metropolitan areas, or metropolitan
counties as they are for the time being called, are narrowly defined to
include for the most part the main areas of continuous development;
in the division of powers between counties and districts the former are
to have all highways, traffic and transport powers, and the latter
education; plan-making is to be a county responsibility but most
development control will be the responsibility of the districts, as
will all housing powers except for certain limited powers, eg for
overspill, reserved to the counties. Six metropolitan counties are
defined, the three identified by the Redcliffe-Maud Commission
together with West Yorkshire, South Yorkshire and the Tyne/Wear
area.

Greatly as these various proposals differ from one another, they
are alike in making the district authorities much more variable in
size and on the whole larger than the London boroughs.

One of the major differences, however, between the three sets of
proposals which have been put forward for the provincial conurbations concerns the nature and powers of the metropolitan area or
county authorities. At one extreme the Labour Government's
proposals would not only have made the area authority a strong
planning and transportation authority, but would have given it
important personal functions in education and housing. At the other
extreme, the Conservative Government's plan will put the metropolitan counties in an almost identical position to the GLC, as indeed
is its intention.

What emerges from these proposals is that so far as the metropolitan areas are concerned the Labour Government's White Paper
drew the right conclusions from the experience of the new London
system during the first five years of its operation, whereas the
Conservative Government's White Paper appears to be either
unaware of, or indifferent to, the need for a stronger metropolitan
authority which has been clearly demonstrated by the experience of
Greater London, and revealed by our own investigations, which
have been in no way politically motivated. It is a curious reflection
on what may be called 'the learning process' in public life, that the

Labour Party, which opposed the London reforms root and branch, should now be ready and anxious to improve on them, while the Conservative Party which initiated those reforms, should now appear to be unable or unwilling to adopt an improved model of the system for the provincial metropolitan areas.

WILLIAM A. ROBSON

Acknowledgments

The research on which this study is based was made possible by a grant from the Leverhulme Trust. The Greater London Group gratefully acknowledges the assistance given by the Trustees.

Many people in or connected with local government in Greater London gave help in providing information and answering questions; some generously agreed to address Group meetings on subjects of which they had particular expert knowledge, such as planning or education. To all these people the Group is deeply indebted and wishes to express its gratitude.

The Greater London Group

The Group was formed in 1958 under the Chairmanship of Professor W. A. Robson from among members of the teaching staff of the London School of Economics and Political Science. It offered substantial written and oral evidence to the Royal Commission on Local Government in Greater London and has since continued its researches into the problems of government within Greater London as well as in the surrounding areas of South East England that constitute the Metropolitan Region. The membership of the Group is as follows:

Professor W. A. Robson, Professor Emeritus of Public Administration (Chairman)

Mr A. J. L. Barnes, Lecturer in Political Science

Mr B. P. Davies, Lecturer in Social Administration

Professor A. C. L. Day, Professor of Economics

Mr D. R. Diamond, Reader in Geography

Mr C. Foster, Director of the Centre for Urban Economics

Mr J. B. Goddard, Lecturer in Geography

Professor J. A. G. Griffith, Professor of English Law

Professor Emrys Jones, Professor of Geography

Dr G. W. Jones, Lecturer in Political Science

Dr P. H. Levin, Lecturer in Social Science & Administration

Miss A. A. Nevitt, Reader in Social Administration

Dr R. A. Parker, Professor of Social Administration, University of Bristol

Mr W. J. L. Plowden, Lecturer in Government

Mr G. J. Ponsonby, formerly Sir Ernest Cassel Reader in Commerce

Dr D. E. Regan, Lecturer in Public Administration

Professor P. J. O. Self, Professor of Public Administration

Mrs E. P. Tate, Lecturer in Social Administration

Mr J. M. Thomson, Rees Jeffreys Research Fellow in the Economics and Administration of Transport.

Professor M. J. Wise, Professor of Geography

Research Staff
Mr Gerald Rhodes, Senior Research Officer

Mrs E. Wistrich, Senior Research Officer
Mr Michael Collins, Senior Research Officer
Mr T. M. Pharoah, Research Officer
Mr D. W. D. Southron, Research Assistant
Mr K. G. Young, Research Officer, Government Department

Secretarial Staff
Mrs E. Steinhart
Mrs E. M. Bedford
Miss P. Margison

Miles 0 — 10

Kilometres 0 — 10

Enfield

Barnet

Harrow

Haringey

Waltham Forest

Redbridge

Havering

Brent

Camden

Islington

Hackney

Hillingdon

Ealing

Hammersmith

Kent & Chelsea

West-minster

Tower Hamlets

Newham

Barking

Hounslow

Richmond

Wandsworth

Lambeth

Southwark

Greenwich

Bexley

Lewisham

Richmond

Kingston

Merton

Sutton

Croydon

Bromley

— Inner London
⬤ City of London
········ London Boroughs

Figure 1. Greater London: administrative areas.

1
Introduction

The Government of London: The Struggle for Reform described how the London Government Act of 1963 came about. In the present study an examination is made of the new system of local government in Greater London with a view to assessing the effect of the reforms. The period covered is roughly the first five years of the new system, that is, from 1 April 1965 when the GLC and the London boroughs assumed their responsibilities, until 31 March 1970, but this has not been taken as a rigid requirement since detailed discussion depends to some degree on the availability of data and the particular circumstances of individual local government functions.

A study of this kind poses one major question, namely, the importance of structural reform compared with other factors in influencing or determining the work of local authorities. We need to consider how far changes which have taken place are the result of political and economic factors whose influence would have been felt even without the change in structure. Where, for example, fresh legislation since 1965 has put new or different responsibilities on local authorities generally and not simply in London, this may have had a significant effect on the performance of functions quite apart from the effects of structural reform.

What the question emphasizes is that local government reform does not take place in a vacuum. Change in structure is one event in a whole series of events which can or do have a bearing on what is happening in local government. Without prejudging the later discussion it seems helpful in this introductory chapter to try to indicate the background of the five years with which this volume is primarily concerned and to draw attention to the situation within which the new London authorities had to operate. The object is not to write contemporary history but to help the understanding of the following chapters. There is, for example, reference in those chapters to individual events or groups of events with a particular bearing on the subject matter of those chapters – the Seebohm Committee for the welfare and children's services, the Town and Country Planning

Act, 1968, in its relation to planning in London, the Local Government Act, 1966, in its implications for local government finance.
What is attempted in this chapter is to connect these disconnected
events and to show what was happening generally in areas of concern
to local government.

The political background

We consider first the political situation. The London Government
Act of 1963 was the occasion of a fierce political debate. It was a
measure introduced by a Conservative Government against the
strong opposition of the Labour Party. By 1 April 1965 the situation
had been transformed. Not only was there by then a Labour Government in power (even though with an overall majority of only 5
in the House of Commons), but the Greater London Council was
firmly in Labour hands as well as 20 of the 32 London boroughs.
Indeed, throughout the whole period covered by this volume and
until the general election of June 1970 a Labour Government was in
power at Westminster, the general election of March 1966 giving the
Party a comfortable overall majority of 97 seats in the Commons.

For the first two years, then, after the new authorities in Greater
London had assumed their responsibilities, the same party was in
power on both sides of the river, at Westminster and at County Hall.
But by 1967 the standing of the Labour Party in the country had
begun to decline; it suffered heavy losses in the local government
elections in that year. A major factor in this decline of the Labour
Party was the economic situation, referred to in the next section. In
the GLC elections of April 1967, Labour was heavily defeated; a
Labour majority of 64 elected councillors to the Conservatives'
36 was converted into a Conservative lead of 82 to 18; only 6
boroughs (Hackney, Islington, Newham, Barking, Southwark and
Tower Hamlets) returned a full total of Labour councillors.[1]
Taking into account the aldermanic position, Labour's 1964 lead
of 75 to 41 was replaced by a Conservative lead of 91 to 25.

The swing to the Conservatives was so strong that they captured
the ILEA, thus ensuring that, as in 1964, the same party controlled
both the GLC and the ILEA. Harmony at County Hall, however,
only increased the chances of disharmony between County Hall and
Westminster. Education and housing, both sensitive political issues,
and in both of which the GLC was involved, were now subject to the

increasing risk of dispute arising from opposed philosophies of the Government on the one hand and the largest local authority in the country on the other. Even more important were the basic questions about the future of London, what transport policies it needed and how its planning and development were to be carried through. These longer-term questions only began to emerge as the Greater London Development Plan was prepared. As a result of the general election of June 1970 the final verdict on the GLDP will have to be taken by a Conservative rather than a Labour Government. Even so the possibility of party political dispute is not by any means removed yet.

Loss of the GLC was not, however, the last of the Labour Party's setbacks. In spite of the provisions of the 1963 Act, elections were not held for the London boroughs in 1967. The curious story of how the borough elections were postponed from 1967 to 1968 is told in more detail elsewhere in this volume.[2] But during 1968 and 1969 the Labour Party's fortunes in the country continued to decline; hitherto safe parliamentary seats were falling regularly to the Conservatives at by-elections. Majorities of 10,000 or more as at Dudley (March 1968) and Swindon (October 1969) disappeared and were replaced by Conservative majorities. Altogether Labour lost 15 seats at by-elections in this period, 12 of them to Conservatives, and held a number of others hitherto regarded as safe (eg Gorton, Newcastle-under-Lyme) by only narrow majorities.

Swings of 10 per cent or more to the Conservatives were general in these by-elections and reached over 21 per cent at Dudley. The situation was reflected in the local elections of 1968 in which the Labour Party fared even worse than in 1967. In the London borough elections of May 1968 no less than 28 of the 32 boroughs were won by the Conservatives; even boroughs like Hackney and Islington, where on the previous councils there had been 100 per cent Labour representation, fell to the Conservatives. Only the East End strongholds of Barking, Newham and Tower Hamlets together with Southwark remained in Labour hands.

The economic background

The political situation at the national level in the first five years of the new London system was thus characterized by the ascendancy of the Labour Party whereas at the local level there was much more variation in political control. At the same time much of the period was dominated by a major economic crisis which not only stole the newspaper headlines but was a potent factor in influencing the policies of the Labour Government which took office in March 1966.

The period of the first Labour Government from October 1964 to March 1966 was, by contrast, a period of relative calm in economic affairs. It included the setting up of the Department of Economic Affairs under Mr George Brown to be responsible for the long-term aspects of economic policy and for regional planning; the establishment of the National Board for Prices and Incomes in April 1965; and the publication in September 1965 of the National Plan.[3] The latter in particular seemed to sum up the economic atmosphere of the time as viewed by the Government. The Plan was based on an assumption of an increase in gross domestic product of 25 per cent between 1964 and 1970 to which increased industrial investment and technology were to make important contributions.

A severe balance-of-payments crisis in 1966 soon made it clear that such a rate of growth was unattainable. There had been a deficit on the balance of payments account since the latter part of 1963. This in combination with a number of other factors brought sterling under severe pressure in July 1966. On 20 July the Prime Minister announced a series of restrictive measures, of which the most politically sensitive was a standstill on prices and incomes for six months to be followed by a further period of 'severe restraint'. Of more immediate importance to local authorities was a cut in proposed public investment for 1967–8 of £150 million; this, it was said, would not affect the education and housing programmes but would affect amenity projects such as swimming baths.[4]

The situation did not improve in 1967. As the Government struggled with its prices and incomes policy in an attempt to limit inflation, it became clear that the economic recovery on which it placed much hope was not happening. In November 1967 the pound was devalued by just over 14 per cent. This was followed in January 1968 by a further series of restrictions on public expenditure in the years 1968–9 and 1969–70. So far as local government was concerned, these included specific items such as reductions in the number of planned approvals of new houses and cuts in expenditure on road maintenance, and also a general reduction in the rate of growth of local authority expenditure amounting to approximately 2 per cent on the total planned expenditure for 1968–9 and over $2\frac{1}{2}$ per cent on that for 1969–70.[5]

1968 saw a slow recovery in the country's external trading position but it was not until 1969 that the balance of payments at last showed a surplus on current account. Apart from the direct effects of the financial crisis on local government through Government restrictions on public expenditure, the years 1965–70 were difficult ones for local authority treasurers because of the very high interest rates prevalent in those years. Bank rate was raised to 7 per cent soon after the Labour Government took office in November 1964 and thereafter did not fall below 6 per cent (except for a short time in 1967) and on occasions reached as high as 8 per cent.

Other consequences flowed from the situation and the Government's attempts to deal with it. The prices and incomes policy pursued from 1966 onwards, particularly under the vigorous leadership of Mrs Barbara Castle as Secretary of State for Employment and Productivity, increasingly brought the Government into conflict with the trade unions. The publication of the White Paper *In Place of Strife*[6] in January 1969, with its proposals for legislation on industrial relations, marked the beginning of a period of extensive discussion and dispute between the Government and the trade unions which was partially resolved when in June 1969 the Government dropped from the proposed legislation the clauses relating to the enforcement of industrial agreements. It was also towards the end of 1969 that the Government tacitly abandoned the attempt to enforce an effective prices and incomes policy at least for the time being, with the consequence that at the end of 1969 and in early 1970 local authorities, in common with other employers, found themselves having to meet increasingly large wage and salary demands. Dustmen and teachers were among those who benefited; and in June 1970 the National and Local Government Officers' Association secured a $12\frac{1}{2}$ per cent rise for local government white collar workers from 1 July 1970 following a 3 per cent increase in August 1969.[7]

The economic situation, particularly in the period 1966–9, had strong repercussions on the political situation, as was suggested in the previous section. For the newly-established London authorities it was chiefly notable for the additional difficulties which it created for their attempts to put services on a new footing; in particular, ambitious capital programmes had in many cases to be postponed.

Local government

Apart from the effects of the general political and economic situation, local government was also faced in these years with a number of specific proposals for change. These included not only proposals affecting the whole structure of local government, but also some which would strongly affect the approach to a number of important functions, notably planning and the social services (personal health, welfare, children).

Although it did not affect Greater London directly, one of the most important developments bearing on the future structure of local government was the setting up in 1966 of two Royal Commissions, one for England (excluding Greater London) and one for Scotland,[8] with wide-ranging powers to investigate the areas, status and functions of local authorities. The reports of these Royal Commissions were published in 1969,[9] so that for a large part of the period when the London authorities were settling down to a new system of local government, authorities elsewhere lived under the hope or fear of radical change. The relationship between the London reforms, the Royal Commission proposals and subsequent Government proposals for local government reorganization are discussed elsewhere in this volume.[10]

While the two Royal Commissions were at work two other official committees made reports which seemed to be equally significant for the ways in which local authorities conducted their affairs. In March 1964 the Minister of Housing and Local Government, at the request of the four local authority associations, set up two committees. The first, under the chairmanship of Sir George Mallaby, was concerned with the staffing of local government, with the ways in which local government officers were recruited, trained and deployed, and with suggesting improvements. The second, under the chairmanship of Sir John Maud (now Lord Redcliffe-Maud) was concerned with the management of local government, that is, with committee and departmental structures, relations between officers and members, the role of committee chairmen and similar matters.

The reports of both these committees were published early in 1967;[11] both made a large number of detailed recommendations; and although their impact was somewhat lessened by the fact that the Royal Commission was sitting and considering even more fundamental changes in local government, they did nevertheless exert an influence on local authorities' behaviour. This was particularly true of the report of the committee on management which provoked a considerable debate within local government.

The Maud Committee argued that there were serious defects in the internal organization of many local authorities and that, even before any changes were made in the structure of local government, local authorities should review their internal organization. Among their main criticisms were, first, that the division of responsibility between elected members and officers was often distorted by councillors' excessive concern with details of administration and their unwillingness to trust and devolve sufficient power to the paid officers; secondly, that there was often too great a fragmentation of the work of local authorities among numerous separate departments.

The Maud Committee recommended drastic reductions in the number of committees. This recommendation has been widely followed. Much more controversial and unacceptable was their view that local authorities should establish management boards of five to nine members, as the principal policy-formulating body subject only to the over-riding control of the council itself, with a consequent lessening of the responsibilities of standing committees which should become deliberative rather than executive bodies making recommendations to the management board. One member of the Committee, however (Sir Andrew Wheatley, Clerk of Hampshire County Council), dissented from his colleagues and thought that local authority committees should continue to have executive functions.

To strengthen the organization of local authorities at officer level the Maud Committee recommended that the Clerk should be recognized as the head of the council's service with specific responsibilities for the organization and co-ordination of the council's activities; that the post of Clerk should be open to all professions and occupations; and that the principal officers should be responsible through the Clerk to the council for the efficient running of the services for which they were responsible. The number of separate departments should, in the Committee's view, be drastically reduced.

The Committee's recommendations covered, in addition, a wide range of possible changes, from giving local authorities discretion to pay members of management boards part-time salaries to the setting up of a Local Government Central Office by the local authority associations. Although no local authority has gone as far as the

Committee wished in re-casting its internal organization, a very large number of authorities have made some move in the direction of streamlining their committee and departmental structure and some (eg Newcastle and Basildon) have gone much further in experimenting than would have seemed possible a few years ago. In Greater London the Committee's report was important in stimulating a new approach to management questions just at the time when these authorities had begun to deal in earnest with the problems which faced them.

The report of the Committee on staffing attracted rather less attention than that on management. This was not because the subject was of less importance but because what the Committee had to say was less controversial and unexceptionable. They recommended, for example, a number of measures designed to overcome shortages of professional staff by such means as the fuller use of technical supporting staff. Among their major recommendations was that local government should provide a career for the administrative officer without professional qualifications, that senior officers should have formal training in management, and that the local authority associations should consider setting up a Central Staffing organization which would, among other things, keep staff training needs under review and co-ordinate resources in the provision of management services.

Nor was it only the structure of local government and its internal arrangements which were under scrutiny in these years. The question of the place of local government in the general organization of government received both implicit and explicit attention. The creation during 1965 and 1966, of Economic Planning Councils on a regional basis implied that some local government planning powers might ultimately be transferred to regional bodies. Although the Councils were only advisory bodies and their sphere of interest did not directly impinge on local authorities' functions it became increasingly evident that economic planning could not be sharply divorced from land-use planning and that with developments in local authorities' planning functions[12] there was bound to be close interaction between the two.

This broadened into a discussion of whether there was a role for regional government and, if so, what form it should take and what implications it had for the future of local government. The Redcliffe-Maud Commission, clearly influenced by the establishment of the regional Economic Planning Councils, recommended that eight provincial councils should be created, responsible mainly for the preparation of provincial plans and replacing the Economic Planning Councils. These provincial councils would be indirectly elected by the new local authorities proposed by the Commission but would also include some co-opted members. Much the largest would be the South East Province, including Greater London, with a population of over seventeen million.

It seems doubtful, however, whether anything on the lines of the Redcliffe-Maud Commission's provincial councils will be set up. The Government's 1971 White Paper deferred any decision on the grounds that it would be necessary first to consider the findings of the Commission on the Constitution.[13] The Commission on the Constitution was appointed early in 1969. Among its tasks it was to 'examine the present functions of the central legislature and government in relation to the several countries, nations and regions of the United Kingdom' and to consider whether changes were desirable either in those functions or in existing constitutional and economic relationships. The election of a Conservative Government in June 1970 with a somewhat different view of local government reform introduced further uncertainty into the whole question.

With all these major structural and internal management questions under discussion one would naturally expect that the subject of local government finance would also be prominent. A White Paper of 1966[14] was the prelude to changes in the system of assessing and financing Government grants to local authorities.[15] The White Paper also suggested that a new local government structure might provide the occasion for a major change in the basis of financing local government. The Redcliffe-Maud Commission, however, in their brief discussion of this topic concluded that whatever additional sources of revenue might be made available, rates would continue to be the main local tax. This conclusion appeared to be endorsed both by the Labour Government and by its Conservative successor; further developments, however, await the publication of a Green Paper on the subject which has long been promised but at the time of writing, has not yet appeared.[16] Apart from the sources of local revenue, the issue which has occasioned most discussion is the financial control exercised by central over local government. Acknowledgment that control over borrowing 'is more restrictive and laborious than it should be'[17] has led to some modification of the procedures for loan sanction but it is not yet clear whether the reform of local government will lead to any real relaxation of central control.

To many people in local government it thus seemed that in the period covered by this volume not only was local government under minute scrutiny from all angles but the whole question of the role of local government in the latter part of the twentieth century seemed far from clear. The avowed aim behind the setting-up of the Redcliffe-Maud and Wheatley Commissions of reforming and strengthening local government seemed to be contradicted by other proposals put forward by the Labour Government of 1964–70. In particular, the decision announced in February 1970[18] to transfer responsibility for the National Health Service to area health boards directly responsible to the Minister and the explicit rejection of arguments for unifying the National Health Service under local government control, means the loss to local government of many services for which it had hitherto been responsible including the ambulance service, health visiting, maternity and child health care and the school health service.

It was not only the health services which were under examination in this period. The most effective organization for a number of other services also received attention for a variety of reasons. This was especially true of the personal social services provided by local authorities, essentially those provided by children's and welfare departments. Growing concern with the fragmentation of these services through the usual local government practice of separate departments and committees[19] was a major factor in the setting up in December 1965 of a committee under the chairmanship of Mr (now Sir) Frederic Seebohm to review the organization and responsibilities of these services and to recommend any changes which might 'secure an effective family service'. The report of the committee published in 1968, recommended that a new social service department should be established in each local authority and be responsible to a single committee.[20] These recommendations were endorsed by the Government and carried into effect by the Local Authority Social Services Act, 1970; in mid-1970 the local government journals were full of advertisements for Directors of Social Services to be responsible for these new departments. Quite a few authorities, including some London boroughs, eg Lambeth, had begun to reorganize on Seebohm lines even before the Act was passed.

Underlying the appointment of the Seebohm Committee were emerging ideas about the way in which the needs of people for certain services might best be met. It is interesting in this connection to note that, independently of the Seebohm Committee, the Children and Young Persons' Act of 1969 made several modifications including the possible transfer of approved schools from the Home Office to local authority control. Other structural changes in services arose from a response to the view that some services now need to be organized on a much larger scale both to take account of changing conditions and to take advantage of the economies possible through specialization. Police and water supply are examples of services where amalgamation and the creation of larger units had begun to operate before the London reforms took place and which continued during the period covered by this volume. But no change has been made in the special organizations of the Metropolitan Police and the Metropolitan Water Board, which are peculiar to the capital city.[21] The latest of these services to be examined has been the fire service. A departmental committee was set up early in 1967 under the chairmanship of Sir Ronald Holroyd and reported in May 1970. Their main conclusion was that there should be larger and more uniform fire authorities, with about 40 brigades in England and 5 in Wales against the existing 131; they recommended no change in the position of the GLC which is by far the largest fire authority in the country.[22]

Finally, in this brief account of changes, actual and proposed, affecting the organization of local authority services, should be mentioned a somewhat different development resulting in a different approach to an important part of the planning powers of local authorities. In 1967 the Ministry of Housing and Local Government referred to the 'revolution in planning thought' which had occurred in the previous few years, partly as a matter of techniques and partly from a realization that planning needed to be carried out in regional and national and not simply local terms.[23]

The feeling that the development plan approach and procedures embodied in the Town and Country Planning Act of 1947 were no longer adequate was put in positive terms by the report of the Planning Advisory Group set up in May 1964 by the central government to conduct a general review of the planning system.[24] The group had representatives both of central government departments (Ministry of Housing and Local Government, Ministry of Transport, Scottish

Development Department) and of local authorities (clerks and treas-urers as well as planning officers), together with two outside consul-tants. It recommended briefly the replacement of existing develop-ment plans with a new type of plan dealing only with the broad physical structure of the area and the principal policies and priorities for its future development. These 'structure' plans as they are called would be submitted to the Minister for approval, in the same manner as the existing development plans; but in addition local authorities would be able to prepare their own more detailed local plans which would have to be in conformity with the policies of the structure plan but would not require Ministerial approval.

These proposals were eventually embodied in the Town and County Planning Act, 1968, along with other changes designed to secure greater public involvement in the planning process and to improve the arrangements for development control. One effect of the new system, when it is fully in operation, should be to broaden the planning process; a structure plan

> will embrace not only policies for the use of land, but also for the management of traffic and for the improvement of physical environment. It must be drawn up with regard to regional economic policies and to the resources likely to be available to put it into effect.[25]

It should also simplify the procedure for Ministerial approval of plans which under the old system had led to great delays.

The 1968 Act came at a time when the Greater London Develop-ment Plan was actively under preparation. For the purposes of the Act the GLDP was to be treated as a structure plan, although the 1963 legislation had been framed on the basis of the previous town and country planning legislation. However, the GLC must have been fully aware of developments which were taking place;[26] how far these planning developments affected the way in which planning has been dealt with under the new system is discussed at greater length in the planning chapter.

Although we have been mainly concerned here with the major de-velopments affecting the responsibilities and organization of local authorities and local government generally, it is clear from the planning example that such developments cannot be entirely separ-ated from changes in government policy which may have profound repercussions on the activities of local authorities. A good example whose impact on the London authorities is more fully discussed in the education chapter, is provided by the Labour Government's aim to make comprehensive secondary education the normal pattern.[27] But perhaps one of the most significant developments of all for its effect on the future of London was the increasing attention which was paid to the problems of South East England at the very time when the new system of government was establishing itself.

Before the London reforms took place the Government in 1961 initiated a study of South East England by officials mainly 'to seek solutions for London's problems over a wide front'[28] at least in terms of land requirements. The area covered by this study was large, the whole of England south east of a line from the Wash to Dorset. The basis of the study's proposals[29] and of the Government's endorsement of them was 'a fresh and much larger programme of new and expanded towns'[30] which would eventually alter the distribution of population growth in the South East. The important point about these proposed new towns is that they were to be much further away from London than the first post-war new towns. The areas of Southampton/Portsmouth, Bletchley and Newbury, all about fifty miles or more from central London, were suggested for major new cities. Large expansion schemes were suggested for even more distant places such as Ipswich, Northampton and Swindon. In the event, not all these proposals were acted upon but in 1965 it was announced that a new city, ultimately of 250,000 people, was to be established in north Buckinghamshire (Milton Keynes), together with major expansions of Northampton and Peterborough. Subsequently, it was also decided to go ahead with a major expansion for the Southampton/Portsmouth area.

These government decisions had, of course, major implications for the new London authorities, which took over their responsibilities on 1 April 1965, through their effect on population and employment trends in Greater London. Furthermore, they were followed by other inquiries. The South East Economic Planning Council's first report[31] raised the possibility of a further approach to the development of the South East by channelling new development along the corridors of growth and existing radial routes, many of which led to the new major growth areas already decided upon. These proposals were referred by the Government in 1968 to a study group under the chairmanship of Dr Wilfred Burns, the chief planner at the Ministry

B

of Housing and Local Government. Their report[32] was published in 1970.

These developments had repercussions most immediately and obviously on the production of the Greater London Development Plan which could not be considered as an isolated exercise but one which could only make sense in the context of the South East's future as a whole. This aspect is further developed in the planning chapter. But even more widely both the GLC and the London boroughs could only meaningfully plan the future development of services for which they are responsible, whether for new motorways or for old people's welfare, within a realistic framework of knowledge about London's future size and relationship to the remainder of the South East. Hence the importance of divergences of view about the future population pattern of Greater London even within the short term up to 1981.[33]

Many acts of Government policy between 1965 and 1970 had their effect on local government, including Greater London. If anything, the situation was more marked in this particular period because of the coincidence of the setting up of the new authorities in Greater London with the accession to power of a Labour Government after thirteen years of Conservative rule at Westminster. Thus an early act of the new Government was to impose severe restrictions on new office building in Greater London;[34] much emphasis was laid on the need to increase the building of new houses by public authorities;[35] and in 1968 a new 'urban programme' was announced of additional government expenditure to assist areas of special social need.[36] These, as well as other measures which have been referred to (eg the greater emphasis on comprehensive secondary school reorganization) were perhaps largely the consequence of a change of government. But developments in other directions (eg planning, the welfare services) might well have taken place whichever party was in power, although not necessarily in the same form as actually happened.

These are, however, speculative questions. What is abundantly clear is that the early years of the new authorities in Greater London were years in which local government, both generally and in relation to its specific activities, was subject to close examination and affected by what in total amounted to a considerable shift in attitudes and policy, partly the result of developments initiated within local government and partly as a result of central government policies.

The significance of the situation for the present study is that we cannot assume that without the London reforms the provision of local government services in London would have gone on after 1965 much as it had before. The London government reforms came into effect at a dynamic rather than a static period for local government. Not only did this affect considerably the problems facing the new authorities, but it increases the difficulty of assessing the effects of the reforms on the provision of services. To see how the different influences operated on the London authorities, specifically and generally, is the aim of the following chapters.

It remains briefly to note the structure of the volume. The following two chapters deal respectively with the effects of the London reforms on the organization of the two main political parties in Greater London, and with the GLC and London borough elections between 1964 and 1970. Thus these two chapters provide an essential link between the earlier historical study of the London reforms[37] and the main theme of the present study, and emphasize the political framework within which the new system has operated.

Next follows a series of six chapters each dealing with an important local government function or group of functions. The main aims of these chapters are to assess the effect of the reforms on the performance of the functions, and to examine how the new authorities have met their responsibilities.

These chapters are followed by three chapters of a more general nature which discuss, first, the financial consequences of the reforms, and, secondly, the nature, role and functioning of the GLC and the London boroughs in the new system.

The final chapter seeks to evaluate the reforms as a whole.

2
Political Party Organization

(a) *Introduction*

One effect of the London Government Act, 1963, was to render the existing party structures in Greater London obsolete. Neither party was organized upon a metropolitan basis, but upon a basis of Areas or Regions broadly co-terminus with the boundaries of certain groups of the then existing Counties.[1] Thus on the Left, the London Labour Party co-ordinated activities within the areas of the London and Middlesex County Councils, beyond which the constituencies fell within the boundaries of three Labour Party Regional Councils: the Eastern (13 constituencies and parts of two others); the Southern (16 constituencies and parts of two others) and the Northern Home Counties (28 and parts of one other). A further complication was that Middlesex was in fact covered both by the London Labour Party and the Northern Home Counties Organizing Area. As a report of a Labour Party committee of enquiry was to explain:

> 'radical changes in the machinery of local administration make necessary equally radical changes in the structure of the Labour Party.'[2]

The same could have been said of the Conservative Party. The Home Counties North Area contained the whole of Essex and Middlesex; the Home Counties South East Area contained the whole of Kent and Surrey; parts of Greater London were the responsibility of the Eastern Area; and in the London Area responsibility was shared between the London Conservative Union and the London Municipal Society. Whereas in the LCC area the London Labour Party had unchallenged local supremacy, having no rivals upon its territory, the official Conservative machine co-existed in a unique duplication with the extremely influential, ostensibly non-Party, London Municipal Society.

With the creation of the Greater London Council these problems of overlap, duplication and fragmentation had to be tackled. The first round of GLC elections was scheduled for the spring of 1964 and neither campaign planning nor the selection of candidates could be left to the existing organizations, but necessitated new machinery which had to be operating by the autumn of 1963 at the latest. Local constituency parties had to be co-ordinated in order that GLC candidates could be selected, and the subsequent London borough elections contested. But the most sweeping changes involved the creation of an organization for the Greater London area to endorse the locally-adopted GLC candidates, to settle electoral policy and effect a single publicity campaign.

What follows in this chapter is an account of the way in which the two major parties faced the problems and evolved new organizations with responsibility for the area of the entire metropolis. But it is more than a story of a readjustment of boundaries and responsibilities: the 1963 Act had other repercussions within the parties, both upon the distribution of power between the national and local organs, and upon the politicians themselves. The London Government reforms brought to a close a distinctive era of London's political life, and in order to convey something of the nature of that era the parties are considered within the context of their own separate histories.

(b) *The impact upon the Conservative Party*

The origins and nature of the London Municipal Society. The original aims of the London Municipal Society were promotional and limited, it having been founded in 1894 to 'extend and complete the policy . . . for the reform of London government by the establishment of District Councils or Corporations and their endowment with adequate authority'.[3] This aim was achieved following the Unionist victory in the General Election of 1895. The Society lobbied Salisbury's Government and claimed to have been instrumental in persuading it to bring forward the London Government Bill, which, when passed in 1899, created the Metropolitan Borough Councils. As *The Globe* commented at the time, 'the success of the Bill is very largely due to the energy and determination which the society has displayed on its behalf'.[4]

At an early meeting of the Society, Lord Salisbury exhorted his listeners not to be 'shy of using all our political powers and machinery for the purpose of importing sound principles into the

Government of London'.[5] The 'sound principles' hardly needed to be identified as Unionist principles; but the Society itself continued to retain the right to interpret these principles as it saw fit, and to translate them into action accordingly. It was in no sense an organ of the Conservative Party; indeed, the brief official history speaks of the Society as 'an independent organization outside the party machine and mainly concerned with promoting London's interests'.[6] The Herbert Commission and the chairman in particular subsequently equated the Society with the Conservative Party,[7] whilst Smallwood echoes this in his bewilderment at the Society's claim to be 'an independent body that tends to look after Conservative interests in London'.[8] The confusion is understandable, but the Society never claimed to be politically neutral. Rather it was partisan, in the sense of being Unionist by outlook and membership, whilst having no formal links with the Central Office machine.

Its partisanship was initially displayed in its support for the Moderate groups upon the LCC and the Metropolitan boroughs around the turn of the century. The Unionist-inclined Moderates were un-co-ordinated, and lacked a coherent municipal philosophy; this the Society gave them. The Moderate label was dropped in 1907 and the more specific Municipal Reform title adopted, the LCC candidates then fighting elections as the Municipal Reform Party. They were subsequently to identify themselves as Conservatives. Their political opponents were initially the Progressives, identified with the Liberal Party, but containing not only such great Liberals as Lords Rosebery and Crewe, but Fabian socialists such as Sidney Webb.

In its opposition to Progressivism and, later, to Socialism, the Society fought on many fronts, and several distinct roles. The story of the years subsequent to 1899 is largely the story of the way in which the Society became, at different times and to different degrees, the co-ordinator of ratepayer organizations, both in the metropolis and nationally; the leading purveyor of anti-socialist doctrine on local government matters; a research and intelligence agency for the conservative group at County Hall; election campaign organizer for Municipal Reform, later Conservative, candidates for LCC and Metropolitan Borough Council elections; and organizational adviser and substantial helper to Conservatives in municipal election campaigns in provincial cities.[9]

The changing roles of the Society, 1905–59. A subsidiary aim listed in the memorandum of objects circulated by the Society at the time of its inception was the promotion of 'efficiency, economy, and consistency' in local administration. This was pursued by the careful scrutiny of public expenditure, by campaigns for greater Parliamentary control of municipal financial legislation, and by the organization and articulation of the 'Ratepayer' interest. Between 1905 and 1920 these activities engaged an increasing proportion of the Society's energies. By the latter year the Society was operating on a national, rather than a merely metropolitan scale, and in 1921 its title was changed to The London Municipal Society and National Union of Ratepayers' Associations. In 1937, however, a separate National Union of Ratepayers' Associations was formed, and in 1953 the term Ratepayer was finally dropped from the Society's title.

With the increase and subsequent diminution of the role of national spokesman for the ratepayers went a similar rise and fall of the Society's function as anti-socialist propagandist. Between 1907 and 1909 in particular the Society published a very considerable body of fervent polemic aimed at local electors and local government activists and this activity continued to occupy a great part of the Society's time at least until the mid 1930s.[10] Smallwood has described the Society in its latter days as 'the intellectual spokesman for Conservatism in London',[11] but it had been more than this. From its foundation the Society was the intellectual spokesman for Municipal Conservatism nationally. When, however, the Society's functions contracted during the early 1950s the propagandist function also declined. In 1946 the Conservative Central Office, with help from the London Municipal Society, established a Local Government Department to serve the entire country, thus relieving the LMS of a rather demanding duty. In this respect at least, Sharpe is probably correct in his judgment that the Society 'diminished in importance since its heyday during the inter-war period'.[12]

This diminution, it should be noted, did not apply to the remaining functions of servicing the Conservative group at County Hall and running the LCC and Borough Council election campaigns. Indeed, it might be said that whilst the LMS declined in national significance, it greatly increased in local importance.[13] During the years of opposition upon the LCC from 1934 onward there was a manifest need for a research and intelligence organization to provide the minority party members with information and political argument. The efficacy of the London Municipal Society in this respect varied with the personalities of the staff, as might be expected in an activity that extended to the drafting of speeches and the preparation of briefs. In the early 1950s the Conservative group and the Society gained enormously from the services of Iain Macleod MP and J. Enoch Powell MP, who directed the Society's affairs.

It is also significant that the post of Director was itself first created in the aftermath of the 1949 LCC election when the two major parties dead-heated in the election of Councillors. The need for organization was clearly never greater if the Conservatives were to maintain their favourable position. There was a strong case for a top-calibre executive director, and the necessary restructuring of the Society was carried through by the Hon. John Hare MP,[14] who as Chairman of the Society, had experience both at County Hall and at Westminster. The years of Mr Hare's chairmanship were years of adaptation to a more limited, but more specialized and perhaps more effective role.

The service and research functions of the Society increased in importance but it also continued to play the major part in the organization of election campaigns in London. Its experience in this respect pre-dated that of the Conservative Party machine itself. At the Society's first meeting in 1894 a resolution had been carried committing the LMS 'to assist in securing and to support suitable candidates at municipal elections'.[15] This it managed to do with remarkable energy, and with initial success. As early as 1898 the Society was employing an organizer of canvassers who was sufficiently astute to mobilize the party helpers in safe divisions and draft them into the marginals – an early use of a now common technique. Such activities soon led to friction with the official party organization. By 1900 talks were being held between the Society's representatives and those of the Metropolitan Division of the National Union – talks at which the Party's Principal Agent 'Captain' Middleton was also present. The outcome was that a joint campaign committee was formed to conduct the LCC election campaign, and was reconvened for successive LCC elections.

The early electoral successes – the LCC had been won from the Progressives in 1907 – were reversed in the inter-war years, the Labour Party capturing the LCC in 1934, winning 28 seats from the Municipal Reformers. In the Borough Council elections of that year, control of twelve Metropolitan boroughs passed to the Labour Party. These losses were a profound shock to Conservatives in

London, and the London Municipal Society became immediately vulnerable to criticism. When the Municipal Reformers failed to recover lost ground in 1937, but lost a further six LCC seats and about 60 Borough Council seats to Labour, the critics became more voluble, and Neville Chamberlain, the then Prime Minister, ordered investigation of Conservative electoral organization in London. The London Municipal Society submitted written evidence to the investigating committee, and claimed to have been completely vindicated of the charges laid against it.[16] The Prime Minister took a close interest in the inquiry, the result of which was to bring the Party's Metropolitan Area organization and the LMS into a very much closer relationship.

The Society, nevertheless, continued to participate in the running of the London election campaigns, jointly with the Metropolitan Division, and with that body's successor, the London Conservative Union. There seems little doubt that the Society, by virtue of its close connection with the LCC Conservative group, was the dominant partner in this relationship. The Joint Campaign Committee was chaired by the Conservative leader at County Hall, who at all times had the closest relationship with the Society's officers. As the main aim of the electoral effort was to ensure the election of Conservative candidates to the LCC, the Leader had the considerable moral authority necessary to ensure the continued predominance of the LMS in electoral matters. This predominance was real; the Society drew up and interviewed a panel of suitable candidates, produced the campaign guide, and handled the publicity. Had a constituency association at any time adopted an unsuitable candidate, it would have been from the Society's officers that the pressure to reconsider would have come. The London Conservative Union was left to provide the agents and organizers in the constituencies, and look after the details of meetings, and campaign expenses.

A significant consequence of the 1938 inquiry into London political organization was the decision to bring about a deliberate identification of local candidates with the national party, by the adoption of the Conservative label for future local government elections – a decision which, due to the war years, came into effect only in 1946. The dropping of the 'Municipal Reform' label perhaps symbolizes the waning power of the limited, Metropolitan appeal to which the LMS was committed. Following the adoption of the Conservative label, local government elections came to be increasingly the concern of the party's headquarters. By the time the Society's official history appeared in 1954, a slightly defensive note could be discerned: 'In London, no less than in the provinces, the party machines have extended their activities beyond their primary function and engage almost as energetically in local as in parliamentary elections . . .' But the need for the LMS to survive remains – 'the one place whereby common policies on the County Council, the Borough Councils and in Parliament can be hammered out and given expression, is the London Municipal Society . . .'[17]

In 1958–9, the Society gave evidence to the Herbert Commission, both orally and in writing, submitting two separate memoranda of evidence. The first memorandum featured criticism of the LCC on grounds of remoteness and inflexibility, and called for strengthened boroughs, although, as Rhodes has observed, 'in view of this conclusion the Society at first came to a suprisingly indefinite conclusion about how local government should be reorganized'.[18] In this memorandum, the Society, above the signature of one of its oldest retainers, Miss Phyllis Gelli, was simply elaborating the objections to the existence of the LCC with its wide range of powers, objections that had provided the impetus to the Society's establishment in 1894.[19] It was the Society's traditional view.

Some months after the first memorandum the Society submitted a second one, this time drafted by the new director, Mr Roland Freeman,[20] and this came out in far more definite terms in favour of a single strategic authority for Greater London and a number of 'County Boroughs' of around 200,000 population. This sudden hardening of view led to accusations of undue influence on the part of the Conservative group at County Hall, for the Society's line at this later stage was certainly in accord with the evidence submitted by the minority party on the LCC.[21]

The second memorandum, with its more radical arguments for boundary reform, implicitly challenged the continued existence of the London Municipal Society itself; were the LCC area to disappear as a local government unit, the Society could hardly hope to exercise its existing electoral functions within that area, and a take-over of the peripheral local parties was out of the question. As the Society's spokesmen emphasized during their appearance before the Commission, the Society had no contact with or support from the Conservative organizations outside the County of London. And no body analogous to the LMS with which it might combine, existed in Middlesex – or indeed anywhere else. Paradoxically then, as in 1946 when the Society collaborated in the setting-up of a Local Government Department at Central Office, it was attempting to create circumstances that would throw its own future into question.

The response to the creation of the GLC. The Herbert Commission's proposals and their acceptance by Mr Macmillan's Government raised the problem of the future of the party organizations in the new Greater London area. It remains to trace the course of events by which the Society passed away, to be replaced by the centrally-controlled Area Organization of the Conservative Party. The first steps were taken uncommonly quickly by the officers of the National Union. The White Paper on London Government appeared on 29 November 1961; on 4 December, Mr Cyril Norton, Area Agent for the North of England, arrived in London to act as secretary to a Planning Committee which was to be established to review the Party machinery in Greater London.[22] The chairman of the committee was to be Sir Eric Edwards (later Lord Chelmer) an important figure in the National Union.[23]

That the party headquarters were able to act with such speed was largely due to the initiative of Sir William Urton, the General Director of the party, who was himself to serve upon the Planning Committee.[24] On the publication of the Herbert Report some eleven months before, a committee was set up under Sir Toby Low, deputy chairman of the Party Organization, to bring together the representatives of the areas affected by the Report. This committee held preliminary discussions of the problems created for the party structure, but could go no further until the Government made its intentions clear. By the time the White Paper appeared, the General Director was ready to take immediate steps. He reconvened the Low Committee on a more formal basis under Sir Eric's chairmanship, and summoned Mr Norton to London. These two key appointments he had decided upon in advance, in anticipation of the Government's intention to legislate. Local Conservative leaders, in contrast, were in no position to make an appraisal of the probable fate of the report. Sir Percy Rugg could not draw the Government on the content of the White Paper, nor would they even give him a firm indication of a possible publication date. And the minutes of the Executive Committee of the LMS reveal that as late as 22 November

1961 – seven days before publication – the Executive had no fore-knowledge of the Government's intentions.

The Planning Committee when it first met on 8 February 1962 had ten members, others being co-opted at later stages. Apart from Sir Eric, the chairman, and Sir William, the original members were Sir Toby Low (now Lord Aldington); Sir Percy Rugg, Conservative leader on the LCC since 1949; Sir Dan Mason, Chairman of the South-Eastern Area of the National Union since 1957 and a member of the National Executive; Lieutenant-Colonel (later Sir) Edmund Joly de Lotbiniere, Chairman of the Eastern Area since 1961; Mr (later Sir) Neil Shields, Chairman of the London area since 1961, deputy leader of the Hampstead Metropolitan Borough Council, and a member of the National Executive; Mr (later Sir) Theo Constantine who was later to become Chairman of the National Union; Sir Samuel Salmon, LCC member and Chairman of the London Municipal Society, and Mr Rex Bagnall, Chief Organization Officer at Central Office.

This very high-powered committee met to consider the future shape of the party in London almost every month from February 1962 to June 1963, occasionally meeting more than once a month and in January 1963 in three successive weeks. From the very first it was accepted that some new body would need to be operating by the autumn of 1963 at the latest, and with responsibilities for the whole Greater London area, in order to select candidates in time for the GLC election campaign. For reasons that merit full discussion elsewhere, the Committee decided in favour of a conventional Area Organization of the Conservative Party, under the control of Central Office.

The Committee's main task, after settling some important issues of principle, was to draw up the constitution and rules for this new body. This complex and perhaps tedious business occupied much of the time of the full committee; to facilitate the swift conduct of more clearly political decisions a sub-committee was established. This sub-committee fulfilled the functions of the Area Organization during the long gestation period of the latter; it prepared a list of approved candidates for circulation to the constituency Associations; it set up several policy study groups to prepare the ground for the campaign itself; it exercised pressure upon certain MPs during the passage of the Bill, and constituted deputations to the House of Commons; and it split into two working parties, one to consider arrangements for the GLC elections, the other for the borough elections.

The full planning Committee agreed the final draft constitution for the Greater London Area Organization at a meeting on 6 May 1963. This constitution was then considered and approved at a representative conference called for 31 July. The conference was addressed by Mr Brooke, who had been the Minister responsible for setting up the Herbert Commission, and was himself a former leader of the Conservative Group at County Hall. There was a certain amount of opposition to the Planning Committee's proposals, especially from the peripheral areas in suburban Surrey, where the metropolitan sentiment had never been strong. Old loyalties died hard – but they died more quickly than in the case of the London Labour Party. It was perhaps historically apt that the conference should have been held on 31 July, for this was also the day upon which the Royal Assent was granted to the London Government Bill.

When the new Area Organization came into being on 17 October 1963 – taking over from the LMS which was thereafter dissolved – several of the Planning Committee members figure prominently in it. Lord Chelmer, who had chaired the Committee, became the chairman of the Area Council, in order to oversee the difficult transitional period; he was thereafter elected to the Presidency, where he continues to play considerably more than a figurehead role. Sir Neil Shields became a member of the Executive and Mr Cyril Norton the Chief Central Office Agent for Greater London. This last post, interestingly, has no parallel in the Provinces, where the chief party officials hold the rank of Central Office Agent – the equivalent of Mr Norton's deputy. The final issue of *The Londoner* reviewing the staff appointments to Mr Norton's team described it as 'one of the most powerful political teams ever seen in London'. It certainly contained few of the staff who had served the London Municipal Society in its last days.

The voluntary liquidation of the London Municipal Society. The London Municipal Society was the victim of its own aims. Committed to the notion of reform of local government boundaries, it achieved – in so far as it had any influence upon the Commission and the Government – not only the redundancy of the LCC, but its own redundancy as well. The final issue of *The Londoner* illustrates this paradox in its headline: 'Act becomes Law – LMS achieves its aims:

London Municipal Society to close down'.[25] But views differed on the necessity of the Society going into voluntary liquidation upon the creation of the Greater London Council. Sir Samuel Salmon himself only reached this conclusion 'after much heart-searching'. Others never reached it. They agreed that the electoral responsibilities for Greater London would be too demanding for the LMS, but believed that there remained a useful job for it to do. Even Sir Percy Rugg, first Conservative leader on the GLC, would now perhaps incline to this view, although he had been less sympathetic towards it during the years of planning from 1962 to 1964. The Local Government section established within the Area Office seemed to some a poor substitute for the LMS, at least during the life of the first Greater London Council.

From 1894, as the preceding pages show, the London Municipal Society underwent several changes of role. Why did the creation of the Greater London Council prove a mortal blow to it? Apart from the impracticability of extending the area of the Society's electoral operations, the reasons for its liquidation seem to be threefold. In the first place there was the question of finance.[26] The Society had traditionally been financed from private sources, and when funds were low an appeal would be launched among the Society's many City friends. The key man in these fund-raising activities was Sir John Musker, Honorary Treasurer of the Society from 1936 to 1963, and a distinguished merchant banker. But Sir John's services would not be available indefinitely, for he was nearing retirement. With the Society's financial future in mind Sir Samuel Salmon approached the senior officials of the National Union shortly after the 1958 LCC elections, and a series of meetings were held to consider whether or not the Society should launch another major appeal. The outcome was that Lord Poole, Deputy Chairman of the Party organization, offered to find a substantial proportion of the Society's requirements from Party sources.

Again, views differ on the inevitability of this development. Some feel still that the Society could have continued to find its own funds. Others maintain that the ever-increasing cost of election campaigns, and the decline of the Society's wider 'civic' functions dictated a shift of financial obligation from the individual or corporate well-wisher to Conservative Party headquarters. In the event, the support of Central Office, once accepted, placed the Society's future in the hands of its benefactors – it had become a dependant.

And in this vulnerable position it was to discover that some if its best placed friends – the late Iain Macleod in particular[27] – were strongly against the Society continuing its work.

The second factor undermining the Society's claim to survival concerned its role as adviser to the Conservative group at County Hall. There was an assumption among some prominent Conservatives in London that following the integration of Conservative suburbia with the Labour-controlled LCC area, their party would control the new Greater London Council as a matter of course. If this were to be the case, they felt that the need for the research and briefing activities of the LMS would be lessened, for, as the majority party, the Conservatives would have a closer relationship with the Council's chief officers, and would be thus better informed on factual and policy questions. As this was not in fact the case during the first Council, the support of the Society was sorely missed.

The third factor concerned the long-standing tensions between the Society and the London Conservative Union. The acrimonious disputes that followed the election defeats of 1934 and 1937 were not forgotten. Central Office felt concern about the duplication of responsibilities in electoral work and believed that the LCU would produce better results if it had a clear field, and did not have to work in harness with the LMS. Certainly the detached observer might have been struck by the committed, enthusiastic yet fundamentally amateurish flavour of the Society's outlook in its final years; it was perhaps tinged with nostalgia for the 1920s when the Society was at its peak. In this it was severely out of accord with the spirit of nattional Conservative politics in the early 1960s. To the party professionals the LMS was if not a downright liability, at least a dubious asset. Such a view, whether right or wrong, was bound to be fatal in the circumstances of the party's re-appraisal of its organization.

(c) *The impact upon the Labour Party*

The origins of the London Labour Party. The Labour Party organization in London came into being in 1914 at a relatively late stage in the politics of the London County Council. In the years immediately following 1906, when Labour first achieved substantial representation at Westminster, Socialism at County Hall was still left in the trust of the Fabians within the Progressive Party, there being no Labour organization for the LCC area as such. As the author of the definitive study of the Left in London, Paul Thompson, observed '. . . it is more remarkable that until 1914 the Labour movement in London remained a diverse collection of separate borough movements than that a special party was eventually formed to unite it'.[28] For this lack of organization, Thompson places the blame upon the London Trades Council which promoted its own 'invariably inadequate schemes for electoral unity in London rather than support the Labour Party's efforts in the same direction'.[29]

Not unnaturally, the National Executive Committee of the Party showed concern. The NEC had put forward proposals for a London Labour Party as early as 1906, yet postponed discussion until after the 1907 LCC election – at which the Moderates for the first time gained control – but the scheme then foundered upon the rock of the NEC's doubts as to the political reliability of such a party's probable leadership. These doubts were engendered by the activities of the Social Democratic Federation, at that time in control of the London Trades Council, but similar fears were to recur, perhaps less reasonably, during the next sixty years until the London Labour Party itself was finally dissolved.

The Trades Disputes Act of 1913 strengthened the financial basis of the labour movement, and the London Trades Council was able to take the vital initiative later that year when Fred Knee – a Social Democrat who had urged close co-operation between the Labour-minded elements in London – assumed the secretaryship of the LTC. In May 1914 the LTC called a conference of 'those eligible for affiliation to the national Labour Party for the purpose of establishing a united working class party on the LCC at the 1916 election'.[30] This inaugural conference of the London Labour Party was attended by 420 delegates the great majority of whom represented trade union branches. It adopted a constitution closely modelled upon that of the national Party. With the death of Fred Knee the following year, Herbert Morrison became secretary of the new party, and thus presided over the period of growth that was to follow.[31]

In the first election to be fought by the London Labour Party in 1919, considerable success was scored; 573 councillors were elected to the Metropolitan Borough Councils, to form majorities in 13 boroughs. Twelve of these were won from the Municipal Reformers (as the Moderates now chose to be known), one from the Progressives, and in one borough a Progressive–Labour coalition was formed. Fifteen Labour supporters were elected to the LCC, where formerly only two had sat. Many of the Borough Council gains of that year were, however, dissipated in 1922, and were not to be regained until 1934. But Labour's representation on the LCC grew steadily – without too severe a setback in 1931 – until the party gained control by winning 69 of the 124 seats in 1934. One interesting feature of these gains was that in the borough elections they were made, initially at least, at the expense of the Conservatives, whereas the rise of Labour on the LCC entailed the annihilation of the Liberals.

In 1951 the Labour organizations in the County of Middlesex affiliated to the London Labour Party, swelling its membership and finances but creating certain anomalies in the relations with the national Party. The Rules Revision that embodied this expansion put the London party's prime aim as 'to unite the forces of Labour in London and Middlesex' – an ambition unlikely to reassure the Transport House officials. Pressure from the LLP to recast the regional boundaries was unsuccessful, and the working of the new situation came to depend upon harmonious relationships between the regional officers of the Party and those of the LLP. As the last Secretary of the LLP was to remark in a retrospective survey of the party 'most people would agree that this arrangement has not worked well'.[32]

The London Labour Party in 1961

The tensions between the London and the national Parties inevitably came to a head in the years following the announcement of the Conservative Government's intention to create a single authority for the London metropolis. That such tensions came to arise, and were not easily resolved, cannot be understood without a grasp of the uniqueness of the LLP within the framework of the Labour Party. For organizational purposes England (with the exception of London), Wales and Scotland were divided into Regions, each with a representative Council acting as territorial advisory committee to the National Executive, and working with a centrally-appointed Regional Organizer answerable to the National Agent in Smith Square. Issues of national policy are not within their remit, nor can the Councils amend their own constitutions; only the National Executive may make such changes, at the request of the Regional Council. They are, in consequence, characterized by political weakness. McKenzie has remarked of them: 'they are so nearly powerless that it is difficult to understand how they succeed in holding the interest of those who attend their meetings'.[33]

The area of the London County Council was, however, excluded from the network of the Regional Councils (the first of which were established in 1938) by virtue of the pre-existence of the London Labour Party. The independence and power of the LLP was in sharp contrast with the subordinate position of the councils elsewhere. It was an integrated political party affiliated to the Labour Party nationally, and responsible for its own finances. Structurally, it was a small-scale replica of the national party, with an executive committee as its highest organ, elected by and answerable to an annual conference. The leaders of the County Council Labour groups sat ex-officio upon the executive committee, which appointed the secretary and staff, whereas outside London the Regional Organizer acted as Secretary to the Regional Council.

The LLP, in affiliating to the Party nationally, gained representation at Annual Conference, as well as at the more specialized conferences on Women's, Youth and Local Government affairs. It also had the right to amend its own constitution, subject only to endorsement by the National Executive, and the 1939 memorandum 'Labour groups on local authorities' and the attached model standing orders did not apply to parties in the LCC area, Morrison having successfully argued for their exclusion as 'exceptional'. Decision-making power was concentrated within the leadership of the LCC Labour group, however, and so great was the influence and prestige of the LCC leadership in the years of Labour control that even Labour politicians could refer casually to the LLP as the 'LCC Labour Party' without greatly distorting reality. This was the perhaps inevitable consequence of Labour's long years of rule at County Hall, which relegated the LLP executive to the status of the LCC leader's personal political *apparat*. Leonard argues that this loss of the LLP's original role followed Labour's capture of the LCC in 1934: thereafter 'important decisions came to be taken by the Labour hierarchy at County Hall, and the London Labour Party was largely reduced to an electoral role'.[34]

This view is somewhat oversimplified. In fact Morrison continued as secretary to the LLP for the next six years, during which time he also led the London County Council. Only upon his entry into the national Government in 1940 was a separation effected between the Council leadership and the Party secretaryship, Charles Latham

becoming leader, and D. H. Daines secretary. It was this separation
that effectively reduced the political significance of the LLP as such,
for after Morrison no major political figure was to act as its secretary. Its future was one of agent or handmaiden to the Labour
group, a position of subordination in which it was kept by a
practice of joint control of Labour Group policy committee and
LLP executive, achieved by means of interlocking membership of
both bodies.

As a result of this interlocking membership, many of the powers of
the LLP executive effectively accrued to the LCC Leadership. Perhaps the most important of these was the control over candidatures.
A single example will suffice to illustrate the manner in which the
LLP executive's power to refuse endorsement to locally adopted
candidates could be used. Six months after Labour's massive victory
in the 1958 LCC election[35] a decision was taken by the leadership to
raise council house rents, despite the fact that the maintenance of
rent levels had been a specific plank in the party's election platform.
Something like one third of the members of the Labour group
objected unsuccessfully in the party caucus meeting. One of their
number Mr Tom Braddock, was so incensed by this *volte face* as to
write a letter to the *Guardian* protesting against 'this undemocratic
procedure'.[36] The LCC leader, Sir Isaac Hayward, was displeased
with Mr Braddock, who, having been unable to reverse the rents
decision, continued to snipe at the leadership during the life of that
council. Mr Braddock's local party re-adopted him for the 1961 LCC
election, but the executive of the London Labour Party refused to
endorse his candidature, and the local party was forced to seek an
alternative candidate.

Mr Braddock's example was soon followed by that of Mr Hugh
Jenkins, who wrote to the Press about certain decisions of the LCC
Labour group on three occasions between his election in April
1958, and January 1959. Mr Jenkins' position was discussed at meetings of the LLP executive, but it was decided to deal with the
matter informally within the Labour group at County Hall rather
than follow the course adopted in the case of Mr Braddock. But on
20 January 1959 the leadership tabled a motion at the caucus meeting to the effect that:

'No member shall write to the Press or make public statements
either orally or in writing, attacking the policy of the party or

the decisions of the group. In the event of any violation of this rule, a member or members render themselves liable to the Whip being withdrawn.'[37]

The motion was passed, but it is significant that thirty members voted against it. The following years saw a campaign against the leadership waged by and on behalf of the dissentients within the pages of *Tribune*, the left-wing Labour weekly. *Tribune* adopted the cause of free expression in Labour groups on local authorities, arguing that 'discussion and argument on vital matters is the life-blood of the Labour movement', and running both major front-page features and series of critical articles.[38]

The latter years of the LCC – and especially the Council of 1958–61 during most of the life of which the threat of dissolution was not yet apparent – were a time of constant criticism. The *Spectator's* Westminster correspondent described the County Hall regime as 'a tyranny so absolute and so unquestionable that it would have appalled Caligula'.[39] Whilst Earl Attlee himself is reported to have referred to the Hayward regime in private conversation as 'the nearest approach to a totalitarian state in Western Europe'.[40] The Sunday *Observer* also joined in the fray, but the nature of caucus rule became a major issue within the party only after the 1959 General Election defeat. At the Labour Party annual conference of that year, Mr Gaitskell launched his attack upon the out-dated nature of the party constitution which also was to become the opening round of his battle for political survival.[41] Mr Gaitskell had familiarized himself with the politics of Labour groups in the larger cities,[42] and launched his attack with the greatest possible diplomacy: 'Another handicap, I'm afraid, was the unpopularity of certain Labour local councils. I do not want to be too critical. I am sure the public often does not know the true circumstances. But the criticisms are too widespread to be ignored, whether it be of the attitude to council house tenants or of excessively rigid standing orders, or more generally of apparently arbitrary and intolerant behaviour . . .'[43]

The impact of Mr Gaitskell's speech was considerable, in that it gave heart to the liberals within the party, at London County Hall and elsewhere. Francis Williams, writing in *Forward* two months later called attention to 'the apparent readiness of far too many members of the Labour group on the [London County] Council to let a small-minded bureaucratic tyranny have its head'. With such encouragement, a liberalizing caucus formed around the person of Mrs Evelyn Denington, drawing much of its total support of thirty or so members from the relatively young, newly elected councillors. What united them was little more than an aversion to what they saw as nepotism, and a claim to genuine contestation of the elections to the offices of the Labour group. There were little or no policy differences between them and the leadership,[44] rather they sought a fundamental change in the *style* of Labour's rule at County Hall.

Sir Isaac Hayward, as Leader of the Council, could not be effectively opposed in the elections for no figure of sufficient status could be induced to stand against him, so great were the Leader's powers of patronage. But the 'young Turks' (as one of their number has since described them)[45] chose to test their strength by opposing Sir Isaac's chief lieutenant, Mrs Freda Corbet, in the elections to the post of Whip. In 1959 Mrs Corbet maintained her position, but in the following year Mr William Fiske was persuaded to stand against her; supported by the 'young Turks', but otherwise having little connection with them, Mr Fiske defeated Mrs Corbet. This was however a limited victory, for Mr Fiske could not be persuaded to challenge Sir Isaac; nor, following his election, was there any marked change in the disciplinary style of the leadership. Immediately after the 1961 LCC election the officers of the Labour Group saw fit to remind their colleagues of the customary channels of decision-making at County Hall, in a circular letter signed by Mr Fiske:

'Whilst on many committees, Party issues frequently do not arise and decisions can be made irrespective of Party, it has been found desirable to follow any lead the chairman feels he should give . . . it is most important to attend the (Party) Group meetings . . . and a member missing this meeting should not come into committee and raise difficulties for the chairman. Every member is entitled to ask up to two questions at a council meeting, but he should ascertain beforehand from the chairman of the committee that the question would not cause the Party any embarrassment.'

The letter also pointed out that any member wishing to raise a policy issue should first give notice in writing to the policy committee, before which he would have to appear in the first instance.[46]

To sum up, this then was the situation within the London Labour movement when the Government published the White Paper setting out details of its Greater London proposals. Labour had ruled the

LCC since 1934 without a break. Herbert Morrison had since gone on to higher things, but the system bore the stamp of his leadership. The seventy-six-year-old Sir Isaac had gained his ascendancy under Morrison's leadership, having been Whip for most of the time between 1932 and 1947. Despite his very great age, he was able to maintain 'a tight hold' (in the words of one of his former committee chairmen)[47] over the decisions of the group, which were taken in the small policy committee, and submitted for approval to the group. Powerful sanctions could be applied against those such as the unfortunate Mr Braddock, who dissented from these decisions, and who failed to recant. The London Labour Party organization served as a power base and headquarters for the leadership, and produced its own monthly journal, *London News*.[48] The dual bequest of Morrison, the personalized, democratic-centralist style of LCC leadership and the LLP as a semi-autonomous 'party within a party' – seemed unassailable. But both were in the event to become casualties of the London government reorganization.

Crisis: the creation of a Greater London Council. The Herbert Report and the Government's White Paper of the following year were anathema to the London Labour Party. The opposition of Sir Isaac, of the London Labour Party executive and of all elements in London Labour politics (short of the Fulham and Hackney CLPs and Mr Hugh Jenkins) was based upon fears that the party would lose control in London following an extension of the boundaries into Conservative suburbia.[49] The 1961 annual report of the London Labour Party expressed it thus: 'We shall continue to oppose any scheme, conceived in political spite, which is calculated to smash up not only the London and Middlesex County Councils but the Metropolitan and Municipal Boroughs in Greater London as well . . .' The opposition of the LLP did not wane during 1962, but the annual report for that year indicated a dawning awareness of the probability of reorganization in its call for 'a sound municipal policy' and 'high calibre candidates' to be chosen in the event of the new authorities being set up. By 1963 the practical problems of fighting these elections were coming to occupy the attention of the LLP executive and secretariat. Then in the spring of that year *London News* ran a front page article by Ben Parkin MP[50] in which he forecast a Labour win of the GLC and of the subsequent General Election following Dr Hill's exclusion of part of suburban London from the GLC area. Such considerations necessarily lessened the degree of opposition to the reform scheme among some elements within the LLP.

Sir Isaac and his closer associates meanwhile steadfastly refused to countenance the possibility of the break-up of the LCC. In this they were not being unduly purblind, for there seemed a good chance of a Labour government coming to power sufficiently early to scrap the reorganization plan. Mr Michael Stewart had given firm public undertakings, notably at a press conference in March 1963, and during the debate on the third reading of the London Government Bill in the following month, that were the General Election to take place in 1963, Labour was pledged to repeal the Bill; were it to take place after the elections to the new authorities, the transfer of services would be halted.[51] This was sufficient to satisfy Sir Isaac, at a time when GLC candidates were being adopted in September 1963 for it seemed to assure the future of the LCC. But two events were to upset his expectations. The first was the unexpected resignation of Mr Macmillan at the Blackpool Conference of the Conservative Party in October 1963, and the accession of Lord Home to the office of Prime Minister, which ruled out the practicability of an election before the spring of 1964; the second was Labour's very substantial victory in the GLC election of April 1964, which left the new Prime Minister with little alternative but to hold on to office in the hope that the fortunes of his party would improve. On 9 March, with just one month to go before the GLC election, the *Daily Telegraph* revealed that Sir Alec (as he now was) would settle the timing of the General Election on the basis of the GLC results; and during the night of 9 April, as the results came in, a statement was released from 10 Downing Street to the effect that the General Election would be held 'not before the Autumn'. The consequence of this was that Labour was returned to office too late in the day to repeal the 1963 Act or indeed to preserve the LCC services in their existing form – this was made clear to the LLP at a meeting with Mr George Brown, immediately after the General Election.[52]

By this time, however, Sir Isaac and some of the other veterans had either excluded themselves or had been excluded from a future in London local government. Sir Isaac and Mrs Corbet, for instance, had not thought it worthwhile standing for the GLC; during early months of the transitional year when the LCC and the GLC co-existed, and both Councils met at County Hall, a degree of tension grew up between the Greater London Councillors, who felt their own future to be assured, and the London County Councillors, who saw them as 'interlopers', whose brief office would inevitably be terminated by the eventual election of a Labour government. Yet Sir Isaac may well have had his private doubts about this view, for in January 1964 he had indicated that he might accept an aldermanic seat on the GLC, were one to be offered. In the event Sir Isaac put himself forward as a last-minute Aldermanic candidate, and – astonishingly, in view of his decades of service and long years of leadership – received only a derisory handful of votes.

But it was not only the disbelief in the future of the GLC on the part of the 'old guard' that brought about the demise of their closed, autocratic form of rule. Greater London with its much wider boundaries was inevitably different in character from the LCC area. The London Labour Party, despite its formal responsibility for political organization in Middlesex, had never achieved there the degree of *rapport* or the sense of identity that had existed between London County Hall and LLP headquarters,[53] whilst the Kent, Surrey, and Essex parties had no connection with their counterparts in Inner London. The County Hall politicians had therefore to accept a sharing of power with others, some of whom had scant respect for the London 'old guard', for what they saw as their authoritarianism, for their identification of 'London' with the LCC[54] or for the bad reputation of Labour caucus rule which some claimed the County Hall regime had done much to foster. To unite the Labour representatives of the whole of Greater London, a leader was necessary who could command the loyalties not only of the elected Labour members, who might number anything between forty and seventy, but the one hundred candidates fighting the election in all parts of Greater London. There was a recognized need for a central, focal figure which the publicity machine could present as the potential 'Mr Greater London'. Only Mr William Fiske could fill this role and, indeed, he had played the leading part in the 1961 LCC election campaign. A key man in Inner London with experience on the LCC, he was not however closely identified with the 'old regime', due to his stand against Mrs Corbet; and he was acceptable to the Outer London candidates, not least for his adoption in a suburban seat for the election. Mr Fiske was chosen unanimously as principal candidate at a pre-election meeting of candidates, and therefore as Leader-elect of the Labour group. When the elated Labour group met for the first time at County Hall, there took place an open election for the leading positions. Mr Fiske (Havering) was elected Leader, Mr Higgins (Ealing) his deputy, Mrs Serota (Lambeth) the chief Whip, and 'backbenchers' were included upon the policy committee of the Group for the first time. A new era had begun in London Labour politics.[55]

The future of the London Labour Party.[56] The creation of the GLC posed an organizational problem for the London parties. The Labour party was slow to face it if only because of the deep-rooted incredulity with which many within the party regarded the Government's plan. This accounts for the temporary, somewhat makeshift arrangements that were put into effect for the 1964 GLC elections; the difficulty of quietly disposing of the London Labour Party accounts for the persistence of these arrangements through to 1968. The elections were organized by the Greater London Co-ordinating Committee of the Labour Party, the conception of which was due to a Working Party under Arthur Skeffington MP, and which was brought into being largely by the efforts of Miss Sara Barker, the National Agent. Originally titled 'Temporary Co-ordinating Committee', this *ad hoc* body had in the first place twenty-seven members. In May 1965 the National Executive Committee reconstituted the Co-ordinating Committee in more permanent form with forty-eight members, having postponed consideration of the regional structure for Greater London. After protests from the LLP the Trade Union representation was increased with the eventual outcome that the Co-ordinating Committee under the chairmanship of Mr Mellish became a more significant body both financially and politically, coming noticeably to eclipse the London Labour Party.

The passing of the London Government Act raised the question of the future of the London Labour Party in acute form, and the creation of the centrally-conceived Co-ordinating Committee did not bode well for the future. In an attempt to avert dissolution the LLP appointed a committee on party structure in November 1963. This included many of the leading figures in London politics, all of them members of the LLP Executive, and was chaired by Mr Mellish, who was to play an ambiguous role in the organizational changes of the next five years. The outcome of this committee's deliberations was a statement of the LLP's hopes for survival, which were summarized in the following points:

a. the London Labour Party would be obsolete in April 1965 – but the functions it exercised would need to be continued by some other political machinery.

b. a Regional Council as existed in the provinces would be inappropriate and inadequately financed to undertake the necessary work.

c. a new Labour Party organization needed to be established to co-ordinate regional local government and local affairs. This should be quite independent of the National Executive for electoral, consultative and administrative purposes.

d. this should also provide a political forum for the co-ordination and formulation of municipal policy.

e. there would need to be a secretariat and staff not directly concerned with organization as such.

f. an annual conference should be held in order to link Labour groups and party, union and co-operative organizations.[57]

The executive of the LLP decided to press the case made out by this committee. In March 1965 the secretary, Mr Peter Robshaw, circulated a memorandum calling for an expansion of the LLP onto a Greater London scale.[58] It claimed that the peripheral constituencies in Kent, Essex, Hertfordshire and Surrey were isolated, having little contact with their County Federations of Labour Parties, these being based as they were upon the outlying rural constituencies. London Labour Party offered these peripheral areas a welcome; its past electoral and organizational successes, its financial resources and its enormous reservoirs of goodwill gave it a prior claim to the absorption of the suburban parties. It claimed that the London Labour Party already co-ordinated one-eighth of the Parliamentary constituencies in the country; the marginal additions would involve no risk of creating an alternative centre of power. In terms of constitutional changes, the proposals involved no more than the offering of seats on the LLP executive to the representatives of the peripheral parties, and thus entailed little disturbance. The funds and functions of the Co-ordinating Committee would be absorbed, and that body cease to exist.

It was indeed vital to the LLP that any proposals should involve the winding up of the Co-ordinating Committee, for, although chaired by the ubiquitous Mr Mellish, the LLP's own chairman, it posed a serious long-term threat to the existence of the London Labour

Party. The electoral and organizational responsibilities of which the Robshaw memorandum had made so much had been 'hived off' to the Co-ordinating Committee, and this loss began an erosion of the substantial reserves of financial and moral support that the LLP claimed to possess. The longer the isue remained in doubt, the weaker the relative position of the LLP would become. If power was seen to lie with the Co-ordinating Committee, and not with the London Labour Party, the Trade Union branches – the dominant element in London politics since 1914 – were likely to act accordingly, and divert both their funds and their attention.

The problem was even more acute in the case of the constituency parties, who had for some time been less than generous in their allocation of goodwill and funds to the LLP. Some Constituency Labour Parties had long been in the habit of understating their total membership in order to avoid paying their full dues to the London Labour Party; but at this point in time some of these parties gave notice that they might disaffiliate. The LLP strove mightily to maintain its support; but the problem of party structure in Greater London was not to be settled for an inordinately long time, and the Party declined into a dangerous limbo. This was perhaps symbolized by the resignation in August 1965 of Peter Robshaw, the LLP's secretary, who left to take up a post with the Civic Trust.[59] It was obvious to many people close to the LLP that no action could be taken on the question of its expansion or replacement in the aftermath of the 1964 General Election; the insubstantial majority of the Labour Government presaged an early General Election, and Transport House could ill afford a running battle with the LLP if the election machinery was likely to be mobilized at an unforseeable moment.

The National Executive Committee of the Labour Party responded to repeated pressure by the LLP from May 1965 onward for an early decision with vague promises of future consideration of the problem. Mr Robshaw's successor as secretary to the LLP, Mr Percy Bell, a GLC councillor sitting for Newham, saw his job as inevitably presiding over the last days of the London Labour Party but fought hard on its behalf. His letter to Mr Len Williams, General Secretary of the Labour Party in June 1966 hinted at the degree of the LLP's debilitation: '. . . uncertainty as to the future pattern of organization in Greater London was already creating difficulties and . . . the difficulties would be cumulative in effect...'[60] The NEC was 'urgently requested to deal with the matter immediately . . .' Mr Bell also submitted a memorandum to the NEC in support of the LLP's case. This latter makes it clear that the issue worrying the constituency parties and the Unions was that of the separation of political responsibility (the Co-ordinating Committee) and fund-raising (the LLP). Although Trade Union support appeared at this stage to be holding, one Constituency Labour Party had reduced its affiliation fee from £200 to £20 while 'other parties have needed considerable persuasion to maintain their support'.[61]

It was against this background that the National Executive took action in the autumn of 1966. At the annual conference of the Party in that year Mr Crossman revealed in the course of debate that the NEC had decided to establish a Committee of Inquiry into the Party Constitution and organization. This was appointed on 26 October and comprised Mr William Simpson,[62] Miss Alice Bacon,[63] Mr Crossman himself,[64] Mr Joe Gormley,[65] and Mr Jack Jones,[66] all members of the National Executive. The Committee decided to present an interim report to the Party Conference of the following year, and chose therefore to select only those problems of apparent urgency for initial consideration. One of these was the creation of a regional organization for Greater London, and the Committee set up a working party under Miss Bacon, to which nine of the more prominent London politicians and organizers (including, again, Mr Mellish) were recruited.

At the end of November 1966 the London Labour Party were invited to submit their views to the working party. Their submission was one of the already familiar case for what was in effect a Greater London Labour Party. Whatever organizational, financial or other benefits might accrue to the Party from this proposal, neither the LLP nor the National Executive had any illusions as to the real nature of the issue at stake: whether or not the national Party could permit the existence, within its structure, of a semi-autonomous regional party. The choice was between pluralism and centralism, and the view of the LLP executive was understandably in favour of pluralism. Their view on the issue had been established at the executive meeting of November 1964, when Mr Walter Fliess, a Middlesex member, had argued in a circulated memorandum (which was not favourably received) against the creation of a Greater London Labour Party on the grounds that '. . . one may feel that a political body, consisting of one-sixth of the whole British Labour Party may in time emerge as a quasi-separate party, with all the embarrassment this may entail for the movement . . .[67] The LLP in the memorandum drafted for Miss Bacon's working party argued, however, that goodwill could continue to obviate any possibility of 'embarrassment' – for 'Throughout the history of the London Labour Party there has never been any occasion on which a dispute has arisen between the NEC and the LLP on a policy issue'.[68]

Whilst it is true that Transport House had not received a direct challenge to their authority from the LLP,[69] there were nevertheless those who felt that the executive, although 'safe' and 'reliable' politically, was vulnerable to pressures from the Left, which could then be used as bargaining points in the dealings with the National Executive. Inquiries into constituency party affairs in London, to cite one example, were the province of the LLP executive, and some members of the NEC were jealous of this limited autonomy. The London Labour Party deputation which appeared before the Bacon working party came away from the hearing feeling that their arguments had been sympathetically heard; but Miss Bacon warned the deputation that the Committee itself, having no connection with London politics, might take a less sanguine view of the possibility of future intra-party conflicts.

In fact, the report of the working party to the Committee on Party organization did not favour the LLP position in any degree. The Committee presented their interim report, the London sections embodying the virtually unchanged findings of the working party, to the Annual Conference at Scarborough in October 1967. The first parts of the section dealing with the London region were devoted to an analysis of the problems, but one crucial passage argued the rejection of the LLP's case on the inevitable grounds of centralism: 'A Greater London Regional Council could become a social and political force of great magnitude but it must in no way be afforded functions that could be a challenge to the National Executive Committee, or to the authority of the Party Conference'.[70]

The National Executive approved the report in principle, and thus sealed the fate of the London Labour Party. It was to be replaced by a Regional Council modelled upon, and with the limited powers of, the Regional Councils in the provinces with a General-Secretary and Assistant Secretary appointed by the NEC. But just as great exertions had been necessary to bring the London Labour Party into existence, so were they necessary to disband it. The supporters of the

LLP were to fight on every possible ground to retain an organization that would not be submissive to the party managers in Smith Square. Some of this opposition, notably that led by Mr Hugh Jenkins MP and vocal in the columns of *Tribune* was undoubtedly ideological.

The first confrontation came at a consultative conference called by the National Executive for 27 January 1968, at which the NEC proposals were to be presented. A sharp conflict of view arose, the LLP executive reiterating their case for a Greater London Labour Party, to have it firmly rejected by Mr Simpson, the Chairman of the Committee on Party Organization, and a declared 'firm advocate' of a conventional Regional Council. Mr Mellish stressed the centralist argument, that it was imperative that the National Executive 'should have no fears that Greater London should wish to usurp the authority of the NEC'.[71] The LLP executive were highly displeased with the January conference, and discussed it at great length at their meeting on 1 February, finally resolving that 'in view of the unsatisfactory nature of the consultative conference, the Executive Committee of the London Labour Party insists that before binding decisions are made about the future of the Labour Party's structure in Greater London those affected by those decisions should have the opportunity to vote on them and make alternative proposals at a delegate conference'.[72]

The delegate conference demanded by the LLP was fixed for the last Sunday in March, but was called by the NEC at short notice, a fact which gave cause for some complaints that insufficient time was available for the organizations concerned to appoint delegates and mandate them. A draft constitution was to be presented to the Conference, but no amendments to it were to be considered. *Tribune* complained of the 'highly centralized' nature of the proposed structure, alleged that Transport House was riding rough-shod over the feelings of the London membership, and advised delegates to maintain the LLP in existence until a satisfactory constitution could be discussed at a more representative conference.[73] The London Labour Party then played their last and strongest card at their Annual Conference on the Saturday, instructing the executive not to hand over the party's assets to the Regional Council until a more satisfactory arrangement for the adoption of the new constitution could be evolved.[74] The result of this was that the Regional Council, though duly established the following day, was unable to function whilst the LLP remained in existence for six months longer than had been anticipated, until a final conference in September 1968, at which amendments to the Greater London Regional Council constitution were permitted.

A major issue upon which the opponents of the NEC scheme took their stand was the right of the new London organization to appoint its own secretariat, a function that was to be reserved to the National Executive. As one of their number, Illtyd Harrington, Chairman of the new Press and Public Relations sub-committee of the GLRC explained: 'We in the London Labour Party guard our rights strenuously and many people feel we should carry on the tradition started by Herbert Morrison and appoint our own secretary. This dispute is one of the reasons we have not got the new Council running already.'[75] The finally agreed position was that the GLRC Secretary should be appointed by the NEC *in consultation* with the Executive of the GLRC.[76]

The other issue to arouse the feelings of the LLP supporters was that of the loss of a voice at Party Conference that would follow from the creation of a Regional Council. Immediately after the Conference on 30 March of the LLP, a special meeting of the executive was convened, to adopt a resolution calling for autonomy for the new Greater London organization, and in mid-June the executive resolved to submit a resolution to the national Annual Conference to provide for Regional Councils to separately affiliate to the Labour Party, and thus gain the right to conference representation.

Although the 'democrats' (as *Tribune* somewhat righteously labelled them) gained some concessions as a result of their opposition, the constitutional superiority of the NEC was bound to ensure eventual victory for the Party managers. That only ninety-seven delegates attended the September 1968 conference of the Regional Council is a measure of the dawning apathy and resignation among the opposition, although left-wing candidates had gained considerable successes in the first elections to the executive in March. In September the London Labour Party finally agreed to go out of existence and hand over its premises and assets, the final executive meeting calling upon 'all associated with the London Labour Party to give their unstinting service and support to the Greater London Regional Council of the Labour Party and to continue their work for Socialism in Greater London'.[77]

(d) Conclusions

This survey of the change within the London political organizations between 1962 and 1968 suggests certain similarities and contrasts, but in considering them one cannot draw simple causal connections between the coming into force of the London Government Act and the changes within the parties. Developments were taking place in both parties, and the important question is to decide to what extent the London Government reforms accelerated, retarded or deflected the pace and course of change.

Because of the incongruities of the new administrative areas and those of the London parties, some recasting of boundaries and responsibilities necessarily had to take place. But in neither case did this take the form of a straightforward extension of the areas covered by the LLP and the LMS. Instead, entirely new bodies were created, as offshoots of the national party organizations. The quasi-independent London Labour Party, and the formally independent London Municipal Society were both victims of the re-organization. This loss of independence means something different in each party. The political impotence of Regional Councils of the Labour Party has been commented upon, and the change in the status of the London organization in the LLP has been severe. But the political realities of the new situation do not quite match the formal position. The LLP, though possessing great freedom to act, and even challenge the national party, never made significant use of the freedom – it remained mere potential, and would not have been realized so long as those on the Left of the party did not gain control. On the other hand, the Greater London Regional Council is now seen to be treated with a certain deference by the National Executive, and almost certainly has slightly greater real independence than the equivalent bodies in the provinces, as the episode of the appointment of Mr Len Sims' successor indicates. This may well be true only in instances of organizational, rather than policy matters, however. Even if controlled by the left, the Regional Council would have no power to embarrass the National Executive.

The Area Organizations of the Conservative Party, on the other hand, are by no means as limited. Indeed, McKenzie asserts that they 'have a certain scope for action that is denied their Labour counterparts'. Ostrogorski wrote in 1902 that the Areas had not succeeded in developing an 'autonomous life' of their own,[78] and McKenzie generally concurs that this judgment is still valid. But there is no guarantee that the representative Area Council, normally effectively controlled by the centrally-appointed Area Agent, would not revolt against the national leadership. Only the unlikelihood of a clear-cut regional issue emerging relegates this to the realm of mere possibility.[79]

The overall result of the 1963 Act can still be described, on balance, as a loss of the pluralistic nature of London party politics. It is imperative that other factors contributing to this process should, however, be recognized. In the case of the London Labour Party the administrative untidiness of the Regional and London boundaries, which resulted in dual responsibilities for the county of Middlesex, constituted a case for the eventual re-allocation of powers. The report of the party's Committee of Inquiry suggests that this would have happened whether the London reforms had taken place or not.[80]

Since Morrison's day the London Labour Party had declined in influence and prestige, and was in a weaker position to fight any attempt by the National Executive to disband it. But would the NEC ever have made a successful frontal attack upon the London Labour Party? The evidence of the preceding pages, which give some indication of the vigour with which the London Labour Party fought – even in circumstances where simple logic dictated its obsolescence – suggests that no such assault would have been forthcoming. True, the National Executive had the power to dis-affiliate the London Labour Party, but there would have been strong reasons for leaving this sleeping dog to lie in peace. The London Government Act provided an unprecedented opportunity to tidy up the organization of the Labour Party in the London region, and draw tighter the reins of power. Without the reforms, this would not have been an easy task; even with the wider boundaries as an excuse, it took almost seven years from the Government's White Paper to the creation of a single Greater London-wide Labour organization.

The position in the Conservative Party was rather different. The London Municipal Society had no constitutional tie to the Party – it was a quite independent association of individuals and local bodies of a broadly anti-socialist nature, and yet it succumbed to the re-organization. Whatever reservations about the wisdom of permitting the Society to be wound up may now be held by London Conservatives, at the time they were quite willing to sink its identity in a

C

conventional Area Organization of the Conservative Party. Why was this so? First there was the questionable assumption that the Greater London Council would remain under Conservative control in normal times. This was an argument that ignored both the possibility of abnormal times and the need for research and briefing at the London Borough level.

Secondly – and this was surely decisive – there was the new financial dependence of the Society upon Central Office. Whatever the Society's notional autonomy, this made it a *de facto* organ of the Conservative Party. It would have been less difficult constitutionally for the London Municipal Society to extend its boundaries to the entire Greater London area than would have been the case for the London Labour Party, but had it attempted to do so, it would have been obliged to relinquish its electoral responsibilities. That the Society was not permitted so to expand is explicable only in terms of the rivalries within the Conservative Party.

In 1937 great efforts had been made by the Party under Neville Chamberlain to assert some degree of control over the Society; its autonomy was seen as an inconvenience. How much more so it must have seemed in the early 1960s, when a new spirit of rationalization occupied Central Office, a period during which the Conservative Research Department also succumbed to a greater degree of Central Office control. This raises the question of whether or not the London Municipal Society would have survived regardless of the London government reforms. Opinions differ on this: some Conservatives astonishingly saw the Society almost as an institutional embodiment of the personality of Miss Gelli, its last Secretary, and assumed that it would be dissolved as a matter of course upon Miss Gelli's retirement. Others saw it as still fundamentally strong and vigorous. It is this latter impression which is supported by a reading of the Minutes of the Executive committee; although these are notably reticent on matters that might allow one to judge the financial strength of the Society in its latter years.

These two contrasted views illustrate the mystery that surrounds the Society, and point up the degree to which it was a creature of a past age, struggling manfully to survive in a harsher world of professional party politics. Whereas the London Labour Party strikes the outside observer as a twentieth-century institution, a product of the rising Labour movement, and a reflection of its new sense of identity (of which Herbert Morrison was perhaps a symbol), the

London Municipal Society on the other hand seems to belong with an earlier generation, a generation pre-occupied with prudence in municipal administration, a product of the late Victorian age; (and the name that might be symbolically associated with it is that of Joe Chamberlain).[81]

Having attempted an explanation of the factors behind the dissolution of these two institutions, and the centralized character of their successors, it remains to try to pick out the reasons why the transition was made so much more readily and so much more successfully in the Conservative Party than in the Labour Party. The new Conservative machinery was operating in October 1963; the new Labour Party Regional Council did not begin work until five years later. Part of the explanation lies in the acceptance, by London Conservatives, of the Greater London reorganization as a certainty from the moment the White Paper was published. And as the preceding pages suggest, some of the top party officials had this expectation before the White Paper appeared.

People in each party – as is often the wont of professional politicians – acted on the assumption that their party would be successful in the General Election that was to come. Given the polarization of party attitudes to the reforms, the effect of this was to discourage Labour Party officials from preparing permanent arrangements for the new Greater London area, whereas Conservative Party officials were encouraged to do so.

This explains why the Conservatives were quicker off the mark in adjusting to the new area. It does not explain why the Labour Party dragged its feet beyond the point when it became clear that the GLC area was more than just a bad dream. The reasons for this delay seem to be twofold. First, there was the real political problem of dealing with a vociferous opposition. For many people in Labour politics 'Transport House' symbolizes the bureaucratic centralism of the party, and these people would be counted upon to fight long and hard against any accretion of power in Smith Square. What emerged was a largely ideological campaign founded upon the sentiments of pluralism and populism, and led by Mr Hugh Jenkins, the left-wing Member of Parliament for Putney. But time was on the side of the National Executive, and delay was therefore initially chosen as a weapon with which to defeat the opposition. This was a successful strategy, but was pursued at the cost of setting up a temporary, makeshift party structure. Few believe that Labour's *ad hoc* arrangements that lasted from 1963 to 1968 helped the party in the GLC election of 1967; indeed, it may be that the Conservative Party gained some electoral benefit from its superior London organization.

The second reason is to be found in the different responsibilities of the London Labour Party and the London Municipal Society. The latter played no part in Parliamentary elections, whereas the former had overall responsibility for this electioneering in the London area. The London Labour Party could not therefore be too severely offended in the aftermath of the 1964 General Election, for it was clear that a second General Election was only months away, and the party's goodwill could be crucial to success. Not until March 1966 could the National Executive begin to consider permanent machinery for Greater London. And because of the Labour Party's practice of consideration by Annual Conference of constitutional changes, it took a further year – the interval between the 1966 and 1967 Party Conferences to consider the possible changes and report. By the time the report of the Committee of Inquiry was available, the 1967 GLC election was a thing of the past.

The question remains of the part played by Mr Mellish in the course of events in the Labour Party during these years. As a staunch Morrisonian and Chairman of the LLP he was firmly opposed to the 1963 Act, and lobbied for its repeal immediately after the Labour Government came to power in October 1964. This, for reasons explained above, was not practicable, but Mr Mellish led the LLP in resisting attempts to create a more permanent organization for Greater London until it became clear that repeal was out of the question. However, once faced with the reality of the new local government boundaries, he began work in his capacity as Chairman of the Co-ordinating Committee on what was – as he saw it – the long and delicate diplomatic process of winning the support of the peripheral parties prior to their absorption into a Greater London political structure. At the same time he still had to pacify the truculent Inner London parties and the LLP executive who were wary of a possible 'sellout' to Transport House.

In a sense Mr Mellish possessed a 'Greater London' perspective, for he was anxious that the suburban parties should not be dominated by the old LCC area, which had been the fate of the Middlesex representatives on the LLP executive. He was not of course an LCC member, a factor which may well have shaped his view of the problem, and fitted him for the role he was to play. He was (as a result of his chairmanship of the Greater London Co-ordinating Committee) acutely aware that the latent tensions between the different elements in the party throughout Greater London might well manifest themselves if the pace of adjustment was forced. And the gradualism that this consideration dictated was also a useful tactic for the wearing down and defeat of the opposition, which wished to maintain the LLP in being for largely ideological reasons.

It was his centralist outlook that inclined him against the concept of a Greater London Labour Party in the discussions on Miss Bacon's Working Party. He also saw the financial advantages of a Regional Council, whereby the expenses would be met in part by Transport House. Mr Mellish was not interested in internal Party democracy but in winning elections, and the support of Transport House seemed to offer better electoral prospects than 'going it alone'. The precise institutional arrangements that finally emerged and the timing of their establishment owed something to the influence of Mr Mellish.

The process of change within both parties is now only of historical interest. The political consequences of the changes are however of contemporary significance. The London Government reforms created circumstances in which the national parties could hope to acquire an unprecedented degree of control over the existing organizations. The new institutions are now more centralized, and in terms of personnel more professional than their immediate forebears. But it should be remembered that the tide was flowing strongly in the direction of centralization and professionalization within the Conservative Party, less strongly but perhaps still discernibly in the case of the Labour Party. The evaluation of this change is a judgement to be left to the reader.

3
The Electoral System and Elections

(a) *The Electoral System under the 1963 Act*

The Commission's proposals. The Herbert Commission concerned themselves very little with the electoral arrangements of their Greater London proposals. They discussed the constitution of the proposed Council for Greater London and argued that the Council should be directly rather than indirectly elected, on the grounds that it had important functions to perform 'upon which Londoners are entitled to make their opinions felt through the medium of the ballot'.[1] They thought that the Council should have about 100 members, one to be elected for the area of each Parliamentary constituency, that they should serve for three years, and that as in County Councils there should be triennial elections at which the whole Council would be elected. They made no recommendation about aldermen.

They were even less specific about the composition of the proposed Borough Councils. It was implied that borough councillors would be elected *en bloc* at triennial elections as was the case in the metropolitan boroughs, but unlike borough councillors elsewhere. Apart from this respect 'The constitution of the Greater London Boroughs will be the same as that of Municipal Boroughs'.[2] The Commission also hoped that 'it would be possible, and, if so, we think it very desirable that elections for the [Greater London] Council and elections for the Greater London Borough Councils should be held simultaneously'.[3] This, it was felt, would simplify arrangements, but would also 'help Londoners to understand better the types of Council for which they are voting and the functions of the Councils to which the candidates are seeking election'. Why the Commission thought this to be so is not clear, and their view would appear to rest upon *a priori* assumptions about voting behaviour rather than upon the findings of any attitude survey.

The 1963 London Government Act gave statutory effect to these proposals, though not all were to take effect in time for the initial round of elections in 1964. The net result was to provide for the whole of Greater London something resembling the electoral arrangements that had formerly obtained within the County of London, namely triennial elections for both first and second tiers, and electoral divisions which followed Parliamentary, rather than local authority boundaries for the election of members to the upper tier.[4] One point of difference was however the provision for elections ultimately to be held on the same day for both the local and the 'county' authority, which contrasts with the year's interval that had, since 1949, separated the LCC and metropolitan borough elections. This provision was however repealed before it came into effect and a one-year interval interposed between GLC and London Borough Council elections by the London Government Act 1967, which is discussed below.[5]

A further point of difference was that the electoral areas for the GLC elections were not to be the Parliamentary constituencies until such time as these latter had had their boundaries re-drawn to correspond with those of the London boroughs, and any necessary re-distribution of seats effected. Until then – and it was hoped that the temporary arrangements need only apply to the first round of elections – the boroughs themselves would serve as the electoral areas, with each borough returning two, three or four members according to the size of its population, and subject to the total allocation of councillors adding up to the 100 Herbert had envisaged.[6]

The basis of GLC elections: delay in implementing the Act. The Conservative and Liberal Parties in London were both anxious that the boroughs should not remain as the electoral areas for the GLC for any longer than was necessary. The Conservatives, after winning control of the GLC in 1967 resolved on 13 May 1969 to request the Home Secretary to 'receive a deputation on the Council's request for the early introduction of single member constituencies for Greater London Council elections'. The London Liberal Party issued a statement setting out their views on the electoral system on 1 December 1966. The Council of the London Liberal Party subsequently passed a resolution exhorting Liberal members of Parliament 'to exert maximum pressure to ensure that the 1970 GLC election be fought on the basis of single-member constituencies', at its meeting on 2 December 1968. Mr Eric Lubbock, its President, waged a vigorous campaign to persuade the Government to ask the Boundary Commission to produce an interim report for London as soon as the local inquiries were complete.[7] The Labour Party had divided feelings at least until the Boundary Commission report was published. The existing arrangements had served Labour well in 1964, when they were able to win 64 of the 100 GLC seats. This arose from the distorting effect of the multi-member basis of election which may provide the winning party with an exaggerated majority, it being necessary to win only a bare plurality of votes in any borough to return two, three, or four members to County Hall. This system therefore exacerbates the already existing defects of the 'simple majority' system.[8]

The numerical effects of using the London boroughs as electoral areas for GLC elections have been quite remarkable. The London Borough of Enfield contains four Parliamentary constituencies, the Labour and Conservative Parties each having held two of these seats at Westminster in every General Election from 1950 to 1970. In the 1964 London borough elections the Labour Party narrowly gained control of Enfield, winning 31 seats to the Conservatives' 29. But in the GLC election just one month previously the Conservatives had polled 2,000 more votes than Labour out of a total of 83,000 cast in the entire borough – and thus won all three of the GLC seats allocated to Enfield.

In the 1964 GLC election as a whole, Labour won 46 per cent of the votes cast and 64 per cent of the seats. The Conservatives won 41·7 per cent of the votes, but only 36 per cent of the seats. The Liberal party won 10·2 per cent of the votes and no seats. In 1967 it was the Conservatives who were to benefit from the distortions of the system: winning 54·3 per cent of the votes cast, they took 82 per cent of the seats; Labour, whose share of the vote had dropped to 34·6 per cent won only 18 per cent of the seats; the Liberal vote remained at around 10 per cent but again the party won no seats. Apart from the exaggeration of electoral success inherent in the multi-member system,[9] it has also deprived the Liberal Party of at least one GLC seat in each year; they gained 24·1 per cent of the votes cast at Bromley in 1964, 25·4 per cent in 1967, the greater part of their strength in both years being within the Orpington constituency. Before looking at the reasons for the persistence of the borough basis of election, the proposed revision of boundaries themselves need to be examined. The existing system provided quite a number of anomalies: borough boundaries divided existing Parliamentary constituencies in the case of 16 of the 32 boroughs;

and as a result of London's population decline the metropolis had the benefit of considerable over-representation at Westminster. The theoretical entitlement of 92·96 – or 93 – seats was exceeded by six seats if one counts only those constituencies that were wholly contained within Greater London, whilst parts of a further five constituencies fell within the Greater London boundary.[10]

The Parliamentary Boundary Commission's report of 1969 recommended the reduction in the number of Parliamentary constituencies in London to 93. This, of course, would serve not only to reduce the representation at Westminster, but also to reduce the number of Greater London Councillors from 100 (the number envisaged by Herbert and specified by the 1963 Act) to 93. Several boroughs would consequently suffer a diminution of their representation at County Hall. Of those that had returned Labour Councillors to the GLC in 1964 Brent, Hammersmith, Hounslow, Lewisham and Southwark would lose one representative; of those won by the Conservatives, Westminster and Kensington and Chelsea would each lose one member.

The Commission in their workings had to satisfy two desiderata: that the total number of constituencies allocated to Greater London as a whole should not exceed the area's theoretical entitlement as judged by its share of the national electorate; and that in all cases the Parliamentary and local boundaries should coincide. These two desiderata could only be met at a cost of some inequality between the boroughs. Southwark, for example, had a theoretical entitlement of 3·56 seats and was awarded three only, whilst Bromley, with a theoretical entitlement of 3·54 was awarded four seats, the discrepancy being justified by possible future population changes in the two boroughs.[11]

When the English Commission's report was published in June, 1969 the Home Secretary, Mr Callaghan, rejected its recommendations pending the forthcoming local government reorganization. But the Boundary Commissions are constituted under the House of Commons (Re-distribution of Seats) Act 1949 (as amended 1958), and this requires that 'as soon as may be after a Boundary Commission have submitted a report to the Secretary of State under this Act, he shall lay the Report before Parliament together . . . with the draft of an Order in Council for giving effect, with or without modifications, to the recommendations contained in the Report'.[12] With this in mind the Opposition moved a motion in the House of

Commons calling upon the Secretaries of State for Home and Scottish Affairs to 'implement in full, and without further delay, the recommendations of the Parliamentary Boundary Commissions'.[13]

In the debate which followed, Mr Callaghan argued the undesirability of introducing boundary changes at a time when local government boundaries were facing a wholesale re-drawing, the Redcliffe-Maud Commission having reported the previous week, on 11 June.[14] But, he continued, 'the Royal Commission on Local Government did not cover London. London's boundaries were fixed some years ago. The objections that I have been raising in relation to England do not therefore apply to London . . . I propose, therefore, to introduce legislation which will give effect to the recommendations of the Boundary Commission for constituencies in Greater London.[15] Such a course would also enable the 1970 GLC elections to be held on a basis of single-member constituencies.

Some local Labour parties were embarrassed by this move. Only five days before the Home Secretary's announcement the four prospective candidates for the Brent electoral division had been chosen by the Borough Labour Party. These selections not only had to be scrapped, and candidates chosen on a constituency basis, but, because the Commission only allocated *three* seats to Brent, one of the four candidates would not be re-adopted.[16] Critics of the Government's action pointed to the expectation that the implementation of the full proposals would cost the Labour Party between six and twenty seats in the following General Election, but that the changes proposed in London might well work in Labour's favour. The day following the debate on the Opposition motion in the House of Commons, a Bill[17] was published to give effect to the London changes. The intention was to carry this into law before the summer recess, in order to give the Home Secretary sufficient time to make the necessary Order under Schedule 2 of the 1963 Act to make each of the new constituencies the electoral area for the GLC election, and to give the political parties time to choose their candidates and prepare their campaigns.[18]

The Bill passed through all its stages in the Commons only to be rejected by Conservative and Liberal Peers in the Lords. The Opposition Leader, Lord Carrington, subsequently made it clear that were the Bill to be returned to the Upper House for reconsideration it would again be defeated.[19] Meanwhile, the Queen's Bench Divisional Court had granted a petitioner, Mr Ross McWhirter, leave to apply for an Order requiring the Home Secretary to comply with his statutory obligation under the 1949 Act to lay the Commissions' proposals before Parliament. The hearing was arranged for 20 October.[20] The Government, therefore, was now facing a possible clash with the Judiciary, in addition to a certain clash with the House of Lords.

On 14 October the Commons defeated the Lords' amendments, but accepted an amendment moved by the Home Secretary promising to introduce the Boundary Commission proposals by 31 March 1972.[21] During this debate the Home Secretary announced that were the Bill to be rejected a second time by the Lords, it would then be withdrawn, and Orders embodying all of the Commissions' proposals – involving some 410 constituencies – would then be laid before Parliament. The Government, however, would ask Parliament to defeat them.[22] This in fact occurred, and the Commissions' recommendations were rejected in their entirety.

As a result of these Parliamentary machinations the 1970 GLC elections took place once again on the temporary, multi-member basis (and the Labour Party at Brent re-adopted their original four candidates). In the election the Labour Party gained 40 per cent of the votes cast and won 35 per cent of the seats. The Conservatives gained 51 per cent of the votes cast and won 65 per cent of the seats. The Liberal share of the vote fell considerably to 5·5 per cent and as before, the party won no seats.

The basis of borough elections.   One important question which arose at an early stage in the preparation for the introduction of the new system was the need to divide the areas of the new boroughs into wards for the purpose of holding borough elections. This, and the related question of the numbers of members which each council should have within the maximum of 60 laid down in the Act, was naturally of intense interest to the political parties involved in contesting elections. These subjects were the occasion of much debate in many of the joint committees of the merging authorities, established to make the necessary arrangements for the transition to the new system.[23] In almost every case where warding proposals were made by the joint committees to the Home Office they were challenged by one or more of the political parties, with the result that no less than 29 of the 32 borough schemes were the subject of an Inquiry held under Home Office Commissioners.

The warding arrangements finally approved in time for the 1964 borough council elections were noteworthy for their great variety. In seven boroughs[24] a uniform system applied throughout the whole borough area, generally of three member wards, although in one case (Enfield) there are 30 two-member wards. In three other boroughs (Barking, Merton and Sutton) warding was almost uniform, in each case only a single ward having a different number of councillors from the rest.[25] In the remaining boroughs there was a mixture of different-sized wards. In Kensington and Chelsea, the extreme case, there were 4 two-member wards, 3 of three members, 3 of four members, 4 of six members and 1 with seven.

The warding arrangements of the London boroughs thus differed strikingly from those found in boroughs elsewhere in the country where a uniform system of three-member wards is general;[26] again, there is here a perpetuation of the system formerly obtaining in the metropolitan boroughs, which were, before 1964, the only boroughs to have triennial elections, and tended similarly to have non-uniform warding systems. Elsewhere in England and Wales the mechanics of the voting procedure for annual elections (where one-third of the councillors retire) necessitate three-, or sometimes six-member wards. There are also, however, special features of the London system arising from the 1963 Act itself. This specified the desirable aims for re-drawing ward boundaries as being, first, to secure equality in the number of electors per councillor and, secondly, to fix easily identifiable boundaries with regard to local ties.[27]

A novel provision in the Act also enabled the Home Secretary to take the initiative in considering changes in wards or the number of councillors. This was a significant departure from the normal procedure of the Local Government Act, 1933,[28] under which the Home Secretary's power is limited to considering representations made to him. The effect of the Home Secretary's strengthened powers in London did not manifest itself during 1962–3, but in subsequent years. Five boroughs carried out re-warding schemes between the borough elections of 1964 and 1968. In Westminster these affected only part of the area of the borough, whilst in Barnet and Richmond upon Thames the opportunity was taken to introduce a uniform ward-scheme.

From the voter's point of view one-member wards have the disadvantage of limiting the extent of the representation of the ward's views upon the committees of the Council, and only three London boroughs now have single-member wards. On the other hand, if the number of councillors representing any one ward is too large the voter's difficulties are increased by having to select from a lengthy list of names when he gets to the polling booth. There are five boroughs having wards with at least five members, so that in these wards electors may well have fifteen or more names to choose from, although the introduction of party labels to local government ballot-papers has done something to reduce the danger of unintentional cross-voting.

The procedures for securing approximate equality of representation can be considered a test of the 'fairness' of an electoral system. Here it is noteworthy that for both the GLC and the boroughs the initiative no longer lies entirely with the local authorities. In the case of the GLC the linking of the electoral divisions with Parliamentary constituencies ensures that changes in population are not ignored, and that the question of re-distribution of seats comes under periodic review, although a repetition of the political dispute over the 1969 Report of the Parliamentary Boundary Commissions remains a possibility. In the case of the borough wards, the Home Secretary's extended powers should help to ensure that disparities in the numbers of electors represented by each councillor do not grow too wide, as has happened elsewhere.[29]

(b) *The London Government Act, 1967*

One change introduced since the 1963 Act has had the effect of bringing the arrangements closer to those in the former County of London. The intention of the 1963 Act was that GLC and borough elections should ultimately be held not only in the same year but on the same day.[30] This departure from tradition was in accordance with the Herbert Commission's view referred to earlier. But in 1966 the Labour Government introduced a Bill to postpone the London borough elections which were due in 1967 to 1968 and subsequently to hold these elections every three years, ie in 1971, 1974, etc. The effect of the London Government Act, 1967, has thus been to make London Borough elections take place one year later than GLC elections, exactly as metropolitan borough elections used to be held one year later than LCC elections.[31]

The electoral provisions of the 1963 Act had not been uncontroversial, having come under attack during the committee stages of the Bill in both Houses. In the Commons, Mr Robert Jenkins, the Conservative member for Dulwich, pressed his amendment, providing for one year's interval between GLC and LBC elections, to an unsuccessful division.[32] The arguments marshalled at this point stressed the greater convenience to the political parties of separate election years, and were repeated in the House of Lords by Lord Morrison of Lambeth, when Labour Peers introduced a similarly unsuccessful amendment.

There the matter rested so far as Parliament was concerned until after the election of the Labour government. On 28 April 1966 the Home Secretary, Mr Roy Jenkins, in a written reply to the member for Orpington, Mr Lubbock, referred to representations in favour of a year's interval that had been made by the London Boroughs Committee, but pointed out that 'to make the change requires legislation, which confronts us with considerable difficulties from the point of view of time'.[33] On 18 July, however, in reply to a further written question from Mr Lubbock, Mr Jenkins announced the Government's intention to introduce legislation 'in the autumn'.[34] Pressed in the House by Mr Boyd Carpenter to give his reasons for the measure, on 4 August, Mr Jenkins cited the administrative difficulties of holding two sets of elections on the same day, and the low polls that would arise from two consecutive sets of elections in the same year.[35]

Accordingly, a measure to postpone the 1967 borough elections to 1968, and the subsequent elections to three-year intervals thereafter was laid before Parliament on 20 October. The Bill, presented by the Home Secretary as 'a simple, and, I would have hoped, relatively non-controversial Bill'[36] provoked a political storm.

The Parliamentary battle.   The opposition's case was ostensibly based upon a suspicion that the Government were postponing the election in order to maintain the Labour Party in power in certain outer London boroughs for an extra year in order to make further progress with the introduction of comprehensive education schemes. The Government's case for making a change in the electoral arrangements was as follows: the 1963 Act stipulated that the elections to both bodies should be held upon the same day, starting with the first set of elections being held on a constituency basis for the GLC. When this came into effect the burden upon returning officers would be considerable. Before that time, the effect of the one-month interval between the two sets of elections would be to confuse the electors, and reduce the level of interest in the borough Council elections.

In the debate on the Second Reading,[37] the Home Secretary revealed that a working party of London borough town clerks (who act as Returning Officers in London BC elections) had considered the practicability of GLC and borough elections being held on the same day and had found against it. A report had been prepared for the London Boroughs Association, the Special Sub-Committee of which had accepted it unanimously. The individual boroughs had been consulted, and, irrespective of political colour, had divided 27 to 5 in favour of postponement. Although the issue would not arise until 1970 at the earliest, the Government saw no reason to delay.

These were the issues then which were to be elaborated and reiterated in the various stages of the Bill. Mr Iain Macleod, the member for Enfield, an outer borough in which the comprehensive schools issue was hotly contested, was the leading critic of the change. On the occasion of the Second Reading debate Mr Macleod alleged that Mr Jenkins had originally overstated the degree of unanimity among the London Boroughs (a point which the Home Secretary was quick to concede) and pointed out that when Mr Lubbock had first raised the question of postponement in April, the Home Secretary had claimed a higher priority for other legislation. Mr Macleod contended that the Bill was the end product of pressure from the London Labour Party. This allegation was to be heard again during its passage, and was at various times categorically denied by Government spokesmen.

When the Bill reached the House of Lords it met very stiff opposition. Lord Brooke of Cumnor, formerly Mr Henry Brooke, opened for the opposition during the Second Reading.[38] He accepted the arguments in favour of staggered elections but argued that this should be done by postponing the borough elections due in 1970, not those due in 1967. His amendment at Committee Stage to achieve this was carried by 86 votes to 61, the Government thus being defeated. The Commons discussed the Lords amendments on 13 February 1967 in a very noisy and protracted debate, disagreeing with them by 232 to 151 votes.[39] The Bill was returned to the Lords for further consideration, but the Conservative peers led by Lord Carrington decided that this Bill did not provide a suitable occasion for a Lords challenge to the Commons and the Bill was finally passed.

The background to the Bill: events outside Parliament.   Much of the Parliamentary discussion of the background to the Bill revolved around the views of the London Boroughs Association, and of the Returning Officers, and there was some dispute as to whether or not the Government had correctly reported the degree of support for the measure among the local authorities themselves. The question of the attitudes of the political parties was shunned by Government spokesmen, although the convenience of the party organizations had been the major plank in Labour's arguments during 1963 in favour of staggered elections.[40] That the Labour Government did not, in 1966–7 resurrect the 'party convenience' argument is not surprising, for to have done so would have been to concede part of the Conservative case – that the initiative for the Bill came from the London Labour Party. Instead the views of the Councils themselves, and of the Town Clerks' working party were quoted at some length.

It is necessary at this point to explore in some detail the moves made by the various participants in this story outside Parliament, in order to try to ascertain as far as possible the motives for introducing a Bill which forced an unnecessary confrontation with the House of Lords, and delayed important social measures due to be introduced in that session by the Home Secretary.

The London Liberal Party and Mr Lubbock, their President, were the most vociferous campaigners for postponement. They had supported the Labour pressure in Parliament during the passage of the London Government Bill in 1963, and issued Press statements calling for postponements of the 1967 borough elections in July 1966 and welcoming the Government's action in November and December 1966, concentrating in each case upon the alleged confusion caused to electors.

On 20 September 1965 the Greater London Co-ordinating Committee of the Labour Party discussed the question of holding the GLC and borough elections in separate years, and agreed that this was preferable for organizational reasons.[41] On 8 November it was agreed to seek an amendment to the 1963 Act, by making the Committee's views known in 'appropriate quarters' – including the National Executive.[42] It was reported to the Co-ordinating Committee at its meeting on 31 January 1966 that the NEC of the Labour Party was in favour of amendment to the Act.[43] Labour councillors believing that there was unlikely to be political opposition to the proposal in the LBC then proposed to table a motion at the earliest possible meeting of the LBC.

From this point the pressure for amendment followed two separate channels. Members of the National Executive of the Labour Party used their influence nationally upon the Government; and the London boroughs entered a period of protracted discussion of the merits of the proposal. This latter process is discussed somewhat fully, as it is important to establish just what degree of support for amendment did exist in London. We return later to the activities at the national level.

On 27 January 1966, Mr A. G. Dawtry, the Town Clerk of Westminster and Honorary Secretary for the London Boroughs Committee (as it then was) received a letter from the London borough of Greenwich asking that the question of a postponement of the 1967 borough elections to 1968 be discussed at the next meeting of the LBCs Special Sub-Committee, due to take place the following day.[44] At this meeting the Special Sub-Committee, which included members from both parties, agreed unanimously that a postponement was desirable, but instructed that the views of the constituent Councils be sought.[45] Subject to the majority agreement, the subcommittee authorized the submission of the proposal to the Home Secretary.[46] It is certain that the Conservative members of the special Sub-Committee, which deals with constitutional affairs concerning the LBC, were unaware that the proposal had been a topic of discussion at the January meeting of the LBC Labour Group policy committee.

The Honorary Secretary then sent a circular letter to the member Councils, asking for their views by the end of February. On 9 February, the Special Sub-Committee's action was put before the full meeting of the LBC for confirmation. Some Conservative dissent was expressed but no motion was made, and the report was accepted.[47] The views of the member Councils were presented to the LBC meeting on 2 April.

The Home Secretary, it will be remembered, had originally claimed a 27–5 majority among the boroughs in favour of the change, although in the Second Reading debate he admitted that not all of the 27 were in favour of postponing the 1967 borough elections. The picture was in fact exceedingly complex – 17 Labour and 3 Conservative boroughs (later reduced to 2 by the withdrawal of Sutton) were in favour of the change; 3 Labour and 4 Conservative boroughs favoured a change but not for the 1967 borough elections; 3 Conservative boroughs were not in favour of elections in different years; to this last group should be added Havering for the perhaps characteristic reason that the Council were unable to arrive at a decision on the merits of the issue, and, not having submitted a written reply to the LBC, indicated informally that they preferred to be recorded as not being in favour of change.

If, therefore, the line-up of borough councils is reckoned on a basis of those definitely in favour of the immediate postponement and those not so in favour, we find the totals to be not 27 to 5, but 19 to 13. This majority of only six boroughs was perhaps a less substantial basis for the Bill than had originally been indicated by the Home Secretary when the decision to make the change was announced. Nor were the party divisions quite as blurred as was claimed, if we consider only the opinions on the question of an immediate postponement. Of the eleven Conservative-controlled boroughs, nine chose to oppose the suggestion. Of the twenty Labour-controlled boroughs, only three were against it.

The Greater London Council also made representations to the Home Secretary. Under the 1963 Act the date of the Greater London Council election was to be fixed by the Home Secretary after consultation with the GLC, although this provision was in fact repealed by the 1967 Act, which gave to the GLC the power, formerly possessed by the LCC, to determine the date for its own election of councillors. Along with the date for the 1967 GLC election first raised in a Home Office letter to the GLC on 31 December 1965, that Council also chose to consider the question of staggered elections. The General Purposes Committee at its meeting on 24 January reported that the holding of elections on the same day 'could be confusing to electors and that it would place too much work, with risk of serious mishap, on the staffs at the Town Halls. We also take the view that a separate year, as well as a separate day should be provided for the two elections in order to keep alive a sense of civic awareness . . .'[48] The report, which went on to call for legislation to implement the change, was adopted by the Council on 8 February – one day previous to the matter coming before the LBC – and forwarded to the Home Secretary. The Home Office replied, noting the Council's views and a copy of the entire correspondence was forwarded to the LBC as a matter of course. It was subsequently claimed in the House by Miss Alice Bacon, Minister of State at the

Home Office, who was quite clearly in error, that it was this correspondence which first brought the question of an election postponement to the notice of the LBC.[49]

The alleged views of the Town Clerks – Returning Officers for the borough elections – were frequently cited. What however were these views? The first discussions on the electoral arrangements had taken place in a working party established by the Home Office as early as the summer of 1962. Some borough Town Clerks on this working party pressed for a year's interval on various grounds, the strongest of which were possible confusion to the electors and a closer following of the existing LCC – metropolitan borough arrangements.

The Home Office view was originally that the GLC elections should be held in the spring, to be followed by the borough elections in the autumn of the same year, as had been the practice in London before the Second World War. But the borough representatives argued that this would not give adequate time after the first elections for the borough councils to prepare for the appointed day. The Home Office accepted this view, and were then willing to give effect to the system suggested by the Herbert Commission, namely that the elections to both authorities be held on the same day. This could, of course, only take place after the revision of the electoral boundaries, and the election of GLC councillors on a constituency basis, which would not be until 1967 at the earliest.[50] The borough officers on the Home Office working party were by no means unanimously in favour of having a year's interval thereafter, and so the Bill, as drafted, embodied the Government's view that the Herbert Commission's proposals in this respect should be followed.

During the passage of the 1967 Act, however, these early discussions were not referred to, and the sole source quoted by Government spokesmen was the report prepared for the LBC by a Working Party of the Advisory Body of Town Clerks in February 1966.[51] The first point that emerges from a reading of the report is that it dealt only with the issue of simultaneous elections, and had therefore no bearing whatsoever upon the round of elections due to follow in 1967. Secondly, the Clerks 'confined (their) consideration to the administrative problems . . . and (have) not considered the matter from the point of view either of the political parties or the electorate . . .'[52] Thirdly, the problems posed, though severe, were related to the existing formulation of the Election Rules. The report concluded that 'the difficulties referred to . . . may not be insuperable, but they show that the burden which combined elections *under the Local Election Rules as they stand at present* would place on Returning Officers is a very heavy one indeed and the resulting procedure may not be very satisfactory to the candidates or to the electorate'[53] (our emphasis).

This survey indicates that the arguments marshalled (with the exception of the debatable assumptions about electoral turnout) provided the Government with little justification for the introduction of a Bill at that point in the life of the Parliament. Indeed the question of the eventual necessity of a one-year interval was not considered in the light of the alternative course – a revision of the Local Election Rules for Greater London, to overcome the procedural problems of arranging simultaneous counts.[54] Such a course would have lightened to some degree the burden of the Returning Officers, preserved the principles of the Herbert Report, and avoided a confrontation with the opposition, as well as allowing other important social legislation to go forward. It remains to ask why the Bill was introduced.

The correct explanation would seem to be two-fold: at the departmental level the Home Secretary felt that the Bill had genuine merit, and could be made ready in the interval before more important Home Office measures were ready for presentation; at the Cabinet level certain Ministers were amenable to political pressure to provide Parliamentary time for the Bill. When the decision to bring forward a bill was made in May 1966 there were few indications that it would become a party issue. The representations from the LBC were equivocal, and the warning signs – particularly the party alignments among the London boroughs on the issue of immediate postponement – were easily misread. And at this stage – before the economic crisis of July 1966 – an approach to the Shadow Home Secretary might well have guaranteed an unopposed passage to the Bill. Had it not been opposed so fiercely in Parliament, it would have taken an insignificant amount of Parliamentary time.

If the Home Secretary was unable to foresee the controversy the Bill would arouse, no more could the other members of the Government. Moreover, they were uninterested in the merits of the Bill. Given that the Home Secretary favoured it, it was necessary for the Prime Minister and the Leader of the House to grant it Parliamentary time. This would not have happened had they not been persuaded that there was political advantage in the measure.

The factor that induced the Government to give the Bill time was the convenience to the Labour Party. But it cannot be argued that the Government was inspired by a desire to maintain Labour in power in the outer boroughs for a further year. At the time when the Greater London Co-ordinating Committee of the Labour Party first began to take soundings the party organizers were speaking of the 'bright prospect' of retaining control of the GLC and of those boroughs won in 1964.[55] The minutes of the Co-ordinating Committee reveal no evidence of considerations of electoral advantage.

Rather, this was a question of *organizational* preference within the Labour Party.[56] In late 1965 and early 1966 constituency Labour Parties were pressing the Co-ordinating Committee to arrange for a postponement of the 1967 elections.[57] Party funds, and, in the former LCC area, familiarity with staggered elections are likely to have been the influential consideration here. A leading figure in the pressures applied to the Government was Sara Barker, National Agent of the Labour Party, and the Co-ordinating Committee subsequently awarded a vote of thanks to her for using her influence. In the spring of 1966 a meeting took place between Mr Mellish, Chairman of the London Labour Party and of the Greater London Co-ordinating Committee; Miss Barker; Mr Herbert Bowden, Lord President of the Council and Leader of the House of Commons; and the Prime Minister. The outcome was that the two Ministers agreed to the measure going forward, the occasion of the announcement being Mr Jenkins' reply to Mr Lubbock on 18 July. The political gain was negligible – a real or imagined easing of organizational and financial problems for local Labour parties in Inner London. The cost in terms of parliamentary time and in terms of political standing was undoubtedly great. It is unlikely that the Government overestimated the gain to local Labour parties, rather that Ministers were largely unconcerned about their problems. It does, however, seem certain that no one foresaw the intensity of the opposition to the postponement. The issue had passed through the London Boroughs Committee fairly quietly, and it was hoped that its passage through Parliament would be similarly trouble-free. In so far as the introduction of the Bill revealed a serious lack of foresight, it had this much in common with the Industrial Relations Bill, the Parliament (No. 2) Bill, and the House of Commons (Redistribution of Seats) (No. 2) Bill, all three of which were however eventually withdrawn.

GLC and London Borough Elections. As the previous sections of this chapter have indicated, the electoral arrangements which have so far operated in GLC and borough elections have not been quite the same as those which were intended by the London Government Act, 1963. Elections to the GLC on a Parliamentary constituency basis will not take place until 1973 and simultaneous elections to the GLC and the boroughs have been ruled out by the London Government Act, 1967. There have so far been three GLC elections and two for the London boroughs,[58] and, as has been pointed out, all except the first two in 1964 resulted in a predominantly Conservative complexion for London's local authorities.

What this has meant in terms of the effect on major policy issues will be discussed in individual chapters later in this volume which analyse some of the main local government services.[59] But the main effects of the elections in political terms may be summarized briefly. From 1964 to 1967 the GLC and 20 of the 32 boroughs were in Labour hands. In 1967 Labour lost so heavily in the GLC elections that if the pre-1965 system had been in operation the Conservatives would have gained control of the LCC, something which, despite a very close result in 1949, they had been unable to do from 1934 until the LCC ceased to exist. This is not simply a matter of historical irony. The effect of the 1967 GLC election was to give the Conservatives control of the Inner London Education Authority in spite of the fact that all but 3 of the 13 borough representatives came from Labour-controlled authorities. Thus one of the possible points of tension in the new system was avoided; in 1967 as in 1964 the same party controlled both the GLC and the ILEA.

The period 1968 to 1970 was a very low point for Labour in London. Following the 1968 London borough elections, only the four boroughs of Barking, Newham, Southwark and Tower Hamlets remained in Labour hands. The GLC, the ILEA and the remaining 28 boroughs were all Conservative-controlled. This situation has to be seen, moreover, in the context of the national political scene; throughout the period October 1964 to June 1970 Labour was the party in power at Westminster.

The implications of the political situation are obvious in such sensitive areas of policy as education and housing. In 1965, 11 of the 21 education authorities in Greater London were Labour-controlled and they included the ILEA, the largest education authority in the country; in 1968, only Newham and Barking remained in Labour hands. This was at a time when the Labour Government was attempting to implement a policy of introducing comprehensive secondary education in the face of a good deal of resistance from some Conservative Councils. Again, in 1965, of the 33 housing authorities in Greater London (excluding the City), 21 were Labour-controlled; in 1968 there were only 4. Yet the Labour Government's policy depended on a vigorous programme of house-building by the London authorities, and particularly on an effective contribution being made by the outer boroughs to help those in Inner London. Conservative councils, however, tend to be much less in favour of extensive building by local authorities than Labour councils. Finally, we may note that as a result of the 1970 GLC election Labour, by winning 30 of the 40 seats in Inner London, regained control of the ILEA, although they failed to capture the GLC itself; thus County Hall was to speak with two voices, Labour on education and Conservative on the rest.

It is difficult if not impossible to compare this post-1965 with the pre-1965 political situation; only education in Inner London offers any direct comparison and even here both the composition of the ILEA and its relationship with the GLC make it a somewhat different body from the LCC as an education authority. Much more to the point is the fact that the marked changes in political control of the London authorities in the period 1964–70 have less to do with the London reforms than with the national political fortunes of the major political parties. This is not to deny that the reforms had important consequences for the parties in adjusting to a wholly new situation, not least in adapting their party organizations, as has been described earlier. Furthermore, there were other consequential changes resulting from the very different political situation in Greater London following the 1963 Act; there were, for example, far fewer uncontested elections under the new system compared with the old, as is discussed below. Yet important and significant as these changes were or as were the variations in the proportion of electors voting under the old and new systems (also discussed below), the results of the elections, whatever their effect on the policies followed by the GLC and the boroughs, acquired their main significance in national political terms.

This national political significance is more clearly and obviously reflected in the three GLC elections of 1964, 1967 and 1970. Both the 1964 and 1970 GLC elections happened to occur within the last year of a Parliament, and much political commentary was devoted to interpreting the results in terms of the probable timing of a general election and who was likely to win it. Indeed, in 1964, as has been indicated above,[60] there seems to have been a direct and unprecedented connection between the GLC election results and the Prime Minister's announcement that a general election would not be held before the autumn. It is not difficult to see why there should be such interest in GLC elections. The electorate of $5\frac{1}{2}$ millions (in 1964) did indeed seem to offer the prospect of a 'mini-general election' as the commentators were fond of calling the GLC election – on the assumption made by practically everybody writing on the election that national rather than local issues would determine the results of the election. In this sense the contrast with the pre-1964 situation was simply that the GLC offered a much larger electoral sample than any of the previous authorities.

Although it might be true that the results of the election would mainly reflect the national political standing of the parties, this did not mean that specific London issues and the general conduct of the campaign were not important, not least in the effect they might have on whether the supporters of the different parties actually turned out to vote. This is obviously an important question in local elections where commonly only about one-third of the electorate vote, even though the first two elections for the GLC did produce a rather higher turn-out in some areas, as is discussed later.

The dual significance of the electoral campaign for the first GLC was emphasized with the publication in January 1964 of Labour's policy document with the slogan *Let's Go With Labour for a Really Greater London*. The document emphasized the importance of housing in London, but also argued that Labour would give top priority to dealing with London's traffic problems; at the same time, Mr Mellish, at a Press conference on its publication, said that Labour would fight the election as a minor general election.

The Conservatives also with their manifesto *Take a Greater London Pride* put the emphasis on housing to which they said they would give top priority; but they also said that they would set up a traffic department and urge the Government to provide better public transport.

The Liberals put particular emphasis on attacking London's traffic problems through the building of motorways and the improvement of public transport.

In spite of a good deal of activity by the two main parties, there was, as the campaign progressed, little sign that the electors either knew or cared very much about the issues. As one commentator put it: 'many candidates seem to be in two minds about whether their first duty is to inform the electors about the new authorities or to plunge into party polemics'.[61]

It is not surprising in the circumstances that, apart from the Liberals, the parties should have put their emphasis on traditional issues like housing rather than on the new powers of the GLC in planning and traffic. In the event Ivan Yates of the *Observer* probably summed up the campaign accurately enough when he said 'local is just what the campaigns have not been'.[62]

The 1967 campaign began on a different note. Early in 1966, Mr Desmond Plummer took over from Sir Percy Rugg, a veteran of LCC days, as Leader of the GLC Conservatives. In September of that year he launched *Time for Change in Greater London,* a pamphlet described as a personal statement, in which he attacked Labour's highways policy and criticized what he described as their wasteful administration. This was clearly designed as the first stage in an effort not only to present Conservative policies to the electorate but also to make a more personal impact by building up the image of the new Leader. Labour's manifesto published in January 1967 laid more stress than in 1964 on new and specific GLC responsibilities especially for highways and traffic, and also gave prominence to measures designed to make public transport easier (eg special bus lanes). The Conservative manifesto *Let's Get London Moving* went even further in claiming that the Conservatives would appoint a traffic commissioner and would examine new forms of public transport such as monorails. At the end of March Mr Plummer proposed a package deal with the boroughs under which the GLC would be willing to let the boroughs have more power in housing and planning in return for strengthened powers in the traffic and highways fields.

But although there was much more emphasis in 1967 than in 1964 on the newer powers which had been allocated to the GLC, the campaign seems to have had little effect on the voters. Labour, which by April 1967 was going through a period of declining popularity nationally, feared a low poll, apathy and abstentions on the part of its supporters. In the event the party lost heavily. This heavy defeat for the Labour Party, followed by the even more devastating results in the

London borough elections of 1968, was a painful shock to the party, particularly when such traditional 'safe' Labour areas as Hackney and Islington elected Conservative councils, and when other parts of Inner London which Labour would normally expect to hold such as Hammersmith and Lambeth, elected Conservative members to the GLC. The defeats occurred while the party was in process of considering its future organization in Greater London, as described earlier,[63] but the magnitude of the defeats made it unlikely that defeats in party organization were the major reason for them. There was much evidence that traditional supporters of the party refused to vote because of disillusionment with the policies followed by the Labour Government, but the size of the Conservative victory especially in the 1968 borough elections indicated too that some, normally Labour supporters, had voted Conservative. Furthermore the party found difficulty in recruiting for the election campaigns the degree of voluntary help which it needed. The 1967 GLC election and the 1968 borough elections were, indeed, a dismal period for the Labour party, and, correspondingly, a great boost to the morale of the Conservatives who now held a commanding position in London for the first time for many years.

By the time of the 1970 campaign the Greater London Development Plan had been published, and this helped to focus attention on one issue at least, the proposals for a new system of urban motorways. The campaign was notable for the intervention of a group of candidates campaigning in 27 of the 32 boroughs under the slogan 'Homes before Roads'[64] in opposition to the motorway proposals. In some ways the differences in policy between the two main parties were more obvious in this campaign than in the two earlier ones, mainly perhaps because each had now had a spell in office. Labour now promised to stop the motorway programme until there had been a full inquiry whereas the Conservatives saw the programme as an essential part of their balanced approach to transportation problems. On housing the Conservatives inclined increasingly to what they saw as encouraging people to help themselves by proposing, for example, to sell council houses and to encourage housing associations. Labour claimed that the Conservative aim was to destroy the public sector of housing.

Planning figured in all the manifestoes but as one commentator put it just before the election:

'. . . parties competing in the Greater London Council elections are finding it little easier now than in 1964 and 1967 to argue about strategic planning. Apart from motorways and their effects on homes and pockets, planning is unlikely to excite the elector.'[65]

On the other hand the recent take-over by the GLC of London Transport did figure in the campaign; in particular, Sir Reginald Goodwin, the Leader of the Labour Opposition claimed that fares would be bound to increase sharply as a result of Conservative policies, and particularly their refusal to subsidize London Transport from the rates. The Conservatives countered by asserting that support for London Transport had to be seen in the context of their policies as a whole which would benefit public transport (for example through traffic measures designed to discourage commuters).

As in earlier campaigns, however, the familiar comments of apathy and ignorance on the part of the electors were heard, and the results were awaited not so much for the verdict they would give on Labour and Conservative policies for London as for the light they would throw on the possibility of an early general election. We now examine in the light of these considerations the figures of electoral turn-out.

(c) *Voting in Elections*

Greater London Council. 44·2 per cent of the electorate voted in the GLC election of 1964, 41·1 per cent in 1967 and 35·2 per cent in 1970. These figures are roughly comparable with those for English county elections in the same years; GLC figures were rather higher especially in 1964 when the number of electors voting in English counties outside Greater London was 38·3 per cent. Moreover, GLC turn-out in 1964 and 1967 was a good deal higher than it was for county elections in the same area in 1958 and 1961 when the figures were 31·2 per cent and 36·2 per cent respectively.

These figures are misleading partly because the GLC is not a county and, more particularly, because although every seat has been contested in all three GLC elections so far, it is common for one-third or more of the seats in county elections to be uncontested. Figures of turn-out for county elections are based only on contested seats and there is no means of knowing whether, if all seats had been contested in these county elections, the total of those voting would have been higher or lower. Furthermore, it is a striking fact that within the GLC area before 1964 uncontested elections were not uncommon whereas they have now disappeared. The distribution of these uncontested elections is also significant. They had practically been eliminated from the London and Middlesex CC elections and from Essex CC elections for areas which were incorporated in Greater London except for Barking. They were common in Surrey especially in those areas of the county which were brought into the London Boroughs of Sutton and Croydon, but also in Merton and Richmond upon Thames; and they were also to be found though less commonly in those parts of Kent which became part of Greater London.

So far we have considered the GLC elections as a whole. But the GLC electorate is huge, 5·5 million in 1964 and 5·3 million in 1967; this compares with electorates of 7·7 million and 9·8 million in all contested divisions in English counties in the same years.[66] Under the system in operation in 1964, 1967 and 1970 the electorates of some GLC divisions were larger than some county electorates and certainly of the electorates in contested divisions of many counties.[67] It is therefore reasonable to examine in more detail electoral turn-out in the GLC divisions.

Table 3.1 compares voting percentages in the GLC elections of 1964, 1967 and 1970 with voting in contested county elections for the same areas in 1958 and 1961.[68] Two striking facts emerge from these figures. First, voting behaviour in areas formerly in the County of London (Inner London) was quite different from that in the areas formed from Middlesex, Essex, Kent and Surrey (Outer London). Secondly, in Outer London there was a very marked increase in turn-out in the 1964 GLC elections which was sustained in the 1967 elections; but in 1970 there was a sharp decrease to figures comparable with pre-1964 county elections. In order to indicate the position more clearly, the percentages voting have been averaged for the two county elections of 1958 and 1961 and again for the two GLC elections of 1964 and 1967, whereas 1970 figures are shown separately.

The consistency with which increases occurred throughout outer London in 1964 and 1967 is remarkable. Only Newham showed a decrease in turn-out, and here exact comparisons are not in any case possible since before 1964 the area had had only county borough elections. Of the other 19 borough areas in outer London all except Croydon (again formed mainly from a county borough) and Barking registered an increase in turn-out of at least 10 per cent,

Table 3.1
*Average turn-out in contested County elections 1958–61
compared with average turn-out in GLC elections 1964–7
and 1970*

	1958/61	1964/7	Numerical increase or decrease col. (2) over col. (1)	1970
	(1)	(2)	(3)	(4)
LCC/				
Inner London	33·9	36·1	2·2	32·7
Camden	35·9	43·7	7·8	34·3
Greenwich	42·3	45·0	2·7	40·3
Hackney	21·5	21·9	0·4	24·1
Hammersmith	37·8	42·8	5·0	45·1
Islington	23·3	24·9	1·6	24·6
Kensington &				
Chelsea	31·9	35·4	3·5	27·0
Lambeth	35·4	39·5	4·1	33·5
Lewisham	44·9	42·7	—2·2	40·4
Southwark	30·8	30·6	—0·2	28·6
Tower Hamlets	24·3	21·8	—2·5	23·1
Wandsworth	44·4	44·1	—0·3	37·2
Westminster	30·0	35·7	5·7	30·1
Middlesex/Middx.				
Boroughs*	36·2	49·2	13·0	37·3
Barnet	37·1	50·4	13·3	35·9
Brent	34·0	47·4	13·4	38·2
Ealing	39·6	50·0	10·4	39·5
Enfield	30·1	46·2	16·1	37·5
Haringey	31·3	41·8	10·5	33·5
Harrow	39·4	51·5	12·1	39·2
Hillingdon	35·4	52·2	16·8	44·1
Hounslow	43·5	53·8	10·3	42·9
Richmond upon				
Thames	35·3	51·9	16·6	40·7

* The boroughs listed are practically equivalent to the area of the old
Middlesex; they *exclude* Staines, Sunbury-on-Thames and Potters Bar but
include Barnet and East Barnet, Barnes and Richmond.

	1958/61	1964/7	Numerical increase or decrease col. (2) over col. (1)	1970
	(1)	(2)	(3)	(4)
Essex (part)/Essex Boroughs	27·4	42·3	14·8	33·3
Barking	22·9	29·6	6·7	25·8
Havering	31·1	50·3	19·2	35·6
Redbridge	28·3	45·0	16·7	33·9
Waltham Forest	25·5	39·9	14·4	35·9
Surrey (part)/Surrey Boroughs	31·9	48·9	17·0	37·0
Kingston upon Thames	30·2	47·6	17·4	37·5
Merton	33·9	51·8	17·9	39·4
Sutton	31·7	46·8	15·1	35·1
Kent (part)/Kent Boroughs	34·0	54·3	20·3	41·3
Bexley	36·3	55·8	19·5	43·8
Bromley	32·4	53·3	20·9	39·7
Croydon†	39·3	44·1	4·8	36·0
Newham†	30·1	27·7	−2·4	18·7
GLC	33·7	42·7	9·0	35·2
Inner London	33·9	36·1	2·2	32·7
Outer London	33·6	46·9	13·3	36·7

† Croydon figures for 1958 and 1961 combine voting figures in Croydon CB elections in those years with voting figures of Coulsdon and Purley divisions in Surrey CC elections; Newham figures for these years combine figures for county borough elections in East Ham and West Ham.

and Bromley registered practically 21 per cent. The most spectacular increases occurred in the areas formed from Essex, Surrey and Kent where, under the old system, there was also many uncontested elections.

Equally remarkable was the fall in turn-out which occurred in the 1970 GLC elections in these areas. In general, turn-out in Outer London in 1970 was a little higher than it had been for county areas in 1958 and 1961, more so in the ex-Essex, Kent and Surrey boroughs than in those formed out of Middlesex. Only in a few areas (eg Waltham Forest, Hillingdon) was the percentage voting in 1970 significantly higher than it had been under the old county system. Eight boroughs had a decrease in turn-out between 1967 and 1970 of 9 per cent or more and at Havering the fall was nearly 16 per cent (from 51·3 per cent to 35·6 per cent).

By contrast, voting patterns remained much more stable in the areas formed out of metropolitan boroughs. A relatively small increase in turn-out in 1964 and 1967 was followed by a slightly larger decrease in 1970. Only Camden and Westminster achieved more than a modest increase in 1964 and 1967 and neither came anywhere near the increases which were common in Outer London.[69] Whereas Inner and Outer London were very much out of step in 1964 and 1967, in 1970, as in 1958 and 1961, they seem once again to be marching together.

Whether subsequent GLC elections will follow the 1970 or the 1964 and 1967 pattern is impossible to predict. It is clear, however, that the level of turn-out in most Outer London areas in 1964 and 1967 was much higher than is characteristic of county elections in urban or suburban areas. Ten borough areas averaged a turn-out of 50 per cent or more in these two GLC elections in large divisions of 125,000 – 200,000 electors each, whereas such levels of voting are more normally associated with small divisions in rural areas. In 1964, for example, Westmorland, with a total electorate of 50,000 achieved a turn-out of 55·1 per cent in the 6 of its 46 electoral divisions which were contested. But before considering the possible reasons for the GLC situation we must also take into account what happened in the London Borough elections.

London Borough elections. Table 3.2 compares the average turn-out in the two London borough elections of 1964 and 1968 with turn-out in the same areas for borough (or urban district) elections under the old system.[70] The position is complicated by the fact that the metropolitan boroughs held triennial elections whereas all the other authorities[71] held annual elections for one-third of the Council.

Figures in the table are therefore based on the two elections of 1959 and 1962 for the Inner London boroughs and on the three years 1959, 1962 and 1963 for the remainder.

Even allowing for the difficulties in making comparisons, Table

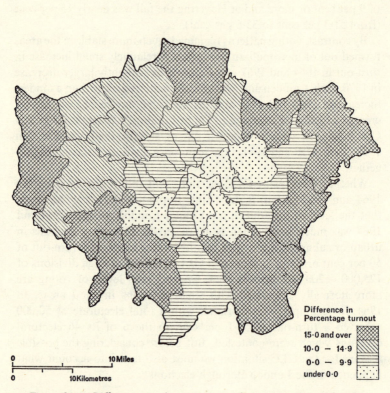

Figure 2(a). Differences in the percentage of electors voting under the new system compared with the old in GLC elections.

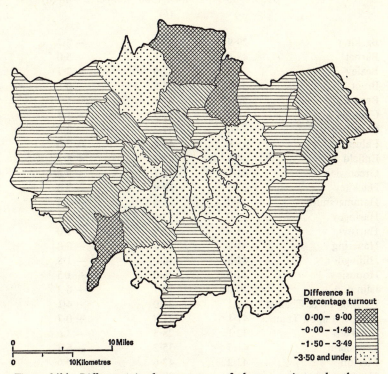

Figure 2(b). Differences in the percentage of electors voting under the new system compared with the old in London Borough elections.

Table 3.2

Average percentage of the electorate voting in contested elections: London boroughs 1964/1968 compared with previous authorities 1959/1962/1963

	1959/1962/1963 (a)	1964/1968 (b)	Increase or decrease (b) compared with (a)
Barking	29·9	27·2	− 2·7
Barnet	48·5	43·5	− 5·0
Bexley	50·3	47·6	− 2·7
Brent	37·6	36·8	− 0·8
Bromley	52·4*	47·1	− 5·3
Camden	36·9	35·2	− 1·7
Croydon	40·0	37·2	− 2·8
Ealing	45·8	43·8	− 2·0
Enfield	34·8	43·1	+ 8·3
Greenwich	41·1	37·1	− 4·0
Hackney	21·7	18·9	− 2·8
Hammersmith	36·0	35·6	− 0·4
Haringey	34·0	33·9	− 0·1
Harrow	45·8	42·8	− 3·0
Havering	41·7	40·9	− 0·8
Hillingdon	45·8	44·0	− 1·8
Hounslow	47·5	46·0	− 1·5
Islington	20·0	19·7	− 0·3
Kensington & Chelsea	29·1	25·0	− 4·1
Kingston upon Thames	41·1	41·8	+ 0·7
Lambeth	32·1	28·5	− 3·6
Lewisham	41·5	35·2	− 6·3
Merton	43·2	42·6	− 0·6
Newham	31·7	27·3	− 4·4
Redbridge	40·4	37·6	− 2·8
Richmond upon Thames	45·4	45·2	− 0·2
Southwark	28·3	23·3	− 5·0
Sutton	47·8	42·1	− 5·7
Tower Hamlets	20·5	15·6	− 4·9
Waltham Forest	33·0	35·2	+ 2·2
Wandsworth	41·7	34·7	− 7·0
Westminster	28·6	27·6	− 1·0

* Excluding Penge UD (elections held in 1958 and 1961).

3.2 is in sharp contrast with Table 3.1. In the London borough elections, unlike the GLC elections, there was no marked change in turn-out patterns dating from 1964. It is true that overall there has been a tendency for the percentage voting to decline. This is particularly true of the Inner London boroughs as the following table indicates:

Percentage of electorate voting

	1953	1956	1959	1962	1964	1968
Metropolitan boroughs	39·9	30·9	32·1	32·3		
Inner London boroughs					27·1	29·5

At the same time, there has been a decline in the number of uncontested elections, particularly in Outer London boroughs such as Richmond upon Thames and Barking.

For the majority of boroughs such changes as have occurred in the percentage of the electorate voting under the new system are thus comparatively small. In 13 of the 32 boroughs the average turn-out in 1964/1968 differed by 2 per cent or less from that for the boroughs under the previous system; in a further seven it differed by more than 2 per cent but not more than 3 per cent. Only in five boroughs did the change exceed 5 per cent and these included Enfield with an increase of 8·3 per cent and Wandsworth with a decrease of 7·0 per cent.

Voting in Greater London. Excluding Croydon and Newham, there are 30 London boroughs which experienced the transition from county and borough (or urban district) to GLC and LB elections. In all 30 boroughs as well as in Croydon and Newham, the percentage voting in the two GLC elections of 1964 and 1967 was higher, and often very much higher than in the two London borough elections of 1964 and 1968.

The remarkable fact is that for the Outer London boroughs this situation was quite the reverse of that under the previous system where voting in borough (or urban district) elections was invariably higher than in county elections.

This phenomenon of higher percentages voting in borough than in county elections is a general one throughout the country. It was not, however, found in the County of London where voting in LCC elections was generally higher than for metropolitan borough elections. Thus whereas all 18 of the outer boroughs listed above reversed previous behaviour, this was true of only 2 of the 12 inner boroughs and for one of these (Hackney) voting in metropolitan borough elections had only been fractionally higher than for the LCC.

On the evidence, therefore, of the first four elections to the new authorities, the level of voting activity in Greater London for the GLC and the London boroughs appeared to be analogous to that found in the old County of London and not to that which is more usually found in counties.

With the sharp decline in turn-out in the 1970 GLC elections in Outer London, however, the position appeared to be reverting to something approximating to the pre-1964 situation. Comparing the 1968 borough elections with the 1970 GLC elections, eleven of the twelve Inner London boroughs had a higher turn-out in the GLC than in the borough election; 18 of the 20 Outer boroughs, on the other hand, had a higher turn-out in the borough than in the GLC election.[72]

Studies of electoral behaviour in local elections in recent years have tended to link turn-out with the degree of marginality of voting areas in party political terms. Thus a study of certain borough elections in 1964 concluded:

'Although there may be other factors influencing turn-out in local elections in urban areas it is clear from the evidence of elections in the study areas that the overwhelming influence is the closeness of the party conflict'[73].

This conclusion was based on a careful ward-by-ward analysis of the elections in the selected boroughs. It is quite likely that a study of the London boroughs on the same scale would yield similar results. But we have here been concerned with looking at the London elections much more broadly, and an analysis appropriate to the small-scale electorate of wards might not necessarily be appropriate to the large electorates in the GLC electoral divisions, many of which are equal to or exceed the total electorate of large towns such as Bradford or Southampton. This is not to say that the closeness of the party conflict might not account for some or all of the differences *between* GLC electoral divisions in particular elections; it might, for example, account for higher turn-out in Bexley than in neighbouring Bromley. It is merely to draw attention to the fact that the interest of the London elections in the period 1964– 70 is in a quite different phenomenon, that is, the remarkable increase in turn-out in GLC elections in 1964 and 1967 in the Outer London electoral areas, compared with previous years.

Here, marginality is no help to us if we are comparing the pre- and post-1964 situations. Many of the Outer London boroughs are by their nature safe Conservative areas. In 1964 the Labour Party, in what was for them a good year, won only 10 of the 20 outer boroughs, and these included Barking and Newham which, unlike most of the outer boroughs, can be regarded as safe Labour seats. Yet boroughs like Barnet, Bromley and Harrow, which could not by any stretch of the imagination be regarded as marginal and which have returned Conservatives in all three GLC elections, all recorded greatly increased turn-out in comparison with the previous county elections.

I remember what Tories are, but what's GLC?

From the Observer, *Sunday, 16 April 1967.*

It would be tempting to seek an explanation in terms of increased public interest generated by the London reforms. If so, one has to explain why only the voters in Outer London showed this interest, whereas those who were formerly in the County of London scarcely seem to have been affected by the reforms. Moreover, it is necessary to be a little more precise about the kind of interest which the voters are likely to take.

The Herbert Commission, for example, seems to have hoped, though not very strongly, that a simplification of structure might lead to a better understanding of the types of Council and their functions and have some effect on voting habits. It might have been thought, for example, that the reason for higher voting figures in LCC than in metropolitan borough elections was that the LCC provided all the major local government services in the County of London, only housing being shared with the metropolitan boroughs. If so, one would have expected some shift in voting patterns under the new system since the London boroughs carry much greater responsibilities than the old metropolitan boroughs; in fact, in the first elections they attracted if anything less electoral interest, not more.

There is evidence, however, both from a survey carried out for the GLC[74] in 1966 and from more general studies[75] that voters are not very well aware of the different authorities and their responsibilities, and it would be unwise to place much emphasis on such knowledge as a factor in voting behaviour. This can be seen better in the outer boroughs. County authorities, which had major responsibilities under the old system, never attracted as many voters as the boroughs with their more limited powers. The GLC in 1964 and 1967 attracted more than the boroughs, in spite of the fact that the latter have concentrated in them those services which are thought to be of the most immediate and vital concern to the electorate, especially education (in Outer London), housing and the personal social services. It is true that the GLC has important powers but strategic planning, traffic management, the fire service and refuse disposal seem unlikely to be the main reason for voters to turn out in larger numbers than before.

We shall probably get nearer the truth by contrasting the effects of the reforms on Inner London and Outer London respectively. In a superficial sense, the 1963 Act merely modified the existing situation for those living in Inner London. In spite of the difference in powers, the GLC might be seen as simply a glorified LCC, and the boroughs as enlarged versions of the metropolitan boroughs; the new system therefore made as much (or as little) impact on the voters as the old. In Outer London, and particularly in areas formed from Essex and Kent, the situation was different. Instead of looking to County Hall in Chelmsford or Maidstone they now had to look to London's County Hall; from being urban fringes of what were still markedly rural counties, they became suburban fringes of a huge urban authority. Moreover, this change was one which had aroused a good deal of local controversy; many of the areas which had high turn-outs in

1964 and 1967 were ones which had resisted incorporation in Greater London, on the grounds that they were not part of London. It would be tempting to argue that the voters were showing that they preferred Greater London to county administration in 1964 and 1967; it is much more likely that the main effect of the controversies had been to bring home to some of the voters that a major change was occurring, and that it was this which was a factor in stimulating increased turn-out. Allied to this was a feeling of uncertainty about what the reforms would mean which may also have helped to increase turn-out initially.

These explanations suggest ways in which the reforms may have had some impact on the voters individually and caused them to cast their votes. But, as we have shown, the reforms also presented a considerable challenge to the political party organizations. The GLC is a political prize of considerable magnitude; at all three GLC elections so far both the main political parties have fielded a full total of 100 candidates for the 100 seats and this in marked contrast to many of the previous county elections, especially in Surrey, which the Labour Party were unable or unwilling to contest. Moreover, it was the Conservative Party which got off the mark first and was quickest to see the potentialities of organizing to meet the new situation. Concentration on getting out the Conservative vote might account for relatively high turn-outs in suburban areas where there was a considerable body of Conservative support. Moreover, determination to win might account for high turn-out in 1964; determination to oust Labour might show why this was sustained in 1967; and realization that much of Outer London was safe for the Conservatives might account for the falling-off in 1970.[76]

All these explanations leave a number of questions unanswered. Why, for example, did ex-Middlesex voters behave more like those from Essex, Kent and Surrey than like those from the old County of London? Middlesex was not a rural county with a remote county headquarters but much more like London. And if strong Conservative Party organization was a major factor in increased turn-out, why was it that in 1964 the Labour Party with its relatively weak organization was able to capture places like Havering and Hillingdon where the turn-out was around 50 per cent?

The data presented here suggest that we still know far too little about what actually influences people to vote in local elections when in normal circumstances only about 30–40 per cent do so. The GLC elections for a time attracted over a wide area of Outer London about half the electorate, but this same electorate showed no greater interest in the London borough elections than before. The explanation is likely to depend on a complex of factors influencing voting behaviour. But it is clear that the reforms under the 1963 Act were the occasion for a remarkable change in the level of voting activity. That this was only a temporary phenomenon is strongly suggested by what happened in the 1970 GLC election. If that election and the 1968 borough elections are to provide the pattern for the future, Greater London will conform to a pattern which has become well established over the country and which indeed was common in the area of Greater London itself before 1964. If, however, 1970 proves to be a temporary phenomenon and 1964 and 1967 are the model for future GLC elections, the London reforms will have led to changes which will be welcomed by those who see increased turn-out as a mark of the strength of local democracy.

4
Personal Health and Welfare Services

The Herbert Commission considered the health, welfare and children's services in one chapter, because as the Report says 'they are so closely connected with one another that we found it impracticable to write about them separately ... The three services also have a common element in that a considerable part of the work consists in giving services to families in their own homes or providing for people a substitute for home when for one reason or another that is needed'.[1]

Nevertheless this chapter will be dealing with the health and welfare services only. The children's service since the 1948 Children's Act had been the responsibility of a separate committee of the local authority and had its own chief officers; its development until 1970 with a few exceptions was distinct from that of the health and welfare departments. Even so it is necessary at the outset to draw attention to the fact that during the period from 1965 the whole question of the relationship of the local authority social services to one another has been the subject of discussion and controversy.

On the one hand there has been the opinion expressed in the Herbert Report that the Medical Officer of Health should lead the personal services; this view received a considerable stimulus with the drive to rationalize local government administration. In London a report of the London Boroughs Management Services Unit advocating a directorate of social services, as one of five major administrative divisions, to include health, welfare and children's departments reenforced this view. On the other hand a trend of opinion was developing which advocated the separation of the medical and social service functions in local government.

The Seebohm Committee was appointed in December 1965 to report on the Local Authority and Allied Personal Social Services. It reported in July 1968[2] to the effect that the local authority personal social services should be amalgamated into one department and separated from the personal and environmental health functions.

This recommendation was enacted in the Local Authority Social Services Act, 1970, which had the effect of amalgamating the existing welfare and children's departments as the new Social Service department and transferring to it the mental health, home help and day nursery services hitherto organized in the health departments. It had been intended to follow this change with a unification of the National Health Service which would have transferred almost all the remaining health functions from the local authorities to new Area Health Boards, leaving behind the public health inspectorate. The change of government in June 1970 has delayed this reorganization.

Thus the attitude of the Ministry of Housing and Local Government changed completely during the first five years of the London boroughs. When the new boroughs were planning their establishments in 1964, they were being encouraged to form joint health and welfare departments; already by 1968 they were being advised not to embark on more farreaching reorganization until the coming patterns of change were clearer. In spite of this advice a number of boroughs made changes in the administrative superstructure of their personal services. This uncertainty about the future and the strongly held differences of opinion about the shape it should take, cast something of a shadow (more obvious in some boroughs than others) over the hoped-for increase in co-operation and co-ordination between the departments of the new authorities.

Before going into more detail it is necessary to establish what is to be evaluated so far as the health and welfare services are concerned. Their future occupied quite an important place in the arguments up to 1963. In the first place, there was dismay at the prospect that services run by the London and Middlesex County Councils which had achieved high standards and some renown for pioneering work would be dismantled, but once the fact of change had been accepted, the argument crystallized around the question of the size of the second tier authorities. There was agreement that health and welfare should be in the charge of the same authority that controlled the children's service, housing and education, if this was possible.

The different opinions which led to such wide divergences – from the 52 boroughs of the Herbert Report to the 5 to 6 London Counties proposed in Plan B of the Greater London Group's evidence[3] – arose from the fact that quite different factors were held to be the most important for the health and welfare services. Thus the Herbert Report laid stress on the need for 'as small units as is consistent with efficiency' because this would facilitate working with people in their own homes and be more conducive to team work; 'the various services should operate wherever possible from one building and there should be a conscious effort to achieve good team work between the various officers concerned'.[4] The Commission saw the major part of the direct work as concerned with admissions to homes, and placed great reliance on joint use of establishments by several authorities and on ancillary services being provided by voluntary organizations.

Ruck in his book *London Government and the Welfare Services* published in 1963 produced evidence to show that the record of the County Boroughs for many key services was to be preferred to that of the counties; he concluded that they were a more suitable size for developing the health and welfare functions. He stressed particularly that the enormous proportions both of the problem of replacing old large institutions for the elderly by small modern homes,[5] and of the increasing numbers of the homeless in the Inner London area, was an inhibiting factor in itself when one authority had to tackle it.

Professor Donnison and his colleagues in proposing Plan B and in his occasional paper[6] paid more attention to the principles underlying the new developments in local authority services which were just beginning to emerge. These may be summarized as a growing emphasis on the family and on greatly increasing the provision of care in the community rather than in institutions; the growing interest in 'human relations' skills' and in preventive work; the increasing demand for training for those entering the health and welfare services. Following from these developments was the need for greatly increased co-ordination both within local authority systems and with outside services such as the mental hospitals, the probation services and the voluntary organizations.

Professor Donnison argued that only very large authorities would be capable of paying for new expensive services and training, and capable of innovation and of providing a wide range of services for the specialized needs of minority groups. Furthermore co-operation with the numerous outside bodies would be difficult if there were a multiplicity of authorities in Greater London and there would be big variations in standards of service which would be unacceptable in an area of 'social, economic and political unity'. When it became clear that the proposal for 5 or 6 London counties was not going to be accepted, Professor Donnison thought that the likely outcome would be either the increased power 'of the Greater London Council, or the informal integration of the boroughs into larger groups, or . . . the increasing intervention of central government'.

In the event the new boroughs were not as small as proposed in the Herbert Report; nor were they the kind of bodies proposed by the signatories of Plan B.

They were modelled on the proposals set out in Plan A of the Greater London Group which produced boroughs approximating in population size to the larger County Boroughs. Although the arguments for Plan A did not include consideration of the most appropriate size of local authority for providing for the social services, it was nevertheless a reasonable assumption that boroughs so large should be fully capable of this task.

In considering the extent to which the health and welfare services of the new boroughs fulfilled the hopes expressed in the Herbert Commission's report or the gloomy prognostications of its opponents, the fact that these boroughs were larger than the Commission would have liked must be borne in mind.

Five years is not a very long time to judge the performance of new local authorities, especially if one takes into account the very considerable upheaval involved in establishing them. Nevertheless it has been a period of rapid development all over the country with some pressure from the Ministry to get departments to plan ahead and to improve standards. Therefore, one looks for evidence of commensurate progress in Greater London. The degree of success achieved in certain specific aspects such as co-ordination, the 'domiciliary team', the programme of small homes for the elderly and the problem of homelessness have been examined; and an attempt is made to evaluate the success of the new boroughs in recruiting and training suitably qualified staff and in raising standards of service.

Attention has already been drawn to the fact that the relationship between the health and welfare services was under discussion during the five-year period from 1965. In 1963 and 1964 when the new Councils were forming their establishments, it had been thought that the majority would set up joint health and welfare departments. This was not so; on the appointed day, 1 April 1965, 20 of the 32 boroughs had separate health and welfare departments, including all but one of the Inner London boroughs. Nearly five years later 15 had not altered this structure, but by this time 11 others had made or were making arrangements for varying degrees of amalgamation between the health, welfare and children's departments. Of these four were Inner London boroughs, one of which within weeks of the publication of the Seebohm Bill had adjusted to the new situation and was advertising for a Director of Social Services to plan the proposed new department. The amalgamations were not always accomplished smoothly. The overall direction was invariably placed in the hands of a medical man,[7] and this caused disquiet, in one or two cases leading to resignations, among the social worker staff who thought that such amalgamations were pre-empting the final decisions on the Seebohm proposals. The quality of the co-operation between departments varied very much from one borough to another.

One or two boroughs right from the very beginning aimed at co-ordinating the policies of their social service departments; for example, Camden, where a Family Service Committee was formed composed of representatives of the children's, health, housing and welfare committees. In many boroughs good working relationships were maintained; Chief Officers meeting at intervals to sort out administrative arrangements required when the responsibilities of departments overlapped or new policy decisions were made. The departments and their committees, however, pursued their programmes and developments independently of each other. In a few boroughs there were signs of lack of co-operation.

Co-ordination of the social services within a borough was further complicated in two ways. First, the GLC was given housing powers and took over the former LCC's stock of housing; in some Inner London boroughs this meant that the GLC managed as much as half of the local authority housing. This difficulty will diminish when GLC housing estates are handed over to the boroughs. Secondly, the formation of the ILEA meant that the education service with its unique and long-established family social service, was the responsibility of a different authority. Indeed in Inner London in many of the boroughs there was almost certainly some initial loss in co-ordination since the LCC had established a practice of issuing circulars over the joint signatures of the Chief Officers concerned (children's, housing, health and welfare) with the result that divisional and field staff were working to identical guidance from County Hall. The new boroughs in the Inner London area adopted a variety of administrative devices for co-ordinating the activities of field workers but for the most part in the early days the departments were still too pre-occupied with their own problems to be able to spare much time for inter-departmental policy making and planning.

The problem of premises was another factor which hindered the kind of development the Herbert Commission had been looking for when it spoke of all the services operating from one Town Hall where the officers could become personally known to one another. In fact, the new boroughs found themselves each in possession of two or three Town Halls, possibly one or two former county divisional offices and municipal buildings and a number of other scattered and frequently unsuitable buildings. The first call on premises was for the Clerk's department which needed to be in whichever building was chosen for the new London Borough Town Hall. Some of these were serviceable fortresses built in the 1920s and 1930s and capable of housing several departments; but not a few of the boroughs inherited magnificent examples of Late Victorian and Art Nouveau municipal architecture dating from a period when municipal staffs were very much fewer in number and their offices planned to be ancillary to the Council Chamber and Committee Rooms. On the appointed day two-thirds of the boroughs were unable to house their health, welfare and children's departments together, often not even adjacent to one another, and the position has not greatly improved since.

This does not necessarily mean that the public found access to the services more difficult. An inquiry was made in 1969 for this study among the organizers of the Citizens Advice Bureau in the Greater London area to try to find out among other things what had been the effect of the change in this respect.

There are over 80 CABs in the Greater London area dealing with approximately 400,000 inquiries annually; an important part of their work consists of being well-informed about the availability of services so that inquirers can be directed to the office they need. Not enough replies were received to give a total picture, but the comments which were made indicate how very differently the 1965 changes affected the public from one borough to another. Thus in three Outer London boroughs, one north, one east and one south, and in one inner south London borough, it was reported that the services were more accessible and liaison with the officers much improved. Whereas in two Inner London boroughs and one Outer London borough, greater difficulties with long journeys to the welfare offices were reported. This partly arises from the fact that quite a number of services important to the elderly were not provided by the counties, but by the smaller lower tier authorities. As the organizers of the CAB inquiry commented:

'Difficulties arise from the necessity for London boroughs to deploy their services over a very large area; this affects the borough resident in tangible ways, in that he may now need to travel some distance to consult the service he needs, but it would appear also that residents of a particular area with a strong local flavour, may still feel a residue of resentment in having to be merged in a larger, more impersonal grouping.'

Discussing the question of easier access for the public, S. K. Ruck[8] has pointed out the scarcity of offices in the counties where the public might go for help or advice. Inasmuch as the LCC Welfare department had only 3 offices and each of the 12 Inner London boroughs has at least 1, the position has obviously been improved. Not much progress has been made in providing area offices; 1 or 2 boroughs have been experimenting with these and others are making plans. However, changes of this kind ought not to be expected too soon.

It would be easy to underestimate the task which had to be faced in getting the new departments operational. Health departments were a patchwork composed partly of pieces cut out of the former county cloth and partly of 2 or 3 smaller but whole cloths – the former boroughs' and urban districts' environmental health services. Each had to become immediately responsible for emergency services formerly provided by the counties; a sphere in which there was and is considerable co-operation between adjacent boroughs. They had the advantage of administering well-established services, the main part of whose staff and equipment was so locally based, that most boroughs had a full range of facilities, and field staff who were working for the same clients from the same bases as before the reorganization; although this was not quite the case with the mental health establishments where there was a good deal of sharing.

The welfare departments, where they were separate, were now establishments employing staff of a great variety of skills and levels of training; the majority were, though experienced, untrained. The departments inherited a share (not necessarily within the borough borders) of the former county residential accommodation, and a heterogeneous assortment of community services. Voluntary organizations whose contribution, especially to old people's services, was considerable, were usually organized on the basis of the former boroughs and districts, so that there might be several such bodies for each welfare department to make working relationships with. Keeping services going and building homogeneous departments with a sense of direction and purpose was no mean task. It made many heavy demands on the senior officers.

Moreover the welfare services, and the home help and mental health sections of the health departments, were operating in a context of increasing powers and duties. They were under pressure to provide greatly expanded and improved services.[9] Interviews, study of annual reports and analysis of some of the key services, creates a definite picture of sharp differentiation between boroughs. Some went ahead with great speed and vigour from the start, while others seemed satisfied to keep services going and some quite quickly fell relatively far below the average. It is not without relevance to the controversy about the relationship between health and welfare that the information given later in this chapter suggests that welfare services were more likely to expand when they were organized by separate departments. Only about a quarter of these seemed relatively lacking in dynamic, while about two-thirds of the services run by joint departments seemed rather stagnant. Of the eleven boroughs which were making more extensive amalgamations up to early 1970, two Inner London boroughs had expanded their services on a considerable scale, 4 had shown some progress, and the other 5 were comparatively stationary. Of the six boroughs which started with joint health and welfare departments and had made no change by 1970, only one gave evidence of a lively and go-ahead policy; these boroughs included among their number two who actually decreased their meals service between 1967 and 1969, completely at variance with the London and national trends.

At this point it might be convenient to consider an objective to which the Herbert Commission, following the advice given to it, attached very great importance, namely 'to achieve the domiciliary team with the family doctor as its clinical leader and one local authority officer as its organizer'. S. K. Ruck, writing in 1963, said 'Those working in this field generally regard this as something of a chimera'. There is no evidence that even the joint health and welfare departments regarded this as an objective attainable within a decade or two. In the face of the growing body of opinion which opposed the concept of social services as only ancillary to medical care, leading to the change of policy at the Ministry already referred to, this particular solution to the problem of offering an integrated social service to families became even less realistic. It is becoming increasingly recognized that the numbers of families and individuals whose social needs might legitimately become the subject of local authority social services (or grant-aided voluntary services) go far beyond those who are in regular contact with their general practitioners, and often have troubles which only a small number of the latter would consider their concern.

That the general practitioner should be able to call on the social services of the local authority when he needs them through 'one local authority officer' is, however, very much the order of the day. The new local authority social service departments will have to organize themselves to achieve this for all who will have cause to approach them.

Before going into the details of particular services, the following tables, using figures taken from the IMTA statistics give a general picture of the rate of expansion of the services, judged by the increase in expenditure, comparing the London boroughs as a group with local authorities in the rest of the country.

In general these figures show the kind of expansion which has been going on in the health and welfare services all over the country, even allowing for inflation; greater expenditure (more than could be accounted for by the higher costs) in the Inner London area remains a feature. There is a suggestion in these figures that the rate of expansion of both health and welfare services was temporarily checked in the Greater London area over the immediate period of the reorganization, but that within three years it had more than regained its impetus on the whole. This did not occur evenly throughout the Greater London area. For example, so far as the former Middlesex boroughs were concerned, while they pulled well ahead of the County Boroughs of England in their expenditure on welfare, they dropped further behind than the former county in terms of health expenditure. In the Inner London area on the other hand, expenditure on health was not so much greater than the rest of the country as it had been before 1965, but in welfare it was greatly increased from about one-third as much to almost twice as much. Of the other 11 London boroughs not included in this analysis, all but one spends considerably less than the average given for the London boroughs for welfare services and 5 spend less than the average for the County Boroughs. The existence of very great varia-

Table 4.1 (*a*)

Average net expenditure per 1000 population on the Health Services
% increases (a) 1966–7 on 1963–4 (b) 1968–9 on 1966–7

	£ 1963–4*		£ 1966–7	£ 1968–9	(a)	(b)
London County Council	2480	Inner London Boroughs	2823	3584	13·4%	26·9%
Middlesex County Council	1817	Former MCC Boroughs	1965	2295	7·5%	16·7%
		All London Boroughs	2294	2812		22·6%
All County Boroughs	1836	All County Boroughs	2408	2783	32·0%	15·1%
Ten large County Boroughs†	1930	Ten large County Boroughs	2601	2985	35·0%	14·7%

Table 4.1 (*b*)

Average net expenditure per 1000 population on Welfare Services
% increases (a) 1966–7 on 1963–4 (b) 1968–9 on 1966–7

	£ 1963–4		£ 1966–7	£ 1968–9	(a)	(b)
London County Council	1164	Inner London Boroughs	1840	2732	58%	48%
Middlesex County Council	740	Former MCC Boroughs	1192	1672	60%	40%
		All London Boroughs	1429	2035		42·5%
All County Boroughs	866	All County Boroughs	1181	1457	35·5%	23·4%
Ten large County Boroughs†	817	Ten large County Boroughs	1185	1453	45%	22·6%

* In 1963–4 the LCC allocated an abnormally large sum as a revenue contribution to capital outlay both in health (1962–3 £40·2, 1963–4 £105·13 per 1000 of population) and welfare (1962–3 £18·19, 1963–4 £415·7 per 1000 of population). This increase has been disregarded for the purpose of these calculations.

† Bradford, Coventry, Hull, Leicester, Newcastle upon Tyne, Nottingham, Portsmouth, Plymouth, Southampton, Stoke on Trent. Populations between 200,000 and 350,000.

tions between the boroughs and the sharp differentiation of the Inner London area from the rest are features which continually appear when different services are examined.

Homelessness. The growing numbers of the homeless and the way in which they were accommodated were very much a matter of public concern at the time when the reorganization of London government was under discussion. S. K. Ruck[10] wrote: 'it can be maintained that the problem of homelessness would not have attained the alarming proportions it has of late in London if its solution had not confronted one single authority'. He also pointed out that homelessness is a problem of housing 'since the only satisfactory eventual solution is a home'.[11]

It was accepted in the Herbert Report that it might not always 'be easy to draw a clear line between the responsibilities of housing and welfare departments. Because of this uncertainty, we do not think that this problem is being satisfactorily dealt with, or that it is likely to be, so long as the responsibility is divided'.[12]

Unfortunately the causes of homelessness appear to have been too deep-rooted to be dealt with by structural reorganization or by reducing the problem which each authority faced. On the contrary as the following tables show there has been a steady increase in the numbers of people in temporary accommodation, which in the past two years has spread out from the Inner London area to a number of other boroughs; temporary accommodation has engulfed even Croydon which had served as a model of how an authority smaller than a county and with welfare and housing powers could deal with the problem.

Inner London area. Numbers in temporary accommodation per 100,000 population

Early 1962	108[13]	Average 1963–4	4,693[17]
31.12.67	204[14]	No. on 31.12.67	6,190[14]
31.12.68	232[15]	No. on 31.12.68	7,051[15]
30. 9.69	257[16]	No. on 31. 9.69	7,627[16]

Outer London Boroughs. Numbers in temporary accommodation

Average number 1965–6	1,700[17]
Number on 31.12.67	1,955[14]
Number on 31.12.68	2,740[15]
Number on 30. 9.69	3,282[16]

This alarming increase in the total numbers did not receive very much attention in the Press. Comment was mostly concentrated on the policies of the new boroughs. The *Sunday Times* and the *Guardian* early in 1967 published articles which referred to the hardship caused by the differing policies of the boroughs; where some continued to treat the homeless with humanity others were described as operating a primitive service, not welfare at all. Families who crossed borough boundaries could be in difficulties and if they came from other parts of the country were sometimes sent back. Social workers' case files contain examples of families who have been passed to and fro between boroughs neither of whom were willing to take responsibility for them. Citizens Advice Bureau organizers gave some information about this in the inquiry already mentioned. The organizer for a central area bureau who advises people from all over Greater London commented particularly on the hardship caused by the differing policies of the boroughs. The majority of her clients were people about to be evicted from furnished rooms and not problem families. The Ministry returns now include headings giving different reasons for admitting families to temporary accommodation; in the quarter ended 30 September 1969, 653 families were admitted to temporary accommodation; 18·6 per cent had been evicted for rent arrears either by a private landlord or local authority; 29·5 per cent were admitted as a result of court orders or other action mostly by private landlords but a few by local authorities. The rest were listed under seven different headings. The difference which a boundary might make to the kind of help a family could expect may be illustrated by the fact that while on average just over a quarter of families applying for temporary accommodation were admitted during the quarter ended 30 September 1969 (the proportion was approximately the same for both Inner and Outer London), the proportion in the individual boroughs varied greatly from approximately three-fifths at one end of the scale to just under one-tenth at the other. A detailed examination of the figures for the different boroughs shows how deceptive average figures can be.

These figures show that the problem was acute in the Inner London area, but they also indicate that it is spreading over the whole Greater London area. On the other hand they show that the problem seems to have become most acute in a few of the boroughs. For example, relative to their size the problems facing Lambeth and Islington by 1969 were far more serious than those which faced the former LCC

Table 4.2
Average number of persons in temporary accommodation

		Average* 1966–7	Average* 1967–8	Average* 1968–9	Actual No.† 30.9.69
	Lambeth	944	1,281	1,456	1,416
	Hackney	711	679	631	747
	Islington	675	839	1,229	1,433
	Wandsworth	541	468	672	694
	Camden	537	513	505	451
	Lewisham	516	480	557	660
Former	Westminster	413	413	436	402
LCC	Greenwich	336	376	295	229
boroughs	Hammersmith	328	326	334	354
	Tower Hamlets	263	212	251	310
	Kensington &				
	Chelsea	232	228	348	282
	Southwark	116	124	130	644
	Total	5,612	6,039	6,844	7,627
	Average	467	503	570	635
	Haringey	145	263	276	412
	Hounslow	144	257	365	365
	Barnet	177	141	128	235
Former	Ealing	158	123	203	195
MCC	Hillingdon	111	96	96	107
boroughs	Brent	109	137	293	374
	Harrow	11	11	11	54
	Richmond upon				
	Thames	38	30	15	105
	Enfield	21	21	22	40
	Total	914	1,079	1,409	1,887
	Average	101	119	156	208

* IMTA Welfare Statistics.
† DHSS Quarterly returns.

Lambeth	Average* 1966–7	Average* 1967–8	Average* 1968–9	Actual No.† 30.9.69
Havering	31	44	7	23
Former ECC boroughs — Barking	37	44	40	31
Redbridge	44	77	72	118
Waltham Forest	57	64	301	114
Former KCC — Bromley	44	67	86	103
Bexley	36	128	231	192
Former SCC — Merton	150	120	166	190
Sutton	51	101	88	115
Kingston upon Thames	2	4	15	24
Croydon	41	30	15	291
Newham	167	214	218	194
All Outer London Boroughs				
Total	1,574	1,978	2,697	3,282
Average for all Outer London Boroughs	78	98	139	164

* IMTA Welfare Statistics.
† DHSS Quarterly returns.

before 1965; between them they were responsible for almost one-third of the people in temporary accommodation in September 1969 in the Inner London area. In the former Middlesex area 4 of the boroughs by 1969 were facing problems of homelessness on a similar scale to some of the Inner London boroughs. Of these Brent had the fastest rate of increase and was the only one of the London boroughs to attract much attention in the Press. On 22 January 1970 *The Times* published an article and picture headlined '5 Homeless Families in one Room'. This was in emergency accommodation where fathers were not admitted and which had been the subject of special criticism in the *Guardian* article of 11 April 1967 already referred to. Southwark and Croydon both experienced a sudden and very great increase in the numbers of homeless in 1969.

Only 6 out of the 32 boroughs had fewer people in their temporary accommodation at the end of September 1969 than their average figure for the year 1966–7. Of these, 4 had managed to keep the figure steady and only 2, Camden and Greenwich, had substantially reduced the numbers. These two boroughs had both given very special attention at the planning stage to the way in which their social services were to work together. Greenwich was the one Inner London borough with a joint health and welfare department and is an exception to the suggestion made earlier that these departments tended to be slower in developing their welfare services. Camden had been able to acquire and convert a warehouse, adjacent to the Town Hall and large enough to house health, welfare and children's and housing departments with a common reception hall.

The nature of the housing problem almost certainly is very different in these boroughs but there is some evidence to suggest that positive welfare policies (achieved not without a good deal of difficulty and differences of opinion) have had some effect on holding the numbers in homeless accommodation down. These would include preventative work with families appearing to be in danger of eviction and supportative work to facilitate the speedy re-housing of families in temporary accommodation, in addition to a co-operative policy on the part of the housing department. Other boroughs may have been making equal efforts with less success, because they have encountered greater difficulties with some of the other factors concerned.

The activities of private landlords and developers, the policy of the housing department on rehousing the inhabitants of development areas, the pressure on the dwindling stocks of privately rented accommodation, the financial difficulties into which many of the London borough Housing departments are now running are among the major factors outside the scope of this chapter which help to explain the growing seriousness of this problem.

How have the new welfare departments looked after the families once they have become homeless? It is difficult to generalize. Much of the accommodation available consists of exceedingly unsatisfactory old tenement buildings not always in the area of the borough responsible for the families. The increasing numbers of homeless have made the problem of replacing these buildings more difficult. This is especially true of the Inner London boroughs. The policy of making some accommodation available only to women and children had been ended by the LCC before its demise; but as has been mentioned one Outer London borough at least did not admit husbands and fathers to some of its accommodation.

There is some evidence from the files of social workers of families being sent from one borough to another within the Greater London area since people in housing difficulty may easily have crossed invisible borough boundaries in their search for a permanent home.

Some boroughs are less willing than others to accept responsibility when a family becomes homeless after only a short while in its area. The help offered to families from other parts of Britain who become homeless in London also varies very much. Some boroughs will only offer the fare back to the place of origin, without taking account of the man's work.

It is clearly an advance that the boroughs who shared Morning Lane with its communal feeding arrangements have been able to close it down. Some boroughs have been able to use sub-standard properties bought for eventual demolition when large areas are to be re-developed; in some cases this may have speeded the passage of families through temporary accommodation.

The real difficulty of finding dwellings at rents which many homeless families can afford to pay frustrates the best efforts of the welfare departments in the Inner London boroughs. In some, homeless people may feel that they are treated with more concern and offered better facilities than before 1965, in others the reverse may be the case. In the last resort, as the Herbert Commission said, the only solution to homelessness is a home. The fact that more people have gone through this bitter experience in 1970 than in 1964 cannot be blamed on the new structure of London Government, but neither has this structure been able to ameliorate the problem as had been hoped by Herbert and Ruck.

Residential care for the elderly and the handicapped. 'Provided there is co-operation in supplying special homes for difficult cases . . . the complexities of running small homes for old people should not be exaggerated.'[18] In making this statement the Herbert Commission recognized there would be initial difficulties because of the numbers of old people at that time still in large institutions. Ruck found that the record of the counties compared unfavourably with that of the County Boroughs in providing small homes and in placing them in the localities needing them and concluded that 'smaller authorities pay more attention to meeting needs locally'.[19] He, too, foresaw difficulties initially when reorganization took place.

In the event the homes throughout the Greater London area were allocated to the boroughs in proportion to their estimated need for places. The large institutions were managed by the boroughs within whose boundaries they fell and beds allocated proportionately to neighbouring boroughs. The old people actually in the beds on the appointed day were not of course disturbed, but became the responsibility of the borough now ruling over the address from which they had moved. This arrangement applied also to the small number of specialized residential establishments. In the former Middlesex area and to some extent in the former Essex area this resulted in a considerable number of homes whose residents had not lived in the borough responsible for the home. In the former LCC area it meant that some boroughs inherited small modern homes outside their boundaries because there were not a fair share of places within their boundaries. There are a certain number of residents who are the responsibility of one or other of the London boroughs in the homes run by Essex, Kent and Surrey.

All boroughs also make use of voluntary homes, particularly for the physically handicapped, and are willing to take responsibility for old people who are offered vacancies in voluntary homes, provided that they are satisfied about the standard of care. The situation outlined was to be expected immediately after the changeover; and could not be expected to alter rapidly.

However there were three things which the new boroughs had to do if they were to show themselves capable of providing modern standards of residential care. The first was to provide sufficient beds in small homes to enable the old, large institutions to be closed; the second was to increase the number of available beds so as to reduce waiting lists and to receive old people from hospitals; the third was, jointly if necessary, to provide homes and hostels for certain small groups of people, for whom this has become an urgent necessity; eg the young physically handicapped; mentally convalescent or subnormal people discharged from hospital; the mentally confused or very frail among the elderly population.

It must be said at once that a major limiting factor on expansion has been the difficulty of obtaining loan sanction due to national economic policy. The programmes which the boroughs sent into the Ministry in 1966 to be incorporated in the plan for the development of community care have fallen behind schedule; to some extent this is due to the fact that the programmes expressed aspirations, rather than plans; this is not surprising since the boroughs were expected to submit their plans within a few weeks of the appointed day. Even more is it due to the fact that projects have been put back from one year's programme to another because of the government's restriction on capital expenditure. It has been suggested that larger authorities are more successful in pressing their claims or resisting this kind of pressure. In particular it has been suggested that the LCC was financially more independent than the London boroughs; it is certainly the case that in each of its last two years of life the LCC Welfare Committee used some £1,250,000 from revenue for capital expenditure. In 1968–9, only 6 of the Inner London boroughs raised money for capital expenditure in this way, the amount being approximately £46,000.

Essex County Council which in 1960 had over 35 per cent of its beds for the elderly in former poor law accommodation, closed the last of these early in 1970, whilst increasing the total number of its beds in council homes from 1,416 in 1963 to 2,095 in 1970. This achievement is of some interest in that the population of Essex after 1 April 1965, at just over one million, approximates to the size suggested in the Greater London Group's Plan B for the proposed 'London Counties'.

It was the Inner London boroughs which inherited the most serious problem regarding the numbers of beds still in old poor law institutions, though they also inherited a considerable build-up of new homes in the pipeline by 1965 as the LCC had been accelerating its programme. All their resources were needed for replacing the beds in the old institutions; by the end of 1969 two of these notorious places (Newington Lodge and Southern Grove) had been closed down and the plan for running down the rest was said to be on schedule. To judge from the frequent references in the annual reports to the plans for starting a building being held back for a year or more, this must be a schedule which is subject to revision at the Ministry. A disturbing feature is the fact that in Inner London the average numbers of old people in residential accommodation had risen slightly up to March 1968, but by March 1969 they were about 250 lower than the previous year and about 50 lower than in March 1964. In view of the waiting lists, the known housing plight of many old people especially in Inner London and the slow progress of sheltered housing and old people's flats, it is most unlikely that this fall was due to a drop in demand. Moreover, it was against the general trend. The rest of the London boroughs, the counties and the county boroughs all reported an increase. It is possible that a coincidence of the ending of the flow from the former LCC's programme and the effects of the restrictions on capital building programmes had a special and temporary depressing effect in that particular year. However, the situation is dominated in the Inner London area by the need to absorb almost all the new accommodation coming into use for the elderly people from the large old homes. There is really no way of judging whether the LCC with its enormous resources or the GLC – if it had been given the task – would have provided modern old people's homes at a faster rate than the London boroughs; what is certain, however, is that the Ministry acting on government decisions about national economic policy, has been able to restrict the London boroughs to a slower rate of growth than they would have chosen, even though priority has been given to building for the replacement of the large old institutions.

Care in the community.　Important as residential accommodation is in the scheme of welfare, especially in terms of expenditure, it only deals with the needs of a small number of old people; far larger numbers of old people are living at home and in need of community services, some of which are supplied by the Health department. At the time when the Herbert Commission reported, these services – for example, meals on wheels, lunch and social clubs, chiropody, home-nursing and home helps – were very underdeveloped, and provided in an unplanned and unco-ordinated way by voluntary organizations, metropolitan boroughs, district councils, the LCC Welfare department and the Counties Health departments. Their importance as part of a general scheme for enabling people to stay at home was only beginning to be widely appreciated. The change of emphasis in recent years can be illustrated from the relative expenditure on residential accommodation as compared with the total expenditure on welfare services, shown in Table 4.3. This table indicates that all over the country expenditure on the welfare services is rising, and the proportion being spent on community services increasing.

It is noticeable that this expansion is going ahead particularly rapidly in the Greater London area as a whole.[20] This may be connected with the fact that some of the London boroughs have made considerable use of their powers under recent legislation to provide services directly. For example, 17 of the London boroughs now directly provide the greater part of their meals service and, with a few exceptions, do so on a much larger scale than had been possible for the grant-aided voluntary organizations.[21] On average these 17

Table 4.3

Expenditure on Welfare Services

| Authorities | | Net Expenditure per 1000 of the population* | | |
| | | Total Residential Homes | Total net Expenditure | Residential Expenditure as a proportion of the total |
		£		%
1963–4	County Boroughs	536	833	64·5
	London County Council	744	1154†	64·4
	Middlesex County Council	477	739	64·5
	Counties (England)	400	631	63·3
1966–7	County Boroughs	720	1142	63·0
	London Boroughs	806	1412	57·0
	Counties (England)	551	910	60·5
1968–9	County Boroughs	871	1419	61·3
	London Boroughs	1033	2018	51·1
	Counties (England)	670	1120	59·8

* IMTA Welfare Statistics.
† Excluding abnormally large sum contributed to capital expenditure.

boroughs now devote *less than half* their total expenditure to residential accommodation.

The increasing emphasis on services which enable handicapped and old people to stay at home makes it important to consider in more detail the performance in this field of the London boroughs. This covers a very wide variety of services from major structural alterations in houses to the loan of articles of housing equipment, from daily nursing care to quarterly visits from a chiropodist, from visits to blind and deaf people by specially trained social workers to the establishment of workshops producing saleable goods. A study of annual reports gives evidence that most boroughs cover this spectrum of services in some degree; what does not emerge is evidence that many boroughs have planned their services with any relationship to need, or have taken steps to find out the extent of need in their areas. There is nothing unusual about this; as the Amelia Harris survey shows, few local authorities have attempted to do this.[22]

A few boroughs give indications in their reports that they have been expanding the whole range of their services in a systematic and planned way and there may be others who have acted likewise but not included this information in their reports. In all areas of the country there has been a steady increase in the numbers of handicapped who are registered with the local authority, and the London boroughs appear to have gone ahead rather faster than any other group of authorities in the past two years, so that in respect of basic data for planning for this group of people they should be in a strong position. Only a very few boroughs, however, describe plans to get to know the true number of elderly people likely to be in need of health and welfare services immediately or in the next ten years; much less to quantify the types and varieties of services required. This is not a particular failing of the London boroughs, as compared with other authorities. It does suggest that the task of identifying and meeting local needs is a more complicated matter than the Herbert Commission suggested. In Greater London the methods vary from a fairly extensive service of geriatric visitors who keep an eye on the changing needs of elderly people to a total reliance on the percipience of neighbours.

Two services have been chosen for detailed examination; the meals' service and the home help service. Both are well-established as a major means of enabling old people (and to a very limited extent handicapped people) to remain fit and cared for whether independently or in the homes of relatives. One is the responsibility of the Welfare Department, the other of the Health Department.

Since these are services which do not involve capital expenditure, their development has not been restricted by central government control. The Health and Welfare committees have been able, within the limitations set by the general financial policies of their councils, to expand these services, or not, as they thought fit. A study of them should be a better indicator of the progress of the London boroughs to deal with health and welfare problems than provision of residential accommodation for the old or the homeless where extraneous factors play so large a part.

In Table 4.4 the figures for 1967 are those supplied by the Working Party of the London Boroughs Association, and for 1969 are taken from the returns to the Department of Health and Social Security.

The quantity of meals supplied has been expressed in an arbitrary way, ie the number served annually per hundred of the population aged over sixty-five, in order to provide a basis for comparison. The meals service includes the familiar meals on wheels to the housebound as well as the provision of meals at clubs of various kinds, the latter being the most rapidly growing section of the service.

Although it is not possible to make comparisons with the meals provision before 1965 the boroughs have been grouped by the former county areas, as this shows clearly the way in which things are developing.

The table shows that it is the Inner London boroughs as a group which have been expanding the meals service most vigorously. Two of the former Essex boroughs, Hounslow (formerly in Middlesex) and Sutton (formerly in Surrey) provide meals on a similar scale. The two Inner London boroughs whose meals service falls well below the general pattern may be assumed to be those with the highest proportion of the upper income group in their population, but it would need a more detailed analysis to ascertain whether their achievement is nevertheless disproportionately low. In the rest of the London area it is difficult to believe that such wide disparities are to be explained entirely by differences in need.

The move towards the local authority assuming responsibility for the meals service has been made possible by legislation which has increased the powers of the local authorities to provide services directly. In almost all cases the voluntary organizations were finding it beyond their powers to extend the service to meet the need they knew existed, and were glad to be relieved of the responsibility. More often than not the services of the volunteers have been retained either in helping with delivering the meals, or running various activities where the meals are served in clubs. This appears to be a classic example of a service pioneered by voluntary organizations yet beyond their capacity because it is meeting such a broadly based social need. It is clear that the London boroughs are perfectly capable of providing this service if they wish to do so.   .

The home help service was originally developed in health departments for the benefit of maternity cases; nowadays the overwhelming majority of people helped (88·2 per cent in the London boroughs for the week ending 21 January 1967)[23] are the elderly. The service is seen as important both for helping the elderly to continue living in their own homes, in spite of increasing infirmities which make many household tasks impossible, and by some, though not all authorities,

Table 4.4

Meals Service

| | 1967 | | | 1969 | | |
	% of population over 65	Meals served Total	Per 100 of population over 65	% of population over 65	Meals served Total	Per 100 of population
Camden	13·1	176,276	54	12·7	272,433*	93
Greenwich	12·7	208,810	71	12·5	414,070*	144
Hackney	12·1	314,673	103	12·0	327,150*	112
Hammersmith	13·5	312,991	108	13·0	338,183*	131
Islington	11·9	177,005	57	11·6	252,062*	89
Kensington/ Chelsea	12·5	80,994	29	11·6	100,261*	41
Lambeth	12·8	N.A.		12·4	362,415*	88
Lewisham	13·2	487,372	127	13·2	541,206*	143
Southwark	12·7	468,194	121	11·9	568,835*	164
Tower Hamlets	11·8	357,675	150	12·0	415,673*	188
Wandsworth	14·0	423,434	92	13·3	483,951*	113
Westminster	14·2	98,386	26	13·4	178,507†	54
Barnet	13·5	61,718	14	13·6	78,624†	18
Brent	11·8	94,297	27	12·0	92,153*	27
Ealing	12·4	122,942	32	12·5	N.A.‡	N.A.
Enfield	13·4	99,249	23	13·2	132,416†	38
Haringey	14·3	31,908	9	13·0	75,871‡	23
Harrow	11·6	100,639	42	12·1	108,786†	43
Hillingdon	9·8	64,127	30		N.A.	
Hounslow	11·9	161,373	64	12·1	200,880*	82
Richmond upon Thames	15·4	39,614	14	15·1	83,236†	31
Barking	10·3	N.A.	N.A.	11·5	219,122§	112
Havering	8·5	193,792	91	8·7	230,032*	104
Redbridge	13·4	37,995	11	12·4	101,185*	33
Waltham Forest	14·5	122,295	35	13·8	118,208*	36

* All or nearly all provided by local authority.
† All or nearly all provided by voluntary organizations.
‡ About half and half.
§ Special arrangement of vouchers in half the borough.

Table 4.4 *cont.*

	1967 % of population over 65	1967 Total	1967 Meals served Per 100 of population over 65	1969 % of population over 65	1969 Total	1969 Meals served Per 100 of population over 65
Kingston upon Thames	13·5	68,506	34	12·4	80,019†	44
Merton	14·8	30,307	12	14·5	70,038‡	26
Sutton	13·7	72,594	34	13·7	163,436†	72
Bexley	10·5	56,270	25	10·9	42,231†	17
Bromley	12·1	74,589	20	11·6	157,688†	44
Croydon	13·4	72,570	16	12·5	74,556†	18
Newham	12·2	159,438	50	12·0	190,564*	62

* All or nearly all provided by local authority.
† All or nearly al' provided by voluntary organizations.
‡ About half and half.

as valuable assistance to families who want to continue caring for their elderly relatives but need some help in doing so.

Since the home help service was provided by the London County Council or Middlesex County Council before 1965 it is possible to make some comparisons. The following table shows the aims set in the ten-year plans and the progress made towards achieving those goals.

Table 4.5

Number of home helps (whole time equivalent) per 1,000 population*

		LCC	Middx. CC		IL Boroughs	MCC boroughs	10 County Boroughs
Actual numbers	1964	0·89	0·45				0·73 (1965)
Actual numbers	1967				0·96	0·46	
	1969				1·00	0·48	
Planned increase to	1972	1·12	0·49	1971	1·09	0·56	0·95

* Population figures from IMTA returns. Plans for DHSS, Cmnd. 1973. Achievements from Ministry returns and LBA working party.

From these figures it can be seen that the Inner London boroughs as a group did not expect to expand the service any faster than the LCC had planned, but the Middlesex boroughs were more ambitious. Neither group however appears likely to achieve the target. Difficulty in recruiting staff seems to be a major factor in restraining expansion, though one borough at least mentioned financial stringency as their reason for halting recruitment for one year. There are frequent references in annual reports to the difficulties experienced in developing the service but very little evidence of a serious attempt to consider what kinds of measures would improve matters. One borough mentioned the need for training schemes and greater recognition of the fact that the job involved more than just cleaning tasks. However this is not peculiar to the London boroughs, as the Government Social Survey of the Home Help Service demonstrates.[24]

These average figures suggest that following upon the structural changes of 1965, only marginal differences occurred, with the Inner London boroughs as a group slightly less successful than the LCC had hoped to be, and the former Middlesex boroughs as a group, starting from a much lower figure, more ambitious but only slightly more successful. This is a service for which the Ministry in commenting on the ten-year plans[25] has suggested a minimum standard by saying that a figure of 0·73 of the population is more likely to be too low than too high in any area in 1972.

However, in assessing the effects of the London reforms, average figures, as has already been mentioned, are deceptive.

The changes discernible in the two years between 1967 and 1969 fall into three groups; of the 13 boroughs which were providing home help well above the minimum recommended figure, only 1 has dropped below that figure, and 7 have further increased their provision; of the 10 boroughs whose scale of provision was very low by this comparison, 7 have made some increase and in 3 boroughs the proportion has remained unaltered; of the 9 boroughs whose provision was about or rather below the recommended minimum, 8 provided a reduced service in 1969, and 1 had considerably increased the proportion.

Comparing the tables for the meals service with that for the home help service, the extent to which certain boroughs have been expanding the community care services at a faster pace than the rest is very marked. A rough and ready estimate would certainly suggest that these boroughs were among those whose need for such services was greatest

Table 4.6

Home help (whole time equivalent) per 1000 of the population

	31.12.67	31.12.69		31.12.67	31.12.69
Camden	0·82	0·81	Barnet	0·30	0·38
Greenwich	1·37	1·33	Brent	0·31	0·31
Hackney	1·68	1·72	Ealing	0·80	0·73
Hammersmith	1·03	0·99	Enfield	0·55	0·52
Islington	0·73	0·64	Haringey	0·58	0·57
Kensington/			Harrow	0·22	0·35
Chelsea	0·29	0·48	Hillingdon	0·28	0·30
Lambeth	0·93	1·15	Hounslow	0·77	0·87
Lewisham	1·20	1·21	Richmond		
Southwark	0·89	1·29	upon Thames	0·34	0·34
Tower Hamlets	0·92	1·05			
Wandsworth	0·71	0·65			
Westminster	0·76	0·69			
Average	0·96	1·00	*Average*	0·46	0·48
Barking	0·82		Kingston		
Havering	0·62	0·58	upon Thames	0·22	0·22
Redbridge	0·95	0·54	Merton	0·64	0·62
Waltham			Sutton	0·22	0·39
Forest	0·85	0·91	Croydon	0·31	0·36
Bexley	0·74	0·42	Newham	0·81	0·99
Bromley	0·33	0·69			

and that these councils were responding as it was hoped they would to the needs of their localities. What is not so certain is that the need for these services is so very much less in the boroughs which make comparatively a very restricted provision. One can agree that boroughs whose social income composition makes services for the elderly less extensively required would also be those where there would be competition from private homes for the services of home helps; but the same difficulty should not apply to the meals service. The suggestion here is that the expansion of this service on a really modern scale is now beyond the capacity of voluntary services. For example, a survey made in 1967 for the whole of England[26] showed only five schemes for a seven-day-a-week service of meals on wheels, whereas most of the Inner London boroughs now provide such a service in cases of need and several take Christmas and Boxing Day dinners to old people without families, using volunteers from the Welfare department staff.

The evidence suggests that in the Greater London area the situation, well-known in the rest of the country, has been reproduced; that is, the existence of differences in service which are greater than can be explained by differences in social composition and the proportion of old people in the population. For instance, there is no evidence that the boroughs have made use of the Government Social Survey Study 'Social Welfare for the Elderly', which was specifically designed with 'the object of trying to assist local authorities to measure need for given services and to develop a basic method which might be of use to those wanting to survey their own areas',[27] in order to validate their assumptions about the level at which the needs of their older citizens shall be met.

There is a wide range of other services for the elderly and the handicapped which local authorities are empowered to supply. Home nursing, chiropody and many requisites for the care of sick and incontinent people are made available by health departments; clubs, workshops, transport, aids and modifications to the home are the responsibility of welfare departments. The available evidence suggests that most if not all the boroughs carry out these tasks but it would be difficult to discover to what extent the need is met. This is an area in which the voluntary organizations are very active and the extent to which they supply or supplement services such as chiropody varies very much from borough to borough. This particular service is one of great importance to the mobility and comfort of the elderly. The Townsend/Wedderburn study indicated that an increase of about fourfold was needed in this service. There are frequent references in the annual reports to the need for expanding this service and the difficulty of doing so, as well as a very definite trend for the local authorities to assume the major responsibility. It seems that an important limitation, as with a number of other services, is the difficulty of obtaining staff.

Clubs, workshops and day centres are provided by welfare departments and by voluntary organizations; visiting may be organized by health and welfare departments, or by voluntary organizations, and often by a combination of two or more. There are great variations in the numbers of clubs and the facilities they offer; the evidence suggests that the initiative for a systematic coverage of a borough is possible only under the overall supervision of the local authority departments and that only a few of the boroughs are at present planning in this way. These same boroughs all appear high in the tables for the meals and home help services.

The arrangements for home visiting show the same variations as the other services; the frequency of visits and the extent to which the departments keep in touch with old people whose circumstances are likely to change with the passing of time varies enormously. This was to be expected since the boroughs took over a great variety of services. It is again an area where the voluntary organizations and volunteers working with departments play a very large role. Miss Slack's recent study[28] shows how much the organization of the Old People's Welfare Committee and its relationship with local authority varies from one borough to another, and the impossibility of generalizing about these services as provided by the London boroughs.

Obviously such variations are not necessarily to be deplored, since they may conform to differences in local needs and in the resources of the voluntary organizations. However they did produce critical comment from some of the organizers of CAB who answered the inquiry already mentioned; one called special attention to 'the different standards of health and welfare services between boroughs, ie some provide the minimum according to statutory requirements, others provide far more than the statutory requirements'. Examples were given; lack of a Family Planning Service in some boroughs; extreme variations in the treatment of the homeless and in the standards of service for the elderly; in Inner London, different arrangements, eg for domiciliary work with old people, which make referrals difficult for voluntary organizations, whereas the LCC divisions had worked to a similar pattern. In some boroughs, the organizers considered that relations between voluntary organizations and local authority departments had improved, while in others the reverse appeared to be the case.

Staff qualifications and training. While the quantity of service available to meet many and various personal needs is obviously of prime importance, the quality of the service is also a matter for concern and indeed the two aspects are inter-dependent. A high level of administrative skill is required to plan, organize and deliver a series of services, some complementary and some substitutes for one another, in co-operation with other departments and organizations both inside and outside local government. Each service must be acceptably offered, if it is to lose all trace of its Poor Law origins. The calibre and training of staff is critical and was one of the specific causes for concern raised by critics of the proposal to set up a large number of London boroughs. 'The recruitment and training of staff for the health and welfare services could not be carried out by the Boroughs unaided.'[29] There are several points to be considered here. The metropolitan area exerts a great pull on trained staff, especially marked in respect of fully trained social workers. The setting up of many new departments, also, offered social workers opportunities for advancement and hopes for more professional influence on the services in which they worked. For these reasons recruitment in the London area probably did not suffer from the changes. On the other hand social worker staff have been in great demand for some time and fully trained staff in this category are in very short supply. There is some evidence too that the numbers of health visitors and public health inspectors are not keeping pace with requirements of establishments, so while London as a whole has continued to attract a high proportion of trained staff, the overall shortages have been accompanied by considerable variations between one borough and another.

Secondly, the kind of work which health and welfare departments are increasingly required to do, especially the rapid development of community care, has called for changed attitudes on the part of staff (and committee members), particularly discarding the Poor Law framework to provide 'services which recognize there are subtle needs beyond the merely physical'.[30] Legislation requiring this kind of change, increasing since the Mental Health Act, 1959, places a great strain on staffs already in posts; for them, in-service training programmes are of the utmost importance. Here the anxieties of the critics may have encouraged the development which has resulted in the formation of the London Boroughs' Training Committee (Social Services) (see below).

Thirdly, there is the question of seconding staff for full-time training, which may include trainee schemes for recruiting young staff with a view to secondment. Schemes of this kind are very expensive, as Professor Donnison pointed out, and the implication was that many of the new boroughs would hesitate to incur expenditure for this purpose on the necessary scale.

In considering the effects of staff shortages, reference has already been made to the annual reports of 1967 and 1968 in which some MOHs call attention to the difficulty of recruiting health visitor staff.

By the end of 1969 12 of the boroughs were reporting vacancies in excess of 23 per cent, while several had no vacancies and a further 9 had only 1 to 3 vacancies on establishments of 20 to 40. Six of the 11 boroughs which had reported difficulties earlier were still in trouble by the end of 1969; while 7 of the 12 boroughs which had not reported vacancies in the earlier years, were among those with few or no vacancies at the end of 1969.

Thus in a situation where many of the London boroughs are having some difficulty in filling their establishments, it appears that nearly a fifth are consistently well below strength, while about the same proportion have no problem. It would not be possible to draw any further conclusions without more research into the social composition and problems of the boroughs, the adequacy of the establishments, the numbers and suitability of supporting staff, the duties laid on the health visitors and their reasons for choosing to work where they do. It is, however, arguable that fewer and larger authorities might have resulted in a more equitable distribution of this important resource.

Health visitors must by definition be qualified; the position for social workers is different. Since an agreement made with NALGO in 1966, social workers without the National Certificate in Social Work[31] (a two-year course taught in technical colleges) or a more specialized professional qualification (often though not exclusively obtained after a postgraduate university course) should not proceed beyond a bar in the salary structure. Staff in post over thirty-five years of age and with five years' experience had had opportunity to qualify for a declaration of recognition of experience. This gives an incentive to staff to become qualified, but is not in itself an incentive to employers. There is no statutory requirement upon local authorities to employ qualified staff; therefore there is not the problem of unfilled establishments.

In examining the mental health services in 1969 one finds that 209[32] persons were untrained, out of a staff of 420. About a further 70, including a number of part-timers, held a university diploma in social services which is not by itself a qualification for social work. Just over 150[33] were qualified, of whom 68 were psychiatric social workers (or held an equivalent university certificate) and 85 held the Certificate in Social Work mentioned above. Compared with England and Wales as a whole, Greater London, with about 10 per cent of the population, employed about 22 per cent of the total social worker staff working in the mental services, and about 80 per cent of the qualified staff.

As with other services, there are some extreme differences between the London boroughs. Croydon (10) and Newham (13) employ psychiatric social workers on a scale which is approached only by Camden (7) and Wandsworth (7), while 13 of the boroughs employ no staff with psychiatric or other university qualifications, and 2 of these have no mental health social workers trained for social work at all; though it is fair to add that these two boroughs had between them 3 members of staff on secondment in 1969. Twenty-two of the boroughs had 34 members of staff on secondment for full-time training including the 4 most fully staffed boroughs mentioned, but not including 4 of the boroughs with no university-trained staff. There were wide divergences in the ratio of staff to population; thus Barnet and Croydon with very similar population figures employed respectively 15 and 32 mental health social workers; Havering and Newham, 6½ and 19 respectively.

The Inner London boroughs had staffs ranging from 12 to 18 (except Hammersmith falling below at 10, and Southwark much higher at 24), the variations being roughly related to population size; but they showed extreme variations in the qualifications of staff from one borough with no trained staff at all to one with only 1 totally unqualified. Five of the boroughs had more than half of their staff qualified: the other 7 a very much smaller proportion; all except 1 (with a high proportion of qualified social workers already) had staff away on full-time training. Almost half of the qualified staff in the Greater London area were employed by the Inner London boroughs; if Croydon and Newham were included, only one-third of the qualified staff were spread over the remaining 18 boroughs.

The former LCC had a staff of psychiatric social workers which had been growing in numbers; there is no doubt that the Inner London boroughs have continued to attract these workers and to increase the proportion of highly skilled staff among their mental health social workers.

The position is rather different in the former Middlesex area. The MCC had built its staff up to 72 by the appointed day with the object of making sufficient staff available to the new boroughs, and this number has increased to 95. But, on the other hand, the former MCC had a staff of at least 14 psychiatric social workers qualified also to do statutory work. Whereas in 1969, only Ealing had 3 such officers and a medical social worker, and Harrow had the part-time services of a psychiatric social worker.

One borough had no qualified staff and the other eight had 17 holding the Certificate in Social Work. Four of the boroughs had staff seconded for full-time training. The proportion of staff to population and of trained to untrained staff was roughly the same as the average for the country as a whole. Since the MCC had prided itself on a community mental health service which was rapidly expanding and carrying out forward planning until the changeover,[34] these figures suggest that there has been a check to development. Further consideration will be given to this question when considering the ability of the new boroughs to experiment and initiate new services.

Mental health social work services were chosen for detailed examination, because this is a field with fairly well-defined objectives, in which a social work training has been long established, and where comparisons were possible with the former counties. In 1967 a survey was carried out for the London Boroughs Training Committee (Social Services) by the GLC Research and Intelligence Unit of the qualifications of social workers in the personal social services of the London boroughs; some general findings were published in the December 1968 issue of the Unit's quarterly bulletin. This survey showed that in welfare departments as a whole 67 per cent held no qualifications in 1967, but that this figure was weighted by the high proportion of staff in residential establishments who hold no qualification; a problem which is not peculiar to London, but which has been receiving some attention in the last two or three years following the publication of the Williams Report.[35] The London boroughs were employing just over 700 social workers, just under a quarter of the number being employed in the welfare departments of England and Wales, of whom 130 held appropriate qualifications, or just over a quarter of the total for the country. In other words, both in numbers and qualification the London boroughs as a whole were more than holding their own.

With rapid and radical changes in the duties and responsibilities of the local authority personal services and a high proportion of untrained staff, in-service training programmes assume a considerable importance. Both the LCC and MCC had an experienced staff organizing courses for health visitors, mental health social workers, teachers of the sub-normal, child care officers and family caseworkers. Those people became the nucleus of the staff of the London Boroughs Training Committee headed by the LCC's chief inspector of child care. The work is financed by an annual grant from each of the boroughs which decide to join the Committee, on which they are represented by a councillor. Twenty-nine of the boroughs joined up to 31 March 1970, when 1 dropped out and 2 became members making the number at 1 April 1970 30.

The staff organize courses from one-day conferences to day-release for long periods – 113 such courses were organized for the period September 1969 to July 1970. They covered a great variety of subjects and were directed at all levels and every type of staff including Chief Officers. Additional work was done by the organizing tutors in individual boroughs assisting in their in-service training programme. An interesting feature is the discussion meetings which have been organized for councillors, especially members of the Training Committee. At an early stage in the life of the new boroughs a conference was organized for councillors who were becoming responsible for personal social services for the first time and wanted to become better informed. Subsequently meetings have been organized to give councillors an opportunity to discuss difficult social problems such as Homelessness or the instigation of new legislation such as the Children's Act, 1969.

The London Boroughs' Training Committee (Social Services) provides courses far wider in scope than the former counties; they are more comprehensively planned and cross boundaries. The institution is interesting in that it is an ad hoc body depending for its continued existence on meeting a need sufficiently felt to be backed by hard cash. Nevertheless there are limitations on the extent to which it can meet the requirements of in-service training.

These limitations are two-fold. On the one hand there is the extent to which the boroughs use the service provided; not only do some boroughs make more use of the facilities than others, but also within the individual boroughs some departments make more use of them than others. It is not unknown for officers to fail to appear for courses because at the last minute their superiors feel unable to release them.

A senior administrative officer may feel that courses arranged by an esoteric body perhaps fifteen miles away in West London are a luxury which can be dispensed with if there is pressure of work in his section. He could not take quite the same line if a course had been arranged by his colleague the Training Officer and staff members were required to attend as part of their in-service development. Yet there is no doubt that the staff concerned suffer from this attitude, both because the London Boroughs' Training Committee calls on the services of experienced tutors and lecturers from universities and technical colleges in the London area, and because meeting and talking with colleagues from all over London is a stimulating and refreshing experience for people whose daily work often makes heavy demands on their personal resources.

On the other hand the staff of the Training Committee can only become involved in staff development in an advisory capacity. A number of the boroughs have training officers who take care of this aspect of the work, and can, if they wish, use the courses arranged by the London Boroughs' Training Committee as required, make use of the tutors' experience by consulting them and act as a channel of communication to the Committee on the need for variations and innovations in the programme of courses. Clearly, there is scope for considerable variations in the extent to which individual staff members will have opportunities for systematic development. The attitude of the borough, or the department, may stand between a staff member and the courses on offer. Where attendance depends more on the initiative of the individual officers than on the staff development programme of the department, there is a marked tendency for the Committee's courses to be used more by those who already have some training than by those with little or none.

In other words the London Boroughs' Training Committee, though it makes a very notable contribution to the London training scene and has greatly enlarged its scope in the past five years, is no substitute for a positive attitude to staff development in the individual boroughs. Here, as elsewhere, sharp differences can be observed between one borough and another.

It is now necessary to consider the difficult question of the success of the new authorities in innovation and experiment.

Attitude to new ideas is the most important factor here and, as other studies have shown, the attitude of the Chief Officer is particularly important. The critics of the proposed London boroughs foresaw more clearly than the Herbert Commission where modern ideas on community care of the mentally disturbed and modern conceptions of social services for the elderly and the handicapped were going to lead. For example, to maintain the mentally disturbed in the community requires a range of hostel accommodation,[36] for those who have no families or cannot live at home, far beyond the capacity of voluntary organizations to supply.

Such provision has to be planned in the light of known or ascertainable medical and social requirements, if the doctors and social workers are to have the necessary resources at their disposal in caring for people. The same considerations apply to the care of the physically handicapped and a range of day care for the elderly, to mention only the most obvious. The provision of hostels and the systematic planning of day-care for a wide range of needs, remain at present in the area of innovation and experiment. It must be said at once that both require capital expenditure, and that government economic policy and Ministry control have acted as a brake.

Could larger authorities have resisted such pressure more successfully or been more able to find their own finance? In the past the LCC had been able to raise some of its own finance; in its last two years the Welfare Committee was contributing considerable sums from its revenue account to capital outlay. At the time of the changeover there were a number of establishments which were and still are showplaces both in the London and Middlesex areas or were in the pipe-line, opened after the boroughs had taken charge. At least one of the Middlesex enterprises, a factory to be adapted for a variety of workshops for people at different stages of rehabilitation, was sold by the borough which inherited it, as being on too big a scale. This raises the question of joint use of establishments. At present there are many such arrangements resulting from the sharing out of former county establishments. Some work satisfactorily; others do not. Admission policies sometimes differ with the result that the borough which controls the establishment will not always admit the client which the neighbouring borough desires to see accommodated. Where populations of around $\frac{1}{4}$ million contain sizeable numbers of people in a particular category of need, this difficulty will disappear in time, as the boroughs expand their range of accommodation.

There is very little evidence, as yet, that where a new facility is needed for a very small group, new establishments for joint use are getting off the ground. For example, a proposal first put forward for serious consideration in 1967 for a hostel for the mentally disturbed to serve the boroughs in the eastern part of Greater London had made little progress by 1970.

Hostels for physically handicapped young adults are another example of a specialized kind of need. At present, as is well known, many such people are spending their lives in the geriatric wards of long-stay hospitals; others are placed in voluntary homes by the borough welfare department, but as one Chief Officer explained, this is very unsatisfactory as such homes are often in quite distant parts of the country and the people concerned are cut off from family and friends.

Obviously local authorities have many calls on their resources and must establish priorities. It is not so easy to see where, under the present local government organization in London, is the place for determining priorities which involve more than one borough. The London Boroughs Association has not this power. It has not even been wholly successful in securing uniform charges and criteria for the assessment of health and welfare services over the London area.

Hostel provision has made slow progress all over the country; the concentration of population in the Greater London area means that it should be possible to provide for highly specialized needs, but the reformed structure of local government does not facilitate such provision.

There is plenty of evidence that the boroughs respond to new ideas for less specialized services, for example, the progress the health departments have made in improving co-operation with general practitioners.

A sub-committee on Child Welfare Centres reported to the Standing Medical Advisory Committee of the Central Health Services Council in 1967[37] in these terms 'we are in no doubt about the continuing need for a preventative service to safeguard the health of children . . . It is our view that in the long term it will be part of a family health service provided by family doctors working in groups from purpose built family health centres'. Traditionally this has been the field, first of voluntary effort and then of the local authorities, whose ante-natal and child welfare, prophylaxis, school health and (very recently) geriatric care programmes have been available to the whole of the relevant populations.

It will be noted that the report envisaged group practice in health centres. In the London area the number of doctors working in group practice is very small; in 1965 the number of health centres was infinitesimal. Out of 80 health centres in operation, 12 were in the Greater London area; 6 in four Inner London boroughs and 6 in three Outer London boroughs. In 1964 the London County Council had 2 health centres and apparently had no plans for more; 5 Inner London boroughs had proposals for health centres in their ten-year plans for completion by 1970, 3 of them on quite an ambitious scale (Camden, capital building programme £675,000; Hammersmith, 6 centres, £659,000; Lewisham, 3 centres, £400,000). The majority of the boroughs in the areas of the North West and North East Metropolitan Hospital Boards had plans for capital expenditure on health centres in the period 1966–7; Ealing and Hounslow planning to spend over £700,000 each, Waltham Forest and Enfield over £300,000 each. In the South East and South West areas only Sutton was planning to spend a modest sum. It is most unlikely that these boroughs will have been able to build on this scale because of the restraint on capital expenditure; but these plans indicate the extent to which the health committees were thinking of providing the basic ingredient of a unified family health service.

There is room for some doubt whether all these plans were for health centres in the true sense; ie to include general practitioners. Even in advance of the building of health centres, however, there has been a move for the attachment of health visitors to group practices; information is not easy to obtain since many boroughs now publish their reports very belatedly. The most recent available reports from the Medical Officer of Health from 23 different boroughs have been studied – 1 for the year 1966, 12 for 1967 and 10 for 1968. Eight out of the 23 boroughs reported that they had arranged for the attachment of health visitors to group practices. The numbers were very small – usually 1 or 2 (though in one borough it was 8) out of establishments ranging from 25–35 health visitors. In addition most boroughs were arranging a good deal of informal liaison between their staff and the family doctors. Several reports mentioned that shortage of health visitor staff had prevented them from increasing this kind of co-operation.

Other developments seem to arise as had been hoped from the more localized character of the departments. In a number of boroughs, for example, there has been a large improvement in the liaison between the mental hospitals and the community mental health services; this has been greatly helped by the concurrent trend towards opening acute psychiatric wards in the local general hospitals.[38] In one borough, at least, the social workers in the local mental section and the local mental hospital have been formed into a joint service.

In two boroughs there is encouragement to the people in the old people's homes and clubs to form committees which take an active part in the day-to-day running of the establishments. In one borough, the Chief Welfare Officer or his deputy, attends all such meetings for the purpose of becoming better informed about the needs of the elderly people as they themselves see them. Another borough has established a day care centre in association with an old people's home where frail old people can be cared for while their relatives are at work, thus avoiding the need for permanent removal from home. One or two boroughs have worked out very imaginative and comprehensive schemes for co-operation with the voluntary organizations in the borough, based on the principle of maximizing the services that each can best provide, for example the Triple S scheme in Southwark.

In conclusion, it is clear that the London boroughs are capable of carrying out the health and welfare functions of local authorities. Whether or not the upheaval involved in making this change was worth while involves a balancing of advantages and disadvantages. Neither the best hopes nor the worst fears of the protagonists in the discussion which took place before 1963 appear to have been justified.

Certainly the prediction that unevenness of standards of service would develop has been borne out; but has this proved unacceptable? The CAB organizers expressed some anxiety on this score. The groups of people affected tend to be small in number or not very powerful or articulate. The concept of territorial justice in the field of the personal social services is discussed at length in Bleddyn Davies' book;[39] but this is a national problem, and perhaps the solution in London is not to be found in alternatives to the structure of London Government. The multiplicity of authorities has been accompanied by an uneven distribution of skilled staff; for example, certain areas of London have consistently found it more difficult than others to recruit and retain health visitor staff. The Middlesex mental health service seems to have been a casualty of the reforms which has not yet been made good.

Central government economic policy has prevented many boroughs from carrying out their plans especially of capital development as fast as they would have liked; this explains to some extent for example why hopes for the rapid run down of large old people's homes have not been fulfilled. Former LCC officers were of the opinion that that body was more independent in many matters and could have resisted government pressure; but once reform had been decided upon, the continuation of the LCC was no longer an option.

The principal advantage gained for the health and welfare functions from the new structure has been to create order over the whole Greater London area out of what had been a multiplicity of large and small authorities. Responsibility for the full range of these services lies clearly on each London Borough Council in its area. This allows the possibility, whether it is taken advantage of or not, for the integration of the many services, and the rational planning of expansion. If co-operation and co-ordination has not progressed as far as Herbert would have hoped, the basis has been laid for a big move forward with the implementation of the Local Authorities Social Services Act, 1970. Without the reorganization this Act would have been far more difficult to apply in the London area.

How much has been gained from the impetus due to more local enthusiasm and interest is difficult to assess. Certain welfare services and the environmental health services were, in fact, incorporated into larger authorities by the change; some officers and councillors were not very happy about this. On the other hand in many departments a sense of *élan* and determination to provide better services can be observed, though in some boroughs this might be true in one department and not in the other. Taking the London area as a whole, there is evidence from the overall figures that its services are progressing at a faster rate than those for the country in general.

One further aspect of the London reforms needs to be considered – the consequence of giving all the health and welfare powers to the London boroughs. This system does not make provision either for the difficulties or for the opportunities arising from the fact that the 32 authorities operate in a continuously built-up area. The emergence of ad hoc bodies such as the London Boroughs Training Committee and the designation of Greater London as one area for the planning of Community Homes under the Children and Young Persons Act, 1969, are a response to the need in some matters for decisions on an all-London level.

The present arrangement does not exploit the advantages of a large conurbation, which, for example, offers scope for the provision of services for numerically small groups of people such as the young physically handicapped, and the deployment of highly specialized skills as in the field of epidemiology. The London Boroughs Association and its sub-committees cannot meet this need. The Social Services sub-committee acts as a channel of communication and can discuss co-ordination of policies and plans for joint provision of services, but it can only advise. It has neither the powers nor the staff to take decisions and carry through new developments.

The reform of London Government was not carried through primarily for the benefit of the health and welfare services, but it presented opportunities for local enthusiasm and knowledge and for more rational organization at ground level. Some authorities seized the opportunity more vigorously than others. A certain price has been paid with the result that to the citizen the success of the change may depend a good deal upon where he happens to be living. This leaves for future consideration the suggestion that there is a need for some place in the local government system of a continuously built-up area where decisions can be taken jointly when they concern highly specialized needs or skills.

Nevertheless the evidence shows that the London borough was a structure capable of sustaining the burden placed upon it.

5

Children's Services

London Boroughs' Children's Departments existed for little more than five years. Child care responsibilities were transferred from the former counties to the new boroughs in 1965, as a result of the reforms in London government. In April 1971 the work of children's departments throughout the country was integrated with other local authority personal social services. This was the direct administrative result of the Seebohm Committee's report,[1] the recommendations of which were largely incorporated in the Local Authority Social Services Act, 1970. Seperate children's departments as they existed since 1948, disappeared. Children's services in London underwent two major upheavals within the short space of some five years.

This, however, was not all. At the same time that boroughs were having to establish their new children's services after 1965 they had also to implement the provisions of the Children and Young Persons Act, 1963. This legislation was of great importance for it placed upon local authorities a general duty to prevent the need for children coming into care or appearing before the juvenile courts. Formerly there was no power for children's departments to take any action of this kind, for example by way of advice or the support of families at risk. The new requirements and opportunities demanded additional resources as well as the development of rather different patterns of work. Priorities also had to be reconsidered.

Added to this at the end of 1965 the Government appointed the Seebohm Committee and, although it did not report until the summer of 1968, its existence throughout the middle part of this period created a growing sense of uncertainty and speculation about the future structure of the child care services. The publication of its report increased the preoccupation with possible reform, especially since the Government did not announce its reaction until almost a year later. During this period social workers throughout the country, and particularly those in child care, pressed hard for the implementation of Seebohm's recommendations and the campaign was especially vigorous in London after a report of the London boroughs' management services team recommended changes along somewhat different lines.

In addition to these developments, the Government introduced a new Children and Young Persons Act in 1969. This was another major piece of legislation which increased the responsibilities of children's departments for young offenders and those in need of care or protection. Informal procedures of consultation became available with the aim of keeping children out of the courts.[2] However, this made greater demands upon social work resources. In addition certain responsibilities for those children who do appear before the courts is shifted to the local authority from the probation service. Moreover, the Act also introduced changes in the residential provision for children 'in care' whereby approved schools will be incorporated into the new general system of community homes, and a range of 'intermediate treatment' provided. These changes in residential care are beginning to be planned on a regional basis by joint planning committees. The full impact of the 1969 Act has not yet been felt, partly because its various sections became operational by stages. Nevertheless, children's departments in London, as elsewhere, had to start planning and building up their staff resources in readiness.

Thus during the short life of the London Boroughs' Children's Departments they have been exposed to continual and radical change. Indeed, throughout the country, the five-year period from 1965–70 saw more change in the work of the children's services than at any time since 1948. Furthermore the changes we have noted are only the formal landmarks, other less obvious shifts in professional thinking took place as well as the development of new kinds of provision. All this occurred against a backcloth of a generally high rate of turnover amongst child care field staff,[3] a considerable programme of professional training for the untrained and severe shortages of residential staff.[4] To the outside observer the remarkable fact about the child care services in London during this period is that they continued to operate without major breakdown, and this is especially true of the Inner London boroughs which normally receive higher levels of 'bombardment' than other parts of the country.

Undoubtedly changes would have occurred with or without the reform of London Government: for instance, the greater emphasis upon prevention; the build-up of reception facilities; the establishment of family advice centres; the developments in training and so on. There were certain national pressures operating which influenced the work of the London boroughs no less than other authorities. Additionally, of course, all the counties which were formerly responsible for London child care services had various schemes in the pipeline at the time of the transfer, and this is especially pertinent in the case of the LCC and Middlesex. The LCC for instance had plans for the closure and replacement of certain large children's homes and for more delegation to each of its nine areas. In some fields, however, especially residential provision, the transfer of responsibility to the boroughs appears to have consolidated or quickened some pipeline schemes. This partly reflected the determination of the new authorities to provide at least as good a service as the counties. Indeed this was specified as a 'primary objective' by several of the new borough children's committees at their first meetings. The acceleration of some schemes can also be explained in light of the kinds of problems the boroughs faced. These will be discussed later.

From this brief description of the environment of change which surrounded the London Boroughs' Children's Departments during their short history it is obvious that a 'before and after' comparison cannot satisfactorily be made. Even if it were possible it would in any case seem of dubious value at a time of further general upheaval. Moreover, as we show later, there appear to be considerable differences between the inner and outer borough children's departments which make it hazardous to generalize about the whole of London. Instead of adopting this approach therefore we will identify and discuss several important features of the London children's services since the reforms. These have implications for the future, not least in drawing attention both to the costs and opportunities of change.

There are certain characteristic features of contemporary child-care work which need to be emphasized if the administrative problems and policies of the boroughs are to be understood. First, the work has a considerable 'emergency' quality which, although varying between authorities, is generally substantial in central urban areas. A comparatively high proportion of children coming into care stay only a short time before being discharged again, usually to their own families.[5] Figures from two inner boroughs will illustrate the point. In both of them 58 per cent of the children discharged from care in the year 1966–7 had stayed in care less than two months.

Table 5.1

Children discharged from Care 1966–7

Length of time in care:	Hackney (738 in care)		Kensington and Chelsea (676 in care)	
	No.	%	No.	%
under 2 months	363	58	283	58
2 under 4 months	39	6	57	12
4 under 6 months	18	3	18	4
6 under 12 months	39	6	22	4
1 year under 2 years	49	8	28	6
2 years under 3 years	30	5	30	6
3 years under 5 years	38	6	27	6
5 years or more	50	8	22	4
Total discharged	626	100	487	100

Although there is a hard core of children who stay in care a long time, the turnover amongst the remainder is considerable. However well developed an 'early warning' system, in areas of high mobility – typically inner urban areas – there continues to be a significant element of unpredictability about requests for admission to care. In most instances when a child has to come into care action cannot be deferred and a placement has to be arranged *immediately*. As a result of high turnover, varying degrees of unpredictability in demand, and the immediacy of service provision, a considerable part of child care work must involve admission, finding suitable accommodation and care, and arranging discharge. The turnover rate is also important in that the time spent on 'new cases' is known to be greater than that spent on those who have been in care for longer periods.[6] Social work with the child in care and his family and prevention inevitably receive second priority and are therefore considerably affected by the extent of the admission and immediate placement problems. These aspects of child care place a premium upon the existence of a pool of available (and hopefully appropriate) placement resources which can be quickly mobilized; that is, which are standing ready and are under the control of the authority. Command over a large number of possible placements is one of the advantages of bigger authorities. This was true of the former LCC, although the advantage had to be offset against the cost to children and visiting relatives of transporting them long distances away from home. If there is no such supply of vacant and appropriate placements or the pool is too small, little regard can be paid to the particular needs of a child since his immediate accommodation *somewhere* becomes of paramount importance. This consumes both considerable time and energy in locating a placement, in persuading other statutory and voluntary authorities to allocate it, or in getting a foster parent to take a child at short notice.

Given therefore that one aspect of good child care practice is the *appropriate* placement of a child, two things are essential: first, the availability of varied placements in sufficient numbers and second, control over their allocation. In our view, this was one of the key problems faced by the London boroughs, accentuated by the dismantling of larger authorities and the division of their resources.

In many authorities the problem of controllable placement resources was further aggravated by two policy preferences. In the first place by the commendable desire of many committees to see all local children who were received into care placed, as far as possible, within the boroughs. The slogan 'local responsibility for local children' appears in several of the first annual reports. The placement resources of the counties had been built up over a much wider area. In the case of the LCC this stretched over a large part of SE England, and was pointed out by the Herbert Commission. 'About half of the children boarded-out by the London County Council were,' they reported, 'in foster homes outside the Administrative County at distances up to 80 miles. About 80 of the 100 homes run by the County Council are outside the boundary of the County; and the County Council also use about 400 voluntary or private homes, the great majority of which are also outside the County.'[7] This remarkably widespread distribution of the LCC's placement resources has to be seen in a historical context. First, many of them would have been developed in the inter-war period when the objective of rehabilitating the child and his family was less emphasized. Indeed it was, in some cases, believed desirable deliberately to sever the child from the 'adverse influences' of his family and local neighbourhood background. Second, in the same period, many of the large LCC 'out-county' estates were built. Several studies have shown that council estates provide a high proportion of foster-parents. Indeed, fostering is pre-eminently a working-class institution.[8] Third, in the earlier part of the century at least, ideas about the importance of open space and fresh air which echoed the general 'garden city' philosophy were influential. Last, land prices were also important. These had to be taken into account by both local authorities and voluntary bodies when building (or acquiring) large residential establishments. These often stood in generous grounds and contained all the services (schools, playing fields, laundries, and market gardens) which were then thought necessary but which are now more often shared with the general community. Some of the large homes stand in as much as 50 acres.

However, where a local care policy was particularly favoured by the new boroughs, the clear implication was that new and additional placement resources had to be found *within* the borough or at least near by. For instance, the third annual report of the Children's Officer for Lambeth pointed out in 1968 'that it becomes more and more obvious that homes are needed actually within the borough boundaries', and went on to say that one of the reasons was because numerous children who came into care had special educational needs which could not easily be met if they were in homes outside the ILEA area.[9]

The second aggravating factor was the fairly general dislike of the various joint-user arrangements which existed for the larger residential homes. Although places were usually divided up on a quota basis, each individual authority exercised only limited control over allocation, standards and staffing. Several children's committees early stated their intention of becoming as independent as possible of these arrangements.

Additionally, several boroughs were clearly dissatisfied with their allocation of residential facilities. Take for example the comments in the Kensington and Chelsea report on the first three years' work of the children's committee:

'At the time of the transfer of the Children's Service to the London boroughs, the former London County Council administered residential establishments capable of accommodating 41 per cent of the children in its care, and had firm building plans which would have brought this figure up to 45 per cent . . . although the Royal Borough assumed the care of 626 children, only four establishments were allocated to it, capable of accommodating 101 children, ie 16 per cent. Three establishments in the planning stage were also allocated to the Royal Borough, and these would ultimately increase the figure to 22 per cent. Strong, but unsuccessful, representations had been made to the Home Secretary in November, 1964, concerning the proposed allocation of establishments, it being emphasized that if a larger number of former LCC establishments were not allocated to the Royal Borough, an undue burden of capital expenditure would have to be borne over the following few years.'[10]

The *balance* between different kinds of accommodation was also criticized, for instance by Lewisham which inherited one of the largest residential nurseries in the country. Not infrequently there was also adverse comment about the age, condition and general standards of some residential accommodation, particularly the lack of central heating. Hence not only were there good reasons for increasing residential provision overall but also to adjust the balance by making provision for specific purposes and to undertake improvements.

The general problem of sufficient appropriate and controllable placements could be surmounted in three main ways: first, by reducing the number of children in care through more stringent admission policies or by engaging in better preventive work. The effect of such policies can take some time to materialize and in any case the London area was exposed to increasing demands for admission and for help and assistance. The only Inner London borough to make an appreciable reduction in the number of children in care was Islington, where 20 per cent fewer children were being cared for on 31 March 1969 than on 31 March 1966. Indeed it was the only inner borough to show a reduction in the number of children in care per 1,000 of the population under 18, and this by 15 per cent. Amongst the outer boroughs Barnet and Richmond upon Thames were the only authorities to show a marked reduction in the number and proportion of children in care: 32 per cent and 25 per cent respectively in the same period. Table 5.2 sets out the numbers in care and number per 1,000 of the population under 18. The second of these statistics indicates changes occurring independently of the population size at risk but does not of course reflect any change in the social structure of the borough which might be crucial in determining that figure.

The second means of overcoming the problem of placement shortage was for the boroughs to increase their foster parent resources. However, most of the inner authorities in particular had limited opportunity to make improvements on this front, especially if they were committed to local care policies. As one annual report put it: 'We are forced to the conclusion that finding good short stay (foster) homes is an unrewarding task.' This was borne out by research undertaken by the GLC Research and Intelligence Unit. Their study concluded that 'most of the difference between boarding-out ratios of boroughs can be accounted for by the characteristics of the areas concerned . . .'[11] Indeed looking at the period 1965–9 it is apparent from Table 5.3 that none of the inner boroughs significantly increased their boarding-out rates except Tower Hamlets, but that borough also reduced the number of children in care during the same period by 7 per cent (this nevertheless still meant some 70 additional children in foster homes). In the outer boroughs the only notable increase in the boarding-out rate was at Haringey where it rose 11 per cent *despite* a 57 per cent increase in the number of children in care (this indicates about 110 extra foster homes found). Richmond upon Thames shows a 12 per cent increase in boarding-out but this has to be seen against a 25 per cent reduction of children in care (ie approximately the same number of foster homes). In general, however, 'trend' figures for the smaller outer boroughs can be unreliable since modest shifts in actual numbers lead to apparently dramatic shifts in percentages.[12]

Table 5.2

Children in the care of London boroughs, 1966 and 1969

	1966		1969			
	(1) No. in care	(2) No. in care per 1000 pop. under 18	(3) No. in care	(4) No. in care per 1000 pop. under 18	(5) % increase or decrease in no. in care	(6) % increase or decrease in no. in care per 1000 pop. under 18
Inner boroughs						
Camden	610	13·3	595	14·5	− 2·4	+ 9·0
Greenwich	494	8·5	510	8·6	+ 3·2	+ 1·2
Hackney	784	11·4	794	12·0	+ 1·3	+ 5·3
Hammersmith	638	13·6	702	16·0	+ 10·0	+ 17·6
Islington	1083	16·4	862	13·9	− 20·4	− 15·2
Kensington and Chelsea	678	17·8	722	23·3	+ 5·6	+ 30·9
Lambeth	1039	12·1	1063	13·1	+ 2·3	+ 8·3

| | 1966 | | 1969 | | | |
| | (1) | (2) | (3) | (4) | (5) | (6) |
	No. in care	No, in care per 1000 pop. under 18	No. in care	No. in care per 1000 pop. under 18	% increase or decrease in no. in care	% increase or decrease in no. in care per 1000 pop. under 18
Lewisham	702	9·6	760	10·4	+ 8·3	+ 8·3
Southwark	848	10·5	895	11·5	+ 5·5	+ 9·5
Tower Hamlets*	1444	26·2	1361	26·7	− 6·9	+ 1·9
Wandsworth	941	12·1	1025	13·3	+ 8·9	+ 9·9
Westminster	697	17·4	699	18·9	+ 0·3	+ 8·6
Outer boroughs						
Barking	178	3·7	199	4·6	+ 11·8	+ 24·3
Barnet	360	4·7	242	3·5	− 32·0	− 25·5
Bexley	109	1·9	128	2·2	+ 17·4	+ 15·8
Brent	579	8·3	592	8·0	+ 2·4	− 3·6
Bromley	238	2·9	294	3·7	+ 23·5	+ 27·6
Croydon	364	4·1	351	3·9	− 3·6	− 4·9
Ealing	433	6·2	435	5·9	+ 0·5	− 4·8
Enfield	257	4·0	226	3·5	− 12·1	− 12·5
Haringey	379	6·1	595	9·0	+ 57·0	+ 47·5
Harrow	110	2·2	97	2·0	− 11·8	− 9·1
Havering	162	2·1	187	2·6	+ 15·4	+ 23·8
Hillingdon	223	3·6	231	3·7	+ 3·6	+ 2·8
Hounslow	292	5·7	276	5·5	− 5·5	− 3·5
Kingston upon Thames	161	4·6	145	4·3	− 9·9	− 6·5
Merton	197	4·4	218	5·0	+ 10·7	+ 13·6
Newham	330	5·0	356	5·3	+ 7·9	+ 6·0
Redbridge	117	1·9	157	2·7	+ 34·2	+ 42·1
Richmond upon Thames	209	5·2	156	4·0	− 25·3	− 23·1
Sutton	161	3·9	167	4·2	+ 3·7	+ 7·7
Waltham Forest	200	3·4	209	3·0	+ 4·5	+ 14·7

Source: Home Office, *Children in Care in England and Wales*, March 1966, Cmnd. 3204, and March 1969, Cmnd. 4263.

* Includes a few City of London children.

Table 5.3

Children in care boarded-out by London boroughs, 1966–9

	% boarded-out*		% increase or decrease in no. of children in care 1966–69 (from Table 5. 2)
	1966	*1969*	
Inner boroughs			
Camden	26	29	− 2·4
Greenwich	43	37	+ 3·2
Hackney	28	25	+ 1·3
Hammersmith	22	25	+ 10·0
Islington	28	28	− 20·4
Kensington and Chelsea	18	19	+ 5·6
Lambeth	25	27	+ 2·3
Lewisham	35	32	+ 8·3
Southwark	28	28	+ 5·5
Tower Hamlets	29	36	− 6·9
Wandsworth	28	23	+ 8·9
Westminster	19	17	+ 0·3
Outer boroughs			
Barking	55	54	+ 11.8
Barnet	51	58	− 32·0
Bexley	47	53	+ 17·4
Brent	28	30	+ 2·4
Bromley	39	43	+ 23·5
Croydon	52	50	− 3·6
Ealing	37	23	+ 0·5
Enfield	46	36	− 12·1
Haringey	22	33	+ 57·0
Harrow	70	64	− 11·8
Havering	57	58	+ 15·4
Hillingdon	38	32	+ 3·6
Hounslow	42	42	− 5·5
Kingston upon Thames	36	34	− 9·9
Merton	47	48	+ 10·7
Newham	59	51	+ 7·9
Redbridge	70	57	+ 34·2
Richmond upon Thames	39	51	− 25·3
Sutton	41	35	+ 3·7
Waltham Forest	54	48	+ 4·5

Source: as for Table 5.2.

* The remainder of children in care were mainly in residential accommodation, and a few in lodgings.

The third and apparently widely adopted solution to obtaining more controllable placement resources was to increase the number of places in the boroughs' residential homes, particularly short-term provision and facilities for reception. Comprehensive figures for residential places are not easily obtainable but Table 5.4 below provides one indicator of this development – that is, the debt and rent payments for residential accommodation provided by authorities themselves. This is based upon the cost per child week.

Table 5.4

Proportionate increases or decreases (grouped) in debt and rent charges per child week for residential accommodation, 1966–9

Increases		Decreases	
100% or more	Bromley Enfield Hounslow Redbridge Sutton Waltham Forest	*0% – 24%*	Barking Hillingdon Lambeth Lewisham Richmond upon Thames Westminster
50% – 99%	Bexley Camden Hackney Kensington and Chelsea Kingston upon Thames Tower Hamlets	*25% – 49%* *50% or more*	Croydon Greenwich Barnet Newham
25% – 44%	Brent Ealing Havering Southwark		
0% – 24%	Hammersmith Haringey Islington Harrow Merton Wandsworth		

Source: Derived from IMTA *Children's Service Statistics,* 1966–7 and 1968–9. (N.B. reception homes are omitted from these statistics.)

Over the financial years 1966–9 therefore half the boroughs increased the debt and rent charges for residential care (per child week) by 25 per cent or more. The most dramatic increases were in some of the outer boroughs. Figures for the year 1965–6 are incomplete but indicate that some boroughs, particularly in the inner area, increased debt charges considerably in their first year of operation, notably Camden and Islington. Annual reports furnish further evidence that the wishes of some authorities to develop this aspect of their provision had been frustrated by various 'external' factors. In Lewisham, for instance, it was stressed that the 'capital programme for provision of children's homes and specialist establishments to counter the imbalance of residential accommodation has met with little success, due entirely to failure to secure the necessary governmental loan sanction'.[13]

Among the inner boroughs another broad indicator of this increasing desire to exercise direct control over their placements was the general decline in the use made of the various voluntary homes. The utilization of these placements reduces the cost per child in residential care considerably: in 1965–6 for instance the *difference* between the average cost per child week in a voluntary placement and that in a London borough home was 157s (£7·85) and in 1968–9 even greater at 216s (£10·80).[14] There is, therefore, no financial incentive whatsoever to cut back these arrangements. However, of the Inner London Authorities all but three (Hammersmith, Lewisham and Wandsworth) reduced the proportion of their children in voluntary homes between 1966 and 1969; 5 of them (Camden, Hackney, Islington, Kensington and Chelsea and Lambeth) by over 5 per cent. The outer boroughs have not used voluntary societies and private homes so extensively and it is difficult to see evidence of a similar trend; indeed half of them increased the proportion of their children placed in this way, a few quite significantly (eg Enfield, Redbridge and Richmond upon Thames by 10 per cent or more).[15]

There is the additional problem of placement provision for certain small numbers of children who present special difficulties and have needs which make them a minority even amongst the minority of children in care: the multi-handicapped; the half-caste; the highly aggressive adolescent boy or the very promiscuous adolescent girl. They are children who cannot, for various reasons, easily be placed in a foster home, who may be unduly disruptive in an ordinary children's home or who need special care and treatment. In a small authority each category may represent only a tiny fraction of those in care.

A high standard of specialized provision demands an organization drawing upon an extensive catchment area. This was one of the advantages claimed for the LCC and indeed its 'special' hostels are one example. No single London borough is likely to develop provision which will largely serve other authorities. So, despite their dislike of joint arrangements, it may well be that ultimately such specialized minority provision will be organized by the London boroughs in this way. Already, for instance, another minority service – that of adoption – is provided jointly by at least two combinations of authorities. Hackney and Redbridge provide a common adoption service, and Hounslow, Richmond and Hillingdon have combined to continue the Middlesex Adoption Service first created by the county in 1962.

The reform of London Government therefore had an effect upon the provision of residential accommodation for children in care. It is likely that more new accommodation was provided sooner than would have otherwise been the case, especially reception facilities; that some highly specialized, experimental, or high risk provision was checked, but that more residential places were provided in the Inner London area than would have been done had there been no reform.

In summary the main reasons for this situation were: first, the fact that the transfer and division of county resources meant that some boroughs (despite joint arrangements) were inadequately equipped with residential places. Second, in several boroughs there was, in any case, a wish to disengage from shared arrangements. Third, in the inner boroughs especially, the pool of foster parents shrank or could not be increased. Fourth, the new children's committees were, in almost all areas, publicly committed to providing *at least* as good a service as the former counties and often anxious to keep provision local. This attitude needed expression in action and its most tangible and understandable form was building. Fifth, especially in certain inner areas, sites were made available by other borough departments (especially housing) which had been denied the children's service when it had been a first tier responsibility. Moreover, in the case of the LCC, it has been claimed that because of high costs in central areas sites for new residential accommodation or property for conversion were, in any case, generally acquired elsewhere.

For possibly the first three years after 1965 therefore, a predominantly generous attitude to capital works for children's purposes prevailed at the borough level, especially, it has been suggested, in the inner areas where there was no direct competition from a school building programme. In addition the Home Office appears to have been generally sympathetic in terms of loan approval although at certain periods and in respect to certain projects restrictions were being imposed. Had the boroughs been more fully allowed their heads the developments in capital works would almost certainly have been more numerous.

The work of the children's departments not only depended upon the kind and amount of placement resources but also upon manpower resources. Child care, like other forms of social work, is labour intensive, and its quality ultimately depends upon face-to-face contacts between the provider and receiver of service. The range of work undertaken by children's departments increased markedly in the 1960s as we have already noted. For example, prevention, advice services and adoption all demanded improved field staff ratios. Work with the families and children 'at risk', but not in care, has grown substantially since 1963. Almost all the boroughs increased their staffing establishments after 1965, although some did so more than others. Despite this, staffing problems remain of central importance. We shall consider three aspects: training, consultation and distribution.

Although social work lays claims to being an emergent profession, it still contains many who are untrained. In child care however the position in London is comparatively favourable as Table 5.5 below shows.

However there is a continuing and sizeable task of providing training (or refreshment) opportunities both for those who are already in the service as well as contributing to the training of the newly recruited. In this whole field the former LCC in particular had major achievements to its credit. Indeed one of the stronger arguments against dismantling their children's service was that the fragmented and smaller borough system could not continue or extend this kind of programme. Partly through its recruitment and training schemes the LCC had tripled its child care staff between 1958 and the final transfer in 1965.[16]

However, the gloomy forecasts about the future of child care (and

Table 5.5

Child care officers – qualifications

	% Qualified	% Part-qualified	% Unqualified	% Total
England and Wales 31.3.66	37	23	40	100
London boroughs 30.9.67	53	13	34	100

Source: for England and Wales, *Committee on the Staffing of Local Government* (Mallaby) 1967. para. 145, p. 48. For London, T. G. Morris, 'Survey of Professional Qualifications of Social Workers in the Personal Social Services in the London Boroughs' in the *Quarterly Bulletin of the Research and Intelligenc Unit of the GLC*, December 1968, No. 5, Table I, p. 19.

other social work) training in London were not fulfilled, largely because of the success of the London Boroughs' Training Committee. This has been able to sustain and develop a variety of training programmes for health, welfare and children's services on the basis of the voluntary membership of individual boroughs. Day release and short courses have been organized for all levels, from chief officers to 'pre-recruits'. Liaison and collaboration have been established with other training institutions and organizations, and in general the overall social service training needs of London have been identified by the Training Committee and provisions made to meet them where no other facilities existed.[17]

Consultation and specialist advice are important aspects of child care practice. There was no comparable London boroughs' organization to fill the gap left in Inner London by the dismantling of the LCC children's inspectorate and in the outer areas the amount of specialist advice which the smaller boroughs could call upon was extremely limited compared with the resources of large counties like Essex or Kent. Some boroughs have tried to overcome the difficulty by arranging to engage part-time consultants. Even so the view was expressed in several authorities that less specialist advice and guidance was available after 1965 than before.

The third aspect of staffing is recruitment and distribution. One of the advantages of the larger authorities, again especially the LCC, was central recruitment and the subsequent deployment of staff on the basis of assumed or calculated local need. The increased number of authorities in London has certainly aggravated the whole difficult question of the equitable distribution of manpower resources over the metropolitan area. At the same time there are certain potential local advantages in the new system: for instance, because less attention has to be paid to questions of overall equity, some authorities can, if they choose, endeavour to capture a large share of manpower resources. The situation appeared to become more openly competitive, by the use of differential salary rates and preferential conditions of service in particular. As the Greater London Group's *The Lessons of the London Government Reforms* pointed out in relation to Hillingdon:

'The disparity of gradings between London boroughs is notorious, and nowhere more so than in the case of Child Care Officers. Hillingdon has been conspicuous in London salary affairs by calls for restraint and for comparability of posts and restrictions on mobility. Until the 'break through' can be achieved in the children's services the field workers shortage must be accepted as the price to be paid for a wider view . . .'[18]

Staffing problems appear to have been relatively most pronounced in some of the small authorities with a dozen or so field staff. One or two unfilled posts presented major difficulties; sickness or secondment for training were added hazards. The acute short-term nature of these problems is dramatically illustrated from the annual report of one of the smaller Outer London boroughs:

'From August until November we were unable to recruit a temporary officer, and almost immediately after her arrival, another officer fell sick and remained absent until late March, 1968. Another qualified officer resigned in December to take up a senior appointment with another borough, and the Principal Child Care Officer resigned in January prior to confinement. For some weeks owing to illness and vacant posts, the field staff was reduced to one Child Care Officer and one assistant Child Care Officer and we were compelled to reduce to a minimum the usual supervisory and preventive work of the department. Boarding-out was halted because of lack of staff, but the statutory frequency of visits in accordance with regulations was met. Fortunately, the position was eased by the end of March . . .'[19]

There is little doubt that considerable problems arose in small departments from sickness, turnover and secondment for training. Adequate supervision and specialist consultation were often scarce as well. The Home Office evidence to both the Redcliffe-Maud Commission and to the Seebohm Committee stressed the desirability of children's departments being at least large enough to require 12 field staff and deal with some 250 children in care, 100 of them in residential homes. Populations of about 250,000 were considered likely to produce these kinds of numbers.[20] As can be seen from Table 5.6, 12 outer boroughs fell below this 'in care' population criterion in 1966 and 13 in 1969.

Table 5.6

London boroughs falling below Home Office
suggested 'in care' criterion

Authority	1966		Authority	1969	
	No. in care	Population 000's		No. in care	Population 000's
Bexley	109	215	Harrow	97	208
Harrow	110	209	Bexley	128	215
Redbridge	117	247	Kingston		
Kingston			upon Thames	145	144
upon Thames	161	146	Richmond		
Sutton	161	166	upon Thames	156	177
Havering	162	250	Redbridge	157	246
Barking	178	171	Sutton	167	165
Merton	197	184	Havering	187	252
Waltham			Barking	199	170
Forest	200	240	Waltham		
Richmond			Forest	209	237
upon Thames	209	180	Merton	218	184
Hillingdon	223	233	Enfield	226	268
Bromley	238	302	Hillingdon	231	237
			Barnet	242	316

Source: as for Table 5.2.
Note: both Tower Hamlets and Hammersmith had considerably more than 250 children in care but less than 200,000 population in 1969.

Of course, as the Commission pointed out, three of the best authorities in the Home Office Inspectors' assessment had populations of

F

under 150,000; but the important conclusion was that *most* of the unsatisfactory authorities had populations of less than 200,000. Problems of staffing and specialist provision are obviously more difficult in the smaller authorities and although similar problems arise in larger authorities they can be more easily absorbed and counteracted. In London it seems more appropriate to concentrate upon the numbers in care as a measure of 'operational size' and with increasingly successful preventive policies this figure is likely to remain fairly steady or decline in many boroughs.

As far as the actual number of child care officers is concerned, in 1969 ten of the outer boroughs had establishments of 12 or less and all had fewer than 25.[21] In contrast *all* the inner boroughs had more than this. The 'in post' position showed 13 of the 20 outer boroughs with 12 or less child care officers. These data are set in Table 5.7 below.

Table 5.7

Child care officers (excluding senior staff)

Establishments and officers in post 31 March 1969

Nos.	Establishment		In post	
	Outer boroughs	*Inner boroughs*	*Outer boroughs*	*Inner boroughs*
12 or less	10	–	13	–
13 – 20	7	–	4	–
21 – 30	3	2	3	3
31 – 40	–	5	–	7
Over 40	–	5	–	2
Total	20	12	20	12

With the amalgamation of the personal social services of the health, welfare and children's departments, some of the difficulties faced by the smaller London children's authorities with respect to training, consultation and 'emergencies' may well be reduced. But in the years *before* 1969, of course, the number of child care staff was smaller in all boroughs. For instance, in 1965, soon after the transfer, 17 outer boroughs had establishments of 12 or less child care officers (4 indeed had fewer than 6).

For all the London boroughs the child care staff establishments drawn up in 1965 had increased by 26 per cent at March 1969 (staff in post during the same period increased more rapidly, by 59 per cent, because authorities took time to recruit their officers). Some idea of the significance of the establishment rate of growth however can be gained from comparing it with that in the surrounding home counties. In Kent during the same period it rose by 38 per cent; in Surrey by 48 per cent; and in Hertfordshire by 50 per cent. Essex had a high establishment figure in 1965 and only half the posts were filled; the period to 1969 saw them broadly meet this establishment. The establishment rate of growth for the London boroughs combined is not remarkable; and compared with the national figure of about 9 per cent their vacancies rate is only slightly better at 8 per cent. Obviously establishment figures can be misleading. Some authorities set them at levels which they felt could be met; others were more ambitious. Some started with generous rates in 1965; others did not, but raised them later. However they are some indicator of the desire of authorities to develop these services on a generous scale, or to maintain standards when new tasks and demands are arising. Relatively the most substantial increases were in the outer boroughs. Seven of them more than doubled their establishments in the five years 1965–9, Bromley and Haringey being most notable. Of the inner boroughs Camden, Greenwich, Hackney, Islington, Lewisham and Wandsworth appear to have made substantial increases.

The proportion of appropriately trained staff is crucial. In 1965 about 35 per cent of the child care staff in the London boroughs were fully qualified; in 1969 the rate was approximately 45 per cent.[22] Progress clearly had been made, both in recruitment of trained staff and in in-service training. Of course, there were and are considerable variations between boroughs. In 1965 at least 6 were without any professionally qualified field staff at all; in 1969 none were in this position. Indeed by then 10 of the boroughs had more than half their staff fully and appropriately qualified, and these mostly included the inner authorities, some of which, like Greenwich and Southwark, had very few trained staff initially.

It is extremely difficult and hazardous to generalize about the impact of London Government reform upon the staffing of the child care service. In this brief review we have omitted entirely the residential staffing problems, partly from want of comprehensive data and partly because the problem is in any case complex. Several things are however fairly clear. The forecasted catastrophic effects of the reforms upon training have not occurred, very largely because of the success of the London Boroughs' Training Committee (Social Services). The small size of some of the outer boroughs has posed difficulties about release for training, adequate cover for sickness and other contingencies, as well as specialized advice. The position in London during this period became increasingly competitive. Despite an overall improvement in the proportion of trained staff certain boroughs recruited more than others but the variation does not seem too large. At March 1969 half the inner boroughs employed about two-thirds of all the professionally qualified child care staff working in the London boroughs. In the outer boroughs however almost half the qualified staff worked for only 4 boroughs. Variation amongst these authorities in qualified staffing is now probably greater than it was between the former county children's services in these areas. Although there is, therefore, no substantial evidence that London Government reform had an adverse effect upon the recruitment and distribution of qualified staff in the inner boroughs, greater disparities appear to have arisen between the present outer authorities and between many of these and the counties of which they were formerly a part.

The question of securing collaboration with other departments and organizations looms large in discussions about the child care service no less than other social work services. This arises because social work is partly concerned with effecting liaison on behalf of individuals and families and because, moreover, the problems with which it deals are often complex and require help from many quarters. The work of children's departments could not satisfactorily be accomplished without the collaboration of many agencies: from the gas boards to general practitioners; from housing departments to the courts; and from voluntary societies to other local authorities. As prevention is increasingly emphasized this becomes even more important. Within authorities moreover collaboration was essential with various other departments; treasurers', architects', health and welfare, engineers' and so on. What kinds of collaborative patterns existed for the London boroughs' children's departments during the five years after the reforms? Where did relationships seem to have been improved? Where did they get worse?

It is of course difficult to generalize but from the evidence of numerous interviews a pattern seems to emerge. First, within each of the boroughs collaboration between children's and other departments improved. In particular officers who had experience of the county system (and especially the LCC) maintained that their departments now received greater assistance, understanding and *speed* from treasurers', architects', engineers' and borough housing departments. Children's officers, unlike the former area officers, felt able easily and quickly to contact their fellow *senior* officers. There was less consensus about changes in relation to health and welfare departments. To some extent the former county area relationships were continued (for good or ill), whilst from the establishment of the Seebohm Committee in 1966 a certain amount of suspicion and strain understandably developed, with health departments justifiably fearing that their social work function would be removed in any reform of the personal social services. Considerable difficulties were reported in the Inner London boroughs in relation to education and those other services (school psychological service; educational welfare and care committees) which remain associated with the ILEA. The problem of securing smooth collaboration appears to have arisen partly because differences in the size of the two kinds of authorities meant that a rather different hierarchy of administrative and professional discretion existed. The senior staff of children's departments, for instance, did not find that local senior officers in education had the same decisional antonomy which they exercised. However, this was not the whole problem. In the case of special education (a service with which children's departments were frequently involved) there was a widespread feeling that boroughs lacked 'leverage' in getting places allocated.

Other difficulties were reported regarding education. For instance those boroughs with residential and foster home placements in the outer boroughs or present county areas (eg Kent or Essex) had to negotiate and work with schools and education authorities outside the area of the ILEA. Of course this was common before the reform but there were then fewer education authorities.

Similar comments concerned collaboration with the GLC housing department. Some former LCC child care staff asserted that relationships had deteriorated since the transfer; that in the GLC's area housing offices a view had developed that children's departments seeking co-operation and special help with regard to housing should, wherever possible, look to their own local borough housing departments.

One of the repeatedly mentioned advantages of the former LCC and Middlesex children's departments was the fact that the separate areas within their authority were subject to common overall policies and administrative practices. This meant that a basis for areas working together was firmly established and that in cases of disagreement or conflict a common superordinate authority existed. At the stage of transfer and initial planning there was clearly a major need for the close co-operation of groups of children's authorities. A Society of London Children's Officers was established in 1963 as well as an Association of Inner London Children's Officers (because of the existence of the ILEA). Many chief officers bore witness to the value of these organizations especially in the early period of the new authorities. Added to which of course, many of the children's officers were former colleagues in the LCC and some in the MCC. Many, both in the inner and outer areas, had been the officers in charge of the counties' area offices. As such they had experience in the control of professional matters but comparatively little experience of administration; and some had also been isolated from direct contact with committees. Hence there were boroughs in which not only had the committee members little or no experience of the work of children's departments but the children's officer had no direct experience of working with committees. There was thus an especially difficult initial phase for several chief officers in which the support and help of their colleagues in other boroughs was invaluable. At a different level the LBA social services committee also had a constituent advisory committee of chief officers.

In general the transfer and the early period of reorganization placed a high premium upon co-operation between boroughs. As time passed it became less imperative and, as we have noted, situations of competition arose which at times made collaboration harder. For instance, considerable co-operation was required in closing down a large children's home with numerous user authorities (nine in one case). Different authorities felt able to progress in planning alternative arrangements at different rates. This created problems associated with the slow run-down of a home with which the authority with primary responsibility had to cope.

Only three aspects of the child care service in London have been examined in this chapter; namely placement resources, manpower and collaboration. The recasting of London government has had both positive and negative consequences on each of these aspects although it is difficult to disentangle them from the more general trends to which children's services were everywhere subject and from the whole question of differences in 'endowment and tradition'. For instance, some boroughs continued an LCC, Middlesex, Kent or Essex tradition where a large proportion of the senior staff had been employed by these authorities. Elsewhere, many committee members *were* the former children's committee members of the counties. In other boroughs senior staff were recruited from outside the London area and new traditions and fresh practices were established from the start.

Although there are exceptions and qualifications which can and should be made about the consequences of the reform of London Government we have drawn a conclusion from this review of children's services. It is that the positives outweigh the negatives in the inner authorities but not in the outer areas.

6
Education

The Education Act, 1944, was a wide-ranging piece of legislation which made significant changes in the status and powers of the Minister, the position of voluntary schools, the education of handicapped children and much else. It would be inappropriate here to summarize all of it. For present purposes two of its provisions are important.

First, it redefined education as a continuous process in three successive stages – primary, secondary, and further – and made the first two compulsory for all. Previously, there had been elementary education for all children and higher education (viz. secondary and further) for a select few.

Secondly, the Act made the councils of counties and county boroughs throughout England and Wales the local education authorities (l.e.a.s); each was required to appoint an education committee and to delegate most educational functions to it. Previously, while the counties and county boroughs had been the principal l.e.a.s, Part III of the Education Act, 1902, permitted boroughs and urban districts above certain population limits to control elementary education. These Part III authorities were swept away by the 1944 Act. The administrative simplification this would have effected was, however, largely vitiated by provision for delegation within counties.[1]

The LCC was a special case. The 1903 Education (London) Act provided for educational administration to remain centralized in the LCC when the London School Board was abolished in 1904; it thus contained no Part III authorities. The 1944 Act also treated the LCC uniquely and exempted it from the requirement to prepare a scheme of divisional administration. The only part which the metropolitan boroughs and the City of London could play in the administration of education was to make representations to the Minister when the LCC submitted proposals for the establishment or

discontinuance of a county or voluntary school in the area. In practice little use was made of the provision.

Within the Herbert Commission's Review Area the administration
of most services presented a certain aura of complexity, but the
variations and intricacies of education administration were particularly daunting. There were in all nine l.e.a.s, nineteen divisional
executives, twenty-five excepted districts and two district sub-
committees.[2] Middlesex had twenty divisions of which no less than
sixteen were excepted districts. A prominent feature of the evidence of
these sixteen authorities was their resentment at what they considered
to be excessive control by the county council. This strongly influenced the Commission against delegation arrangements. The LCC
area was divided into nine divisions,[3] each with an office from which
certain activities, for instance local inspection, were conducted, and
which provided a convenient access point for parents to take complaints or seek advice.

The Herbert Commission's proposals. In respect of education the
great weight of evidence to the Commission – from the Ministry,
most local authorities, most professional opinion – favoured no
significant change. Only the clamourings of some lower tier authorities and the proposals of the LSE's Greater London Group pointed
in the opposite direction.[4] Once, however, it was decided to recommend a major reorganization of local government in Greater London,
largely because of the needs of planning and transportation, the
administration of education would obviously be affected too.

Since the Commission saw their proposed new boroughs as 'the
primary units of local government'[5] they plainly had to have a role in
the largest spending service, education. The Commission claimed
that parents ought to be able to approach their local councillors on
educational matters and added 'We cannot think of anything which
is more likely to ensure a persistence of lively health in the boroughs
of Greater London than that they should be given some active say
in the conduct of their schools'.[6] At the same time the Commission
could not ignore the fears which had been expressed, notably by the
Ministry, concerning the dangers of fragmenting the education
service between a large number of authorities. Moreover, the Greater
London Group had made a case for giving the top tier authority
responsibilities in the field of further education. The Commission
agreed that there would be a great advantage in having 'one well

*Figure 3(a). Administrative areas for education under the old system.
Areas unshaded were administered by divisional executives.*

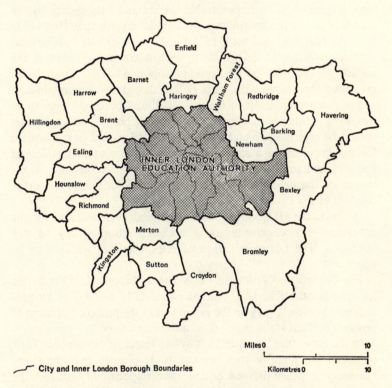

Figure 3(b). Administrative areas for education under the new system.

designed educational plan and one education authority for the area,' but thought that this could be done while leaving 'the control of the day to day running of the schools in the hands of much smaller and more local authorities'.[7] They insisted, however, that delegation should be avoided wherever possible.[8]

At this point one might be forgiven for feeling that the Commission were stuck fast in an impenetrable thicket and held tightly by the thorns of contradictory evidence and conflicting principle. How could both the boroughs and the GLC be given a major role in education without resorting to delegation or to a horizontal division of the service – opposed by professional opinion? The Commission extricated themselves with great skill, although not unscratched.

The competing claims of the new GLC and boroughs were to be met by a subtle two-tier arrangement in which 'the broad division should be that the [Greater London] Council should be responsible for the provision of a statutory standard of education throughout the area and that the boroughs should be responsible for the discharge of the executive work subject to the budgeting and other controls of the Council'.[9]

In short, the GLC was to have an overall responsibility for the education service, whilst largely excluded from its day to day operation. The boroughs' range of responsibilities was not to be enormous.

The Commission's proposals were not well received, either by the local authorities (the LCC reacted particularly fiercely) or by professional bodies, largely on the grounds that the proposed division of power would create as many difficulties as it solved.

In the event, the Government's White Paper of November 1961[10] conceded the force of such arguments. It claimed that the Commission's proposed division of education responsibility between the boroughs and the Greater London Council would be unlikely to work well. Accordingly the boroughs were to be increased to units of 200,000 or more in population so that each could be a viable l.e.a., responsible for the whole educational service in its area – in other words to make possible single-tier educational administration.[11]

Central London was not however, to conform to this pattern and the White Paper proposed a strange authority covering roughly two-thirds of London County. The logic of the latter proposal is difficult to detect. The LCC Education Committee not unreasonably pointed out 'the proposals in the White Paper are demonstrably an attempt to preserve within the framework suggested for the government of Greater London, the advantages of the education system as it at present exists . . . [there have been] no general educational arguments for the change or the exclusion of part of the LCC area.'

One can only suppose that the Ministry of Education urged the retention of at least most of the LCC system partly because of the unique educational history and circumstances of the area and partly to retain a large and progressive l.e.a. for the general benefit of education. In which case, however, why break it up at all?

Attention has already been drawn[12] to subsequent developments. The net result was the creation of that unique body, the Inner London Education Authority covering the whole of the former London County[13] and maintaining the long history of centralized administration of education within that area. In the rest of Greater London twenty new l.e.a.'s replaced part or all of the former educational administrations of five counties and three county boroughs.

THE TRANSITIONAL PERIOD

The general problems confronting the new authorities during the transitional period to the new system have been examined elsewhere.[14] In education as in other services the decisions taken during the eleven months to 1 April 1965 (on staffing, on policy, on administrative procedures, etc.) have affected the service to this day.

The new structure of educational administration contained large elements of continuity. The physical capital (schools, colleges, etc.) was of course the same; the majority of the teachers and many of the administrative staff were taken over by the new l.e.a.s; in a few instances there was even a strong element of continuity in the membership of the education committee. Nevertheless the factors of novelty were equally significant, often more so. Changes in administrative staff and even more elected members were sometimes marked. In most cases two or more previous educational units had to be welded into a single new one. Even where the boundaries of the new l.e.a. coincided exactly with those of the previous administration as in Harrow, the additional responsibilities of a full l.e.a., such as further education, had to be taken on. The ILEA experienced fewest changes in every respect and yet had to adjust to operating in a totally different administrative environment, as a semi-autonomous unit within the GLC rather than as an integrated part of the LCC.

The appointment of senior officers. One of the most important tasks of the new l.e.a.s was the appointment of a chief education officer and other administrative staff. The ILEA inheriting virtually the whole LCC Education Department intact was in a special position. It lost very few people from its administrative staff, nor did many head teachers move to administrative posts in outer boroughs. It therefore enjoyed a very high degree of continuity and prepared early for the 1965 changes appointing Sir (then Mr) William Houghton as c.e.o. of the ILEA at the Authority's first meeting on 5 June 1964. Sir William, like most other c.e.o.s, therefore had a joint appointment with the existing and proposed authority.

The appointment of 20 c.e.o.s in the outer boroughs was less obvious and simple. Where as in most cases several authorities constituted the new l.e.a., the education officer of each often applied for the post of c.e.o. The l.e.a. thus had to decide between them or, if none was thought suitable, to look outside. It was an occasion of much heartache on the part of both staff and elected members. Two authorities (Sutton and Richmond upon Thames) went outside Greater London altogether for their c.e.o.s. Four chose from other areas within Greater London; for example, Waltham Forest appointed Dr Stephens from the North West Division of Kent.

Fourteen boroughs, however, faced up to the difficult decision of choosing between the rivals from the previous authorities. In every case the appointee had served the largest or second largest authority in the new grouping, as in the case of Miss Pratt of Hillingdon who came from Hayes.

The appointment was not always devoid of rancour. In Newham, for instance, the East Ham c.e.o. was so hostile to the reorganization and so confident that it would not take place that he did not even apply for the post of c.e.o. in the new borough; it went to Mr Openshaw, c.e.o. of West Ham. Sometimes too party politics seem to have played a part. On the other hand at Brent one might have expected the Conservatives to promote the appointment of the education officer of Conservative Wembley and for the Labour Party to push the officer from Labour Willesden; in fact the parties agreed on the latter and he was formally proposed by the Conservative minority party.

As regards the appointment of staff other than the c.e.o. – deputy chief education officers, assistant education officers, senior administrative assistants, specialist advisers – there was sometimes the same problem of making a difficult choice between good local candidates. More often, however, there was a different problem, a shortage of applicants. Many of the Middlesex staff, for instance, lived south of the Thames,[15] and when the county disappeared a number of them chose new employment near their home. Sutton seems to have picked up a good number of them. This meant that boroughs like Brent and Barnet were deprived and had some difficulty filling vacancies.

The experience of the boroughs varied widely, even between neighbours. Thus while Hounslow filled its establishment with comparative ease either locally or from other Middlesex staff, Hillingdon had considerable problems and had to recruit extensively from outside. Similarly while Havering had recruitment difficulties, especially in respect of specialist advisers, Redbridge found that the staff of the former county districts provided ample resources; indeed the occasional former officer who went elsewhere made things easier since it left fewer obvious candidates for a particular post.

The extent of recruitment from outside was obviously linked with the number of staff from the previous authorities who stayed on to serve the new. At Croydon for instance the former county borough staff provided the core of the education department of the new l.e.a. At Richmond on the other hand only the deputy chief education officer was local (former divisional officer at Twickenham), most of the other senior staff being outsiders. At Waltham Forest although the former excepted districts of Leyton and Walthamstow brought their rank and file staffs virtually intact into the new borough, it was nevertheless short of certain more senior officers, particularly specialist advisers.

Conclusions. The transitional period was a time of great stress for both officers and members. To keep existing authorities functioning effectively whilst creating the new ones demanded prodigious efforts from all those involved. As has been mentioned, all kinds of problems from accommodation difficulties to staff shortages had to be overcome. Some difficulties were indirect and perhaps, therefore, more unexpected. For instance, when Middlesex's doom became apparent its architects naturally began to look for posts elsewhere. The winding down of the county architects' department greatly hindered school building for the period.

Time and again those who went through all this have claimed that a longer transitional period would have been invaluable. It would have enabled more attention to be given to organizational structure and forward planning instead of being overwhelmed by the daily emergencies. C.e.o.s often had virtually no staff to help them organize their new borough during a twilight period. It would have been eminently desirable to have had a group of qualified and experienced officers working exclusively on preparation for the new service. Anyway the experience of London suggests that more attention should be given to the problems of transition in local government reorganization.

The most enduring decisions taken in London during the transition were undoubtedly the appointments of senior administrative staff; most are still serving with the same authorities. The administrative practice decided on during this period has tended to persist as have many educational policies. It is indeed a danger that these will become entrenched over the years and that new l.e.a.s will be as immobile in this respect as their predecessors.

THE NEW EDUCATION AUTHORITIES

Basic data about the new education authorities are set out in Table 6.1. They may be conveniently divided into two parts. First there is the ILEA, by far the largest education authority in the country, with a total population in excess of three million, a school population of over 400,000 and a penny rate product of over £1,400,000. The ILEA is so important that it is treated by the Department of Education and Science as an association in its own right separate from the AEC.[16]

The size of the ILEA is reflected in its large number of schools (over 1,200) and massive corps of teachers (almost 17,000). The concentration of institutions of higher further education in the ILEA is unrivalled throughout the country. In addition, the ILEA has the largest and most specialized inspectorate of any local authority in the country – a fact noted in the Select Committee on Education and Science's report on HMI's.

The outer boroughs embrace a considerable range of school

Table 6.1
The New Education Authorities – Basic data

	Actual school pop.	No. of organizers	No. of teachers	Number of Schools and Colleges					
				Nursery	Primary	Secondary	Special	F.E. (major Institutes)	Colleges of Education
Barking									
1958–9	32,850			–	71	30	2	1	0
1962–3	29,996			–	71	25	2	1	0
1966–7	28,233	6*	1144	–	66	23	2	3	0
1969–70	27,290	8*	1064	–	63	23	2	3	1B
Barnet									
1958–9	39,362			3	86	33	2	1	0
1962–3	39,599			3	86	33	2	1	0
1966–7	39,828		1640	3	86	33	2	2	0
1969–70	41,910	11*	1748	3	89	33	3	2	0
Bexley									
1958–9	30,705			1	64	24	0	1	0
1962–3	31,304			1	63	25	0	1	0
1966–7	31,517	4	1287	1	64	23	0	1	0
1969–70	32,826	4	1342	1	68	19	0	1	0
Brent									
1958–9	37,111			1	70	35	2	2	0
1962–3	34,640			1	69	32	3	2	0
1966–7	36,504	2	1497	1	71	23	3	2	0
1969–70	39,440	7*	1666	1	72	20	5	2	0

| | Actual school pop. | No. of organizers | No. of teachers | Numbers of Schools and Colleges | | | | | |
				Nursery	Primary	Secondary	Special	F.E. (major Institutes)	Colleges of Education
Bromley									
1958–9	41,843			0	80	32	1	2	0
1962–3	41,935			0	81	33	2	2	1
1966–7	41,558	3	1710	0	82	33	3	2	1A
1969–70	44,726	7	1795	0	85	33	3	2	1A
Croydon									
1958–9	44,150	4 +	1419 +	3	84	40	3	2	0
1962–3	43,347	5 +	1353 +	3	82	40	3	2	0
1966–7	46,405	7	1838	3	90	39	3	2	0
1969–70	50,940	7	2045	3	96	33	3	2	0
Ealing									
1958–9	38,995			1	81	29	4	3	0
1962–3	40,954			1	85	35	5	3	0
1966–7	39,003	8*	1611	0	79	30	5	3	0
1969–70	42,090	9*	1783	0	79	28	6	4	1
Enfield									
1958–9	38,745			0	87	31	2	2	1
1962–3	36,552			0	87	31	2	2	1
1966–7	35,460	7*	1458	0	87	29	3	2	1

1958–9	33,800			3	67	29	3	2	0
1962–3	33,757			3	76	34	3	2	1B
1966–7	34,250	5	1444	3	72	28	3	2	1B
1969–70	35,670	12*	1508	3	73	16	3	2	1B
Harrow									
1958–9	27,374			3	49	18	2	1	0
1962–3	26,067			3	49	21	2	1	0
1966–7	25,835	5*	1046	3	49	20	2	1	0
1969–70	27,830	7*	1103	3	49	20	2	2	0
Havering									
1958–9	41,781			0	73	27	1	0	0
1962–3	42,497			0	75	29	2	1	0
1966–7	41,789	9*	1736	0	77	29	2	1	0
1969–70	43,250	10*	1666	0	84	27	3	1	0
Hillingdon									
1958–9	32,931			2	74	20	1	0	0
1962–3	34,464			2	74	23	1	0	0
1966–7	34,941	10*	1456	2	74	23	4	1	0
1969–70	36,930	10*	1457	2	76	23	6	1	0
Hounslow									
1958–9	27,706			1	61	23	4	2	0
1962–3	27,735			1	61	24	4	2	1
1966–7	28,008	10*	1160	1	63	22	5	2	1
1969–70	30,160	11*	1238	1	63	15	4	2	1

	Actual school pop.	No. of organizers	No. of teachers	Number of Schools and Colleges				F.E. (major Institutes)	Colleges of Education
				Nursery	Primary	Secondary	Special		
Kingston upon Thames									
1958–9	17,980			1	38	14	2	2	0
1962–3	17,890			1	38	14	2	3	1
1966–7	17,634	4	710	1	37	14	2	3	1C
1969–70	17,900	4	752	1	39	14	2	3	1C
Merton									
1958–9	27,870			2	54	32	1	2	0
1962–3	26,065			2	52	32	1	2	0
1966–7	24,227		1019	2	49	27	2	2	0
1969–70	24,920	6	1011	2	38	26	2	2	0
Newham									
1958–9	41,576	6	1627	4	94	30	4	2	0
1962–3	37,801	6	1451	4	93	29	4	2	0
1966–7	37,875	5	1593	4	88	28	4	3	0
1969–70	39,500	5	1667	4	89	27	4	3	0
Redbridge									
1958–9	32,607			0	65	26	1	0	0
1962–3	32,200			0	64	27	1	0	0
1966–7	31,513	7*	1238	0	64	27	1	0	0
1969–70	33,100	9*	1359	0	63	28	1	1	0
Richmond upon Thames									
1958–9	22,184			2	56	22	0	2	0
1962–3	20,940			2	53	22	2	2	0
1966–7	19,605	5*	805	2	51	20	3	2	0
1969–70	20,080	6*	839	2	51	17	3	2	0

Sutton									
1958–9	16,525	2		2	41	15	1	2	0
1962–3	16,674	2		2	38	15	1	2	0
1966–7	19,061	2	735	2	39	15	2	2	0
1969–70	20,455	4	832	2	43	15	2	2	0
Waltham Forest									
1958–9	35,122	1		1	67	31	5	1	0
1962–3	32,291	1		1	66	30	6	1	0
1966–7	29,820	4	1147	1	63	28	6	1	0
1969–70	31,066	3	1270	1	63	26	6	1	0
ILEA/LCC									
1958–9	427,587	53	15,110	25	925	299	89	44	6
1962–3	406,281	61	15,421	23	915	274	90	40	9
1966–7	405,089	69	16,400	28	885	240	93	40	9
1969–70	422,030	79	16,764		884	223	94	50	9

Sources: Department of Education & Science; Association of Education Committees; GLC, ILEA and London Boroughs.

N.B. (i) Where Divisional Executives were divided as a result of the London Government reforms, figures for the periods 1962–3 and 1958–9 are estimates. (ii) Special schools exclude hospital schools.

A. Joint arrangements between Kent and Bromley.
B. A department within a major F.E. establishment.
C. Administered by Surrey.

* Advisers are shared with other outer boroughs.
† Figures relate to Croydon County Borough.

population, ranging from Croydon with over 50,000 and 4 others with over 40,000 down to Kingston upon Thames with 17,900. Similarly Croydon has over 2,000 teachers whilst 3 outer boroughs have less than 850. Only Bexley has no special school of its own, but 6 boroughs contain no l.e.a. nursery school.

Every borough has at least one major institution of further education, and 7 outer boroughs are involved in the creation of 3 new polytechnics, although only Kingston upon Thames is fully self-contained in this respect.

Inter-authority co-operation takes place in a number of fields where the resources of individual boroughs are inadequate, and a particularly good example of this is the sharing of advisers. The total number of advisers has gone up from 87 in 1965–6 to 106 in 1969–70 with 81 in the ILEA at the latter date. The ILEA range of specialists is of course much greater.

Nevertheless, the outer boroughs are by no means small when compared to l.e.a.s in the rest of England and Wales. Only 7 county boroughs and 26 counties are bigger than Croydon; 51 county boroughs and 12 counties are smaller than Kingston upon Thames.[17]

Education priority areas. The Plowden Report introduced the concept of an 'educational priority area',[18] as a deprived, primarily urban area where educational handicaps were reinforced by social handicaps. Plowden wanted 'positive discrimination' in favour of such schools and the children in them to raise their standards above the national average, and the most important step in such a policy would be 'to bring more experienced and successful teachers into these (EPA) areas and to support them by a generous number of teachers' aides'.[19] The Report acknowledged the need for 'objective criteria' to select educational priority schools and area, and suggested 8 criteria. The ILEA's Research Group itself undertook a Pilot Study which further adopted the Plowden criteria and eventually developed sophisticated techniques forming a basis for subsequent negotiations with the DES.

The Government subsequently allocated £16 millions over two years for school building in EPAs,[20] followed by the payment of an additional £75 to teachers working in schools 'recognized by the Secretary of State as being of exceptional difficulty'.[21] Under the former programme the ILEA and 4 outer boroughs benefited as follows:[22]

Special Allocation for EPAs 1968–70

	Major Works £000	Minor Works £000
Inner London	1,630	634
Brent	779	54
Haringey	120	99
Newham	144	55
Waltham Forest	99	72

In all, 57 l.e.a.s in England and Wales received allocation under this programme but the amount received by the 5 London l.e.a.s together represented nearly 23 per cent of the total for England and Wales. Plainly therefore, Greater London contains a disproportionate concentration of problem areas.

Greater London has in addition, a high proportion of immigrants. In June 1965, the DES noted that the 'major educational task (in relation to immigrants) is the teaching of English'.[23] Earlier l.e.a.s had been reminded[24] that such factors would be considered in the allocation of the teachers quota. An ILEA survey revealed[25]

Primary (including nursery)	Jan. 1964	Sept. 1965	Sept. 1966
No. of pupils on roll	233,575	232,780	236,968
No. of immigrant pupils	25,244	34,431	35,035
% of immigrant pupils	10·8	14·8	14·9

Distribution of immigrant pupils was very uneven in the ILEA schools. Around half of the schools had fewer than 10 per cent, whereas the remainder showed varying percentages up to 68 per cent. In only two inner boroughs (Tower Hamlets and Greenwich), however, did the proportion of immigrant schoolchildren to school population fall below 10 per cent. The picture in Outer London was much more varied, as the following table shows.

Borough	No. of immigrant schoolchildren	% of total school population
Haringey	9,145	27·0
Brent	9,543	25·1
Ealing	6,564	16·3
Newham	4,563	11·9
Waltham Forest	2,582	8·7
Hounslow	2,258	7·9
Croydon	3,570	7·4

No other outer borough had more than 7 per cent of immigrants in its total school population, and in boroughs such as Sutton and Havering, the percentage is quite small.

The DES 1969–70 figures[26] for the extra quota allocation of teachers for EPAs and immigrants shows where the emphasis is being placed in Greater London.

Extra Teachers

Inner London	1,113
Barking	0
Barnet	15
Bexley	10
Brent	113
Bromley	7
Croydon	49
Ealing	164
Enfield	29
Haringey	85
Harrow	4
Havering	7
Hillingdon	0
Hounslow	87
Kingston upon Thames	0
Merton	4
Newham	97
Redbridge	27
Richmond upon Thames	6
Sutton	1
Waltham Forest	39
Total	1,857

The total of extra teachers for the whole of England was 5,758. Thus nearly a third were devoted to Greater London, whereas Greater London contains only about one-seventh of the school population of England. Certain provincial l.e.a.s also received proportionately very high allocations of extra teachers, eg Birmingham, Durham. Nevertheless all the indications are that few provincial l.e.a.s have to cope with the incidence of immigrant problems that are the lot of some London l.e.a.s.

Administrative resources. The performance of an l.e.a. is heavily dependent upon the adequacy of its administrative staff. Objective criteria for measuring this adequacy are, however, few. One is forced to rely primarily upon the numbers and range of staff.

Some of the initial difficulties in staffing; notably in relation to c.e.o.s in the transitional period have already been discussed. It was in the middle management areas however that the greatest recruitment difficulties were experienced initially, despite considerable movement of officers to fill these posts. In 1965–6,[27] seventeen assistant education officer (a.e.o.) posts and 22 administrative/professional assistants posts were vacant in the outer boroughs although the position varied greatly between individual boroughs. Both Richmond and Newham had vacancies for six administrative/professional assistants, whilst the former also had two a.e.o. vacancies. Merton had 3 a.e.o. and four administrative/professional assistants posts vacant whilst Enfield had 3 vacant a.e.o. posts. In addition, 37 advisers' posts were vacant – although about two-thirds of these were shared posts. Only four boroughs – Bexley, Brent, Harrow and Hounslow – advertised no vacancies for a.e.o.s, administrative/professional assistants and advisers in 1965–6.

Nevertheless, simply noting the number of vacancies is unsatisfactory. Initially, c.e.o.s had to accept existing staff and to safeguard the salary and (so far as possible) the status of existing officers. The new administrative structures were complicated by particular provisions for individual officers. However, with the combined effects of natural wastage and recruitment to an expanding service, c.e.o.s invariably feel today that a 'good team' has been built up in the years since 1965.

When one turns to the specialist advisers the effect of the reorganization is more complex. There were a large number of vacancies initially because of the shortage of advisory staff relative to borough demands for them, but this situation has eased subsequently. Some effects were more long lasting. On the one hand the range of advisers which any one borough can employ is smaller than the previous counties could. On the other hand, as has been pointed out,[28] the total number of advisers in the metropolis is considerably higher than existed before, and consequently the caseload of schoolchildren for each is considerably lower.[29] We are back here to the old problem of evaluating intensity of provision against range of provision.[30]

Several of the outer boroughs try to reduce the hindrance of a smaller range of advisers than they enjoyed under the counties, by creating a joint corps of advisers with a neighbouring authority. In 1965–6, 6 boroughs were involved in sharing 13 advisers but by 1969–70, 12 boroughs were sharing 31 advisers. Usually the arrangements operate between two boroughs but occasionally as for example between Ealing, Hounslow and Richmond, three are involved. It is interesting that all the ex-Middlesex boroughs (including Richmond) are involved in sharing arrangements, but none of the ex-Surrey, ex-Kent, or ex-county boroughs do so. Sharing of advisers is confined to north of the Thames.

From the expansion of sharing arrangements, it would seem that this has proved a convenient method of overcoming the problems of limited resources, etc. There are, however, problems associated with the sharing of advisers. The combined schools of two boroughs may total around 150 primary and over 50 secondary schools. The adviser is faced with the problem of trying to give an adequate service to all the schools and this can be aggravated where communications are especially difficult, eg between Hillingdon and Harrow. The schools often want immediate advice or attention and some heads seem to feel that advisers are no more readily available than under the previous arrangements.

For the local authority itself, there is the problem of the shared adviser's responsibility. He is appointed to the service of one authority but if schools in each borough want him at the same time, then plainly there are conflicts of interests. This may become particularly acute where he encounters a problem requiring prolonged attention, which may lead to a feeling of neglect in the other borough. Furthermore, not only may the immediate needs of the boroughs differ, but where one borough is significantly larger than the other, the larger l.e.a. will expect to see more of the adviser. Such difficulties seem to have deterred some boroughs from entering into sharing arrangements, and have caused others to limit their 'shared' involvement.

However, other solutions have also been sought. Some boroughs insist that their assistant education officers (a.e.o.s) should have an advisory role; this indeed is possible – up to a point, but as their function is mainly administration and liaison, there are limits on the lengths to which this can be taken. Several boroughs are prepared to pay for specialist advice when it is needed, and they may also cultivate close relations with higher educational institutions in their boundaries. A good example of this is Hillingdon and Brunel

University. Others again are happy to use HMIs for certain kinds of specialist advice.

On the qualitative side, opinions again vary, with some l.e.a. advisers held in considerable esteem but others less well thought of. A certain amount of suspicion still attends l.e.a. advisers. Teaching staff sometimes feel that their career prospects are more likely to be affected by an adviser's opinion than by an HMI's (since the latter is not part of the l.e.a.'s organization).

Finally, perhaps the most beneficial impact of the reorganization in terms of staff, has been that new men with new ideas have taken over.[31] Some of the changes introduced, in most cases on the initiative of the staff, have been mentioned. Almost any reorganization would, however, have produced some changes at the top.

Relationships with teachers. The outer boroughs were aware of the need to foster teachers' loyalties to the new authorities especially where, as in the case of Middlesex and Hertfordshire, the previous l.e.a. had enjoyed a high reputation. One means of fostering this loyalty was the establishment of a borough teachers' consultative committee (t.c.c.) Most t.c.c.s were set up on the initiative of the c.e.o., in 1963 or 1964 to advise on the transition. They are, however, such a convenient forum of consultation between the l.e.a. and its teachers that most have become a permanent feature of the educational scene and are welcomed by most teachers.

Since 21 l.e.a.s replaced part or all of 9 l.e.a.s in Greater London, it may be thought that any problems of remoteness would if anything be eased. Nevertheless many head teachers complain of the greater remoteness of the new authorities. While the 20 outer boroughs are smaller than the previous counties, they are generally larger than the former divisional administrations (and county boroughs).[32] Under the former structure, senior county administrators, especially the c.e.o. would seldom be seen by head teachers, and contact with county headquarters was generally infrequent; day-to-day matters were, however, generally dealt with at divisional level and relationships between the officers there and the head teachers frequently became very close and cordial. The Herbert Commission seem insufficiently to have appreciated this in expecting the new system to 'provide an outlet for that interest administratively closer to the actual schools than the county council'.[33]

It is interesting that our findings based on research throughout

Table 6.2

School Population per Adviser

LEA	1962–3	1965–6	1969–70
LCC/ILEA	6,660	5,650	5,340
Essex	12,010	10,570	13,280
Kent	20,380		14,970
Surrey	14,170	10,860	
Middlesex	26,580	(8,520)*	(6,130)†
Barking		8,070	4,800
Barnet		7,770	4,930
Bexley		7,760	8,210
Brent		17,670	8,760
Bromley		13,870	6,390
Croydon		6,570	7,280
Ealing		13,070	9,700
Enfield		8,890	7,670
Haringey		8,560	5,100
Harrow		5,170	5,570
Havering		5,050	4,550
Hillingdon		4,300	4,920
Hounslow		4,620	5,320
Kingston upon Thames		4,890	4,470
Merton		6,150	4,150
Newham		5,480	7,900
Redbridge		7,840	5,230
Richmond upon Thames		6,540	4,050
Sutton		8,320	5,110
Waltham Forest		15,360	10,360

* Advisers counted as fractions were shared between boroughs.
† Average of nine Middlesex London boroughs.

outer London are substantiated by Deborah Lewis's more intensive study of primary education in Haringey.

'All [primary school heads] stressed the importance of "knowing someone by Christian name" and many felt that there had been a loss of this essential ingredient since the reorganization. This reaction was clearly due to their identification with the smaller District and comments like "we were better off as Hornsey because you could get at people" or "when we were Tottenham we knew everybody" were frequent.'[34]

To some extent head teachers' complaints of remoteness may simply be the result of staff changes disrupting relationships cemented over many years.

Personal relationships can be built up again in a new system. In some cases, however, there can be no doubt that access is genuinely more difficult today – partly because the office is physically more remote and partly because it deals with many more schools than the previous divisional office. Some c.e.o.s showed awareness of this problem and made considerable efforts to tackle it – despite the heavy workload arising from reorganization. At least one c.e.o., Mr Greenwood, of Merton, lunched in a different school each week. Sometimes the assistant education officers were given in addition to their administrative duties an advisory role, as in Waltham Forest; there they expect to spend half their time visiting schools and half in the office.

According to the Herbert Commission 'in general, the English education system . . . allots an enormously important place to the head teacher'.[35] The counties gave their head teachers, especially of secondary schools, rather more autonomy than most of the new boroughs are prepared to allow. The tighter control is greatly resented by head teachers but it is arguable whether it is necessarily harmful educationally. Some would consider that a head teacher can be dangerously omnipotent. In any case the real core of his autonomy has not been affected.

Some views were expressed at the time of the London reforms, that the smaller l.e.a.s would have greater difficulty in attracting sufficient qualified staff. In fact, teacher/pupil ratios[36] have improved steadily over the period in Greater London's primary schools (part of a national improvement).

If staff are being attracted in sufficient numbers, there is still the difficult question of quality. Many outer boroughs adopted a more outward-looking policy than their predecessors had. Kingston for example, advertised nationally for head teachers although, in agreement with local teachers, one-third of the short list is drawn from the borough's applicants. This contrasts with Surrey practice at the time of reorganization which was based on an internal promotion system. A similar situation existed between West Ham and East Ham, with the former's national advertising prevailing under Newham. In fact, national advertising is the general rule in the outer boroughs and the broader range of applicants this policy could be expected to attract would tend to lead to a higher overall standard of head teachers.

Conclusions. The new London l.e.a.s are comparatively large and wealthy in the national context but among themselves they vary considerably. The differences are, of course, especially marked between the mammoth ILEA and the outer boroughs but is also significant among the latter. Some of the London l.e.a.s have graver problems than are faced by most provincial l.e.a.s; on the other hand their size and resources make them better equipped to tackle them.

The administrative resources of each individual outer borough are smaller in size and range than the counties they replaced; in aggregate, however, the administrative resources in Outer London are greater than previously. The most convincing test of the adequacy of the new regimes is an empirical one, namely how well they have coped since 1965. The new l.e.a.s not only had to face the enormous strain of the transition but of the many additional burdens which followed hard on its heels – comprehensive reorganization of secondary schools, government economic stringency, rapidly rising costs and the continued growth of school population. No l.e.a. was clearly overwhelmed. There can be little seriously wrong with administrations which can handle such pressures. On the other hand, it is difficult to determine how far this success was achieved because of, and how far in spite of, the new structure.

In terms of relationships with teachers one's verdict must go marginally against the new l.e.a.s. Far from a significant improvement in relationships being effected there seems to have been a slight worsening. It is difficult to avoid the impression, from our own research, that the drawbacks of the former two-tier system of educational administration were far more apparent to divisional educational officers and elected members of divisional executives than to teachers. Even head teachers who had served under Middlesex did not usually find that county's administrative arrangements awkward or frustrating. The Herbert Commission were not necessarily right in asserting 'that the administrative friction associated with the old system must have in the end some bad effects upon the service'.[37] It is difficult to find teachers, especially head teachers who strongly welcomed the reform. And most have mixed feelings about the way it has turned out. For some, continuity of staff and policies has meant little difference. For others there are both benefits and drawbacks.

PRIMARY AND SECONDARY EDUCATION

By any criterion primary and secondary education is the central responsibility of l.e.a.s. It involves more pupils, more teachers, more money and more buildings than any other part of the education service and generally attracts most public interest. Performance in primary and secondary education must be the first consideration when assessing the success of our educational system.

Self-sufficiency. Any structure of l.e.a.s involves a measure of interchange of pupils. Especially is this likely with secondary schoolchildren, given the larger catchment area of secondary schools and their tendency to be specialist institutions. Obviously, however, an enormous volume of cross boundary movement, especially of primary schoolchildren would not only entail great administrative and financial complexities but would make it hard to establish satisfactory relations with parents. For such reasons and because l.e.a.s are anyway reluctant to pay another authority to educate a child who could be accommodated in one of their own schools, such movements are often discouraged.

The Herbert Commission did not expect much interchange of primary and secondary schoolchildren between their proposed 52 l.e.a.s. Since the Government in fact created larger l.e.a.s and left the LCC system virtually intact,[38] it might have been expected that cross boundary movement would be relatively insignificant. As a 'free trade' safeguard was written into the London Government Act, 1963,[39] it might also have been expected that whatever cross boundary movement there was would take place easily. The second point has been substantiated. Movement seems to take place freely, although initially where a divisional executive was split between two l.e.a.s, a certain amount of friction sometimes occurred, as for example, between Sutton and Surrey. The extent of self-sufficiency, enjoyed by the new l.e.a.s in the provision of primary and secondary school places is, however, much less than the Commission's analysis would lead one to expect. Perhaps it underestimated the importance of the highly urbanized nature of Greater London and the mobility that entails. Moreover, the new l.e.a.s inherited a fabric developed by different authorities which could not be expected to correspond conveniently with the new authorities' boundaries.

The extent of interchange of primary pupils in the nine ex-Middlesex boroughs is shown in Tables 6.3 and 6.4. These Tables do not of course show how much interchange is between the ex-Middlesex boroughs themselves and how much between them and other l.e.a.s. If, however, the total number of pupils which the ex-Middlesex boroughs 'import' (from each other and elsewhere) is subtracted from the total which they 'export' (to each other and elsewhere), the difference is about the same as the number which Middlesex county used to 'export'. This implies that much of the cross-boundary movement is between the boroughs themselves. As this would formerly have been internal to Middlesex, the administrative costs and accounting work now caused by the cross boundary movement is entirely a consequence of the reorganization.

Other parts of Greater London vary in their experience. In Croydon and Newham only 0·7 per cent of their primary children travel over the boroughs' boundaries, whereas Sutton has 8·9 per cent doing so. Some l.e.a.s are major 'importers' of primary pupils, notably Harrow, Redbridge and Merton. In all, 2·8 per cent of primary schoolchildren in the outer borough attend school in another l.e.a.'s area. The comparable figure for the ILEA is 1·3 per cent.

Not surprisingly, greater inter-authority movement occurs in secondary education. Thus in 1968–9, the Outer London l.e.a.s had an average of 11 per cent of their secondary school population attending secondary schools maintained by other l.e.a.s but this ranged from 1·9 per cent in Newham[40] to 19·6 per cent in Redbridge. Furthermore, comparing Croydon as a county borough in 1963–4, with its position as a London borough in 1968–9, reveals that the secondary school population increased from 15,918 to 20,801, whilst the percentage attending other l.e.a.s' secondary schools rose from 2·7 to 8·2 per cent. The average inter-authority movement for the nine ex-Middlesex boroughs is broadly in line with that for the rest of Outer London, but once again embraces considerable variation. The ILEA has less than 2 per cent of its secondary pupils attending other l.e.a. schools, but 'imports' over three times that figure – a position broadly similar to the LCC in 1963–4.

In addition, whereas at the primary level, the number of pupils attending *independent* schools is small, at secondary level it is much more significant, rising in the case of Croydon, to 8·6 per cent of the secondary school population. The proportion of secondary pupils at non l.e.a. schools is much smaller in the remaining l.e.a.s. in Greater London, but taken together with those attending other l.e.a. schools, means that a considerable proportion of the secondary (and a smaller proportion of the primary) school population of a borough is not under its direct control. Where a large part of its school population is not under a borough's direct control, not only may it be confusing for the affected parents but would seem unlikely to encourage an active interest by them in the borough's education service.

Furthermore, the administrative costs arising from the inter-authority payments for these cross boundary movements within what was formerly one l.e.a., are a cost directly attributable to the reform of London Government.

On the other hand, the creation of the twenty outer London l.e.a.s, with the 'free trade' provision applying between themselves and between them and the contiguous counties, has increased the range of choice for parents. Thus, for example, parents strongly in favour of, or strongly opposed to, comprehensive education might choose to send their child to a neighbouring l.e.a. Harrow, in 1967–8, for instance, whilst strongly against comprehensive education for itself,

Table 6.3

Interchange of Primary Pupils in Ex-Middlesex Boroughs

	(a) Number of Pupils in other l.e.a. schools		(b) Number of primary pupils for whom the l.e.a. is financially responsible		% of l.e.a. primary pupils in other l.e.a. schools (a/b × 100)	
LEA	1966–7	1968–9	1966–7	1968–9	1966–7	1968–9
Middlesex (1963–4)	1,320		163,473		0·8	
Barnet	431	481	21,798	23.307	1·9	1·9
Brent	1,353	1,741	23,617	26,675	5·7	6·5
Ealing	692	722	24,549	26,773	2·8	2·7
Enfield	155	364	21,560	23,205	0·7	1·6
Haringey	572	706	20,023	22,131	2·9	3·2
Harrow	516	613	15,753	17,041	3·3	3·6
Hillingdon	541	630	21,427	22,684	2·5	2·8
Richmond upon Thames	283	383	11,446	12,322	2·5	3·1

G

paid £202,304 in adjustment to other l.e.a.s for provision of comprehensive education. Comprehensive reorganization (particularly in Brent) *de facto* provided comprehensive education for many Harrow schoolchildren.

Table 6.4

Interchange of Secondary Pupils in Ex-Middlesex Boroughs

LEA	(a) Number of pupils in other l.e.a. schools		(b) Number of pupils for whom l.e.a. is financially responsible		% of pupils in other l.e.a. schools (a/b × 100)	
	1966–7	1968–9	1966–7	1968–9	1966–7	1968–9
Middlesex (1963–4)	(3,567)		(128,379)		(2·8)	
Barnet	1,675	1,689	17,432	17,892	9.6	9.5
Brent	2,176	2,549	15,134	15,760	14·4	16·2
Ealing	1,911	1,988	16,157	17,087	11·8	11·6
Enfield	1,535	1,493	14,713	15,577	10.5	9·6
Haringey	1,464	1,858	13,031	13,719	11·3	13·6
Harrow	2,115	1,596	11,721	11,462	18·0	13·9
Hillingdon	1,016	846	14,409	14,888	7·1	5·7
Hounslow	1,780	1,578	11,677	12,041	15·3	13·1
Richmond upon Thames	1,263	1,357	8,369	8,634	15·1	15·7

SECONDARY REORGANIZATION

The 1944 Education Act laid down no statutory form of secondary school organization. It was largely historical accident that the tripartite system (grammar, technical, modern) prevailed in most of Greater London – as in the rest of the country.

Comprehensive secondary schools were, however, by no means unknown even in the early post-war period. Middlesex, for instance, decided in favour of comprehensive secondary education in the late 1940s. Changes of political control and economic difficulties, however, made the implementation of the decision slow and by 1965 Middlesex had only two purpose-built comprehensive schools, in Hayes and Potters Bar. Comprehensive schools existed also in Essex, Kent and Hertfordshire, by 1965.

The LCC, however, made the most notable developments in comprehensive organization and is rightly regarded as a pioneer in the field. The need to tackle the damage and decay caused by the war and to cope with the bulge in the primary school population, meant that at first no new secondary schools could be built at all. Nevertheless the LCC established 8 experimental comprehensive schools in pre-war buildings. The London School Plan envisaged a general system of comprehensive education in the secondary sector where the 55 voluntary grammar schools made this impossible.

As new and secondary school building became possible in the early 1950s the LCC was able to erect its first purpose-built comprehensive schools, although not without some friction with Whitehall. The LCC kept the comprehensive schools position under continuous review and all their data, together with the 77 completed comprehensive schools, was bequeathed to the ILEA.

The London Government Act, 1963, required[41] the 20 Outer London boroughs to continue to operate the existing educational development plans applicable to their areas but to submit new ones to the DES by 1 April 1966. The ILEA was allowed to retain the LCC plan. Before the outer boroughs could proceed far in preparing new educational development plans, however, the political context changed.

The general election of October 1964 brought to power a Labour Party pledged to comprehensive secondary education. On 12 July 1965 the DES issued circular 10/65 requesting all l.e.a.s in England and Wales, if they had not already done so, to submit plans for the reorganization of secondary education on comprehensive lines. As a guide, the circular outlined six main ways of going comprehensive although two of them were to be regarded as interim only. It is important to note that the DES decision to require universal comprehensive education was taken with two qualifications; no additional finance would be provided to assist reorganization and the Government would not seek additional powers from Parliament to enable it to *compel* l.e.a.s to reorganize their secondary schools on comprehensive lines.[42]

Although some l.e.a.s in Greater London were already proceeding in a comprehensive direction, all reacted to the circular. Every one set up one or more working parties under its education committee to examine the problems associated with the introduction of comprehensive education. The major working party was usually chaired by the chairman of the education committee.

Of the various possible comprehensive schemes the all-through 11–18 school with its egalitarian outlook and tendency to stress the average child had strong appeal particularly for Labour.

Conservatives, and many grammar school head teachers and staff, generally felt that if comprehensive education were really necessary, then the interests of the most talented children should not suffer as a result. This often led to suspicion of the speed at which Labour wanted to introduce comprehensive schooling, and, frequently, a favourable disposition to schemes of comprehensive education which seemed to allow the possibility of attention being paid to talent, ie these involving sixth form colleges. Of course there was also some outright opposition to the disappearance of the grammar schools.

Since no additional finance was to be forthcoming to assist reorganization it was clear that existing buildings would have to be used largely unmodified. In most cases comprehensive schemes involved the grouping of previously separate schools, often some distance apart, and this implied much movement of staff and students during the course of the day.

Labour councils generally felt that comprehensive reorganization should go ahead despite these difficulties, arguing that to wait for purpose-built schools would mean long delays whilst there were great benefits to be gained from immediate comprehensive reorganization. Conservatives on the other hand, often saw Labour proposals as 'botched up' and 'rushed'; it was feared that the grammar schools would be destroyed without any guarantee that something as good would be put in their place. Where Conservatives were prepared to accept comprehensive reorganization, they usually wished to proceed more slowly and carefully.

Not surprisingly, therefore, the l.e.a.s reacted with varying enthusiasm to circular 10/65. On the one hand certain Conservative strongholds, notably those formerly in Surrey – Kingston,[43] Sutton and Richmond[44] – and the most Conservative of the former Middlesex divisions – Harrow – firmly resisted any real comprehensive reorganization. At the other extreme certain boroughs with relatively small and therefore unsafe Labour majorities, for instance, Enfield, Bexley and Brent, pressed ahead urgently with comprehensive reorganization. Between these two groups a varying degree of favour and hostility was displayed, party control being the most significant factor. There were exceptions. Merton although Conservative-controlled (and ex-Surrey to boot) went ahead with a genuine scheme of comprehensive reorganization because of the moderating leadership of Sir Cyril Black and Alderman Talbot.

A particular problem for the outer boroughs was the production of an educational development plan at the same time as considering circular 10/65. Both items required the widest consultations and careful consideration, which, linked with the problems of launching the new education services, presented considerable difficulties. However, a review of the 1966–67 situation, showed that of the eight outer boroughs which had not submitted comprehensive proposals to the DES, seven were Conservative-controlled.[45] The presence of the political will to achieve comprehensive education could overcome these problems it seems.

On the other hand, enthusiasm for comprehensive education could also lead to problems. Some outer boroughs, eg Hillingdon, submitted schemes quickly which, whilst fully comprehensive, were ruled out by the DES on cost grounds. In such cases, the borough had to go through the preparation process again which inevitably led to longer delays in the introduction of comprehensives.

Consultation and negotiations invariably took place at official level between the c.e.o. and Departmental officials. Indeed the Department proposed a joint committee of civil servants and local government officers to suggest ways of overcoming the deadlock at Kingston; this was rejected by the Council. There were often meetings at a political level too.

The influence of national figures on the course of local events could also be significant as at Enfield where one of the two MPs was a leading Conservative, Mr Iain Macleod. His opposition to the Labour borough council's proposals aroused a storm when he attacked Enfield's c.e.o. in the House of Commons.[46] The impact of these influences is difficult to assess, but it is interesting that the DES rejected some parts of the Enfield scheme and later recanted when Enfield pointed out that similar proposals had been accepted in Manchester's scheme which had been submitted later than Enfield's.

The influence of officers is obviously significant, particularly that of the c.e.o.s. As the professional experts, their views rightly carry weight with members and as they had responsibility for preparing possible comprehensive schemes, opportunities were plainly available to persuade members to their point of view.[47] Where policy is clearly not to go comprehensive, the c.e.o. cannot impose his view on the education committee, but he will no doubt be aware of the possibility of future policy changes. The c.e.o.'s relations with his chairman are particularly important and, generally, the two seem to have worked very closely on comprehensive plans. It is not surprising perhaps that most c.e.o.s got the comprehensive scheme they wanted.

The c.e.o.s personal commitment is important, not only because he and his staff are responsible for implementing and running the new service, but because they represent the permanent element in the situation – particularly noticeable after the 1968 borough elections.

In eight of the outer boroughs, Conservative replaced Labour as the majority party and each had to decide what to do with their predecessors' comprehensive proposals. Most of these boroughs were far advanced in their preparations and a quick decision was vital. In several instances, the new Conservative majority delegated very considerable powers of decision to the chairman of the education committee. The c.e.o.'s own belief in the educational validity of the comprehensive scheme was important, and it was the educational arguments which persuaded several new education committees not to halt the introduction of the comprehensive proposals. To have done so, in Waltham Forest and Hounslow for example, would have greatly disrupted the borough's education service immediately prior to a new school year.

This is not to say that the Conservative majorities had changed their views, although to some the comprehensive scheme perhaps seemed less objectionable when it had to be viewed on its educational merits. The alternative seemed, however, to be chaos. Nevertheless, whilst many Conservative education committees did not 'kill' comprehensive education as Labour councillors had feared, they equally saw no reason to extend its scope and did not therefore incorporate any more schools in the proposed scheme.

Moreover, not every new Conservative majority accepted the *fait accompli*. In Bexley, where comprehensive reorganization posed particular physical problems, the scheme submitted to the DES by the previous Labour council was immediately withdrawn by the Conservatives.

In most cases, however, the 1968 elections had less impact; delays in implementing schemes were due more to financial stringency and physical problems, although, as was mentioned earlier, lack of political enthusiasm could capitalize on such difficulties.

The comprehensive issue was not simply a matter to be settled by the political parties in each l.e.a., it was also a matter of great concern to parents, and authorities went to great pains to involve parents in the decision-making process. Prior to submission of the first ILEA plan, 42 meetings attended by 19,000 parents were held; Merton held more than 30 meetings in its schools, whilst Enfield's public meetings were attended by 5 to 6,000 people. There were more vociferous groups as well, both favourable and opposed to, the comprehensive principle, and these groups made a significant contribution to the whole debate. Local branches of the Confederation for the Advancement of State Education were active in support of comprehensive education, pressing strongly in boroughs such as Sutton, which was little disposed to comply with circular 10/65. On the other side, groups of parents frequently centred on one part of a borough, formed themselves into bodies opposed to the abolition of local grammar schools. These latter groups attracted national attention by virtue of their activities in some of the outer boroughs, for example Ealing. An ad hoc group called the Joint Parents' Committee of the Grammar Schools of Ealing opposed the absorption of the borough's grammar schools into the comprehensive system; they took their case to the High Court but were defeated.

The Ealing parents complained of the lack of consultation and similar protests were heard in another London borough, Enfield. Here the comprehensive issue proved particularly acrimonious, but illustrated most of the main forces operating on both sides. The Labour Party had a small majority after the 1964 borough elections but were none the less determined to press ahead with comprehensive reorganization.

The existence of several well-established grammar schools in Enfield, notably Enfield Grammar School, founded in 1558, provided here, as elsewhere, a focus for the opponents of comprehensive reorganization. In this case the Enfield Parents' Association were the main group opposed to the borough's scheme. As was commonly the case in other boroughs too the objecting parents were seen by the Labour majority as 'a front organization for the opposition'.[48] The opponents of the scheme were confronted by a political majority anxious to press ahead with comprehensive reorganization, because of the uncertainty of their majority after the next borough elections.

Equally, comprehensive opponents could hope for a reprieve if the Conservatives became the majority party.[49] Delay in the introduction of comprehensive education would, therefore, not only postpone the early abolition of the grammar schools, but perhaps preserve them indefinitely. In Enfield, again as in some other l.e.a.s, there was also opposition from some members of the governing bodies of the selective schools, especially the Foundation Governors of Enfield Grammar School.

Without the power to reverse the political decision in favour of comprehensive reorganization, but with a precedent in Ealing, the opposition in Enfield decided to question the legal validity of what the Council proposed. Legal action succeeded in postponing, until January 1968, the incorporation of the schools concerned into the borough's comprehensive system, and some parents continued their protests after that date. Four months later, however, the Conservatives were elected by a substantial majority and their attitude in Enfield, as in several other boroughs, was to avoid disrupting the borough's educational system by trying to put the clock back, but equally not to go beyond what Labour had intended to do.

In the ILEA too the comprehensives issue led to a considerable inter-party conflict. The Labour-controlled ILEA submitted to the DES in 1967 a scheme to eliminate all grammar schools (except the voluntary ones) in the long term, and to reduce their number from 71 to 49 by 1970. By the same date comprehensive schools would increase from 78 to 111. The Conservatives captured the ILEA, obviously to their own astonishment,[50] at the GLC elections in 1967. Their first action was to withdraw the scheme submitted by their Labour predecessors. The revised scheme submitted in February 1968 envisaged the retention of at least 40 grammar schools in 1975, a further 18 being subject to negotiation by that date.

The scheme did not look beyond 1975 but it was intended to make longer term plans in the early 1970s. Some Conservative supporters were disappointed that the scheme was not more substantially altered. On the other hand, the Minister regretted the failure to look beyond 1975 and the holding back of some proposals, and asked the ILEA to reconsider four proposals, but approved the scheme overall. Labour members of the ILEA criticized the retention of 40 or more grammar schools and the failure to introduce fully comprehensive education more quickly. The argument that a heavily populated area with excellent communications, such as Inner London would support grammar *and* comprehensive schools did not convince Labour members. Selection in their view was incompatible with comprehensive education. When Labour recaptured the ILEA in April 1970 they immediately announced their intention of accelerating comprehensive reorganization. Work began on revising the scheme to achieve as quickly as possible a 'fully comprehensive and non-selective system of secondary education'.[51]

The main lesson to be drawn from the comprehensive reorganization issue in Greater London is the variety and complexity of factors involved. Party politics were obviously important but there was no simple divide. Conservative Merton for instance was rather co-operative and in Barnet the conflict was Conservative versus Conservative. Financial and physical factors were often equally or more significant.[52] Hillingdon, for example, although extremely keen to reorganize its secondary schools was hindered by the awkward shape of the borough and the cost of its proposals.

The history of comprehensive reorganization in many cases provides some instructive examples of local pressure group activity. It is also a most interesting case study of central local relations. But one of the most unfortunate aspects of the comprehensive schools controversy was that it followed so closely upon the heels of the reorganization of London Government that the strain upon officers and to some extent members was very considerable. It is difficult to point to hard evidence of evil consequences but it must have made it more difficult for the new l.e.a.s to run themselves in and in some cases (eg Enfield, Barnet) occupied a disproportionate amount of staff time and effort. From the point of view of the outer boroughs in particular, it would have been far better if the comprehensives storm had broken in 1970 rather than 1965.

The Conservative success in the General Election of June 1970 and the speedy withdrawal of circular 10/65[53] did not cast everything back into the melting pot. The main effect was to restore discretion to the l.e.a.s. Those implacably opposed to comprehensive reorganization are not spared the goading of the DES; those in favour can still proceed as fast as money will allow.

SPECIAL EDUCATION AND SCHOOL HEALTH AND WELFARE

The Education Act, 1944, requires l.e.a.s to ascertain which children need special educational treatment and to provide for them appropriately.[54] While schools for mentally and physically handicapped children had existed previously, the requirements of the 1944 Act are both more comprehensive and more flexible.

The field of special education is a complex one. It covers the whole range of handicaps both mental and physical, and Departmental regulations lay down different statutory requirements, eg on class sizes for each category. Moreover, the number of children in each category ranges widely. The commonest are maladjusted children, followed by the educationally sub-normal (e.s.n.);[55] at the other extreme, blind, deaf and physically delicate children are a far smaller proportion of the school population. Provision is also diverse: there are special boarding schools, special day schools, special classes in ordinary schools, home tuition, even ordinary classes in ordinary schools combined with attendance at a child guidance clinic.

Perhaps the most complex feature of all is that special education is a field of constant change. New handicaps are isolated; autism for instance has only been properly recognized in the last decade or so. Medical advances mean that certain kinds of handicapped children now survive in greater numbers, eg the hydrocephalic; on the other hand, the physically delicate have declined in number. Above all changes in educational opinion lead to different approaches. The general trend of thinking today is that wherever possible handicapped children should be treated as normally as possible – living at home and attending ordinary day schools (although perhaps special classes). Boarding schools, except for certain kinds of handicaps, are going out of favour.

A major recent change concerns responsibility for the severely subnormal (s.s.n.) children, sometimes referred to as the ineducable. Broadly they are children with IQs lower than 40. Previously they were the responsibility of local health authorities, not l.e.a.s, and most attended the health departments' junior training centres not special schools. The Seebohm Committee, however, recommended that they become an educational responsibility,[56] and the Government agreed. Both the children and the facilities were transferred to the l.e.a.s in 1970.

Provision in Greater London. The provision of special education in Greater London prior to the reorganization was patchy. Middlesex and the LCC had made considerable efforts but some of, the other l.e.a.s were less advanced. When therefore the schools were shared out among the new l.e.a.s some did rather badly. As may be seen from Table 6.5, Bexley ended up with none and Redbridge with only 1; most outer boroughs had only 2 or 3.

Since 1965, as may be seen from the same Table, there has been some new building in Outer London notably in Brent, Ealing, Havering, and Hillingdon. It is difficult to avoid the impression that much depends upon the particular predilections of c.e.o.s and education committees. Where these are favourable the position has improved since 1965. Nevertheless, the outer borough with most schools – Hillingdon with 7 – still cannot be considered in the same league as the ILEA, at present operating 94.

Table 6.5

Number of LEA Maintained Special Schools

LEA	1965–6	1969–70
Barking	2	2
Barnet	3	3
Bexley	0	0
Brent	2	5
Bromley	3	3
Croydon	3	3
Ealing	5	6
Enfield	3	5
Haringey	3	3
Harrow	3	3
Havering	2	3
Hillingdon	4	7
Hounslow	5	4
Kingston upon Thames	2	2
Merton	1	2
Newham	4	4
Redbridge	1	1
Richmond upon Thames	3	3
Sutton	2	2
Waltham Forest	6	6
ILEA	91	94

One must indeed pay a tribute to the excellence of the special education service in the ILEA. It has a long history; the first resolution about blind and deaf children was passed by the London School Board in February 1872. It has made many advances. The ILEA also has the resources to devote to special facilities; for instance it operates a fleet of buses to take children to special schools and many are adapted to accommodate handicapped children.

Special education is obviously a field in which there is a supreme need for inter-l.e.a. co-ordination, especially in outer London. This is more, however, than a metropolitan need; specialist institutions should be available to children from l.e.a.s all over the South East, indeed, in certain circumstances, all over Britain. Moreover, l.e.a. provision needs to be co-ordinated with voluntary bodies.

There is much bilateral co-operation between l.e.a.s. There is also some wider discussion under the aegis of bodies like the LBA Education Committee and the Standing Conference of Assistant Education Officers for Special Education in the South East. The DES too takes some regional initiatives. Nevertheless there is little gathering of data to determine demand over the whole area, and few agreements are made whereby one l.e.a. builds a special school of a certain kind specifically to meet demand from a wide area. The South East is perhaps behind certain other regions like the North East in positive co-ordination of specialist provision.

Staffing. The ILEA has seven assistant education officers, each responsible within County Hall for a branch of the education service. One of these branches is concerned with welfare and special services and has developed considerable expertise in this field by virtue of its resources and long experience.

In the outer boroughs, only Ealing and Merton designate an assistant education officer solely to deal with special services. Elsewhere, special schools are normally part of the responsibility of the a.e.o. dealing with schools.

Furthermore, the outer boroughs have no advisers devoted exclusively to special education whereas the ILEA has a staff inspector and three ordinary inspectors. Finally, one of the ILEA 'teacher centres' is for special education, a specialist facility not available in Outer London. At present, the outer boroughs have a total of 35 educational psychologists. Eight boroughs have only one, whilst Sutton has none at all.[57] The ILEA has 25 full-time educational psychologists with some part-time assistants.

Ascertainment of pupils needing special education is the statutory responsibility of the medical officer of health. The trend today, however, is to use group ascertainment by a team of doctors, psychologists and educationalists. The London l.e.a.s have considerable difficulty in recruiting educational psychologists, but this is not a problem unique to London; there is a national shortage. Nevertheless the London reorganization exacerbated the position, with twenty new l.e.a.s competing for the small band of psychologists available. The reorganization also brought into office a number of new medical officers of health, some of them without great experience of ascertainment for special education.

Table 6.6

Special Education

Proportion of l.e.a. children in maintained and voluntary day and boarding schools to total l.e.a. school population.

LEA	*School population*	*Malajusted*		*E.S.N.*	
		Total	Proportion	Total	Proportion
ILEA	411,166	1,922	0·47	5,212	1·27
Barking	27,941	31	0·11	210	0·75
Barnet	42,058	79	0·19	212	0·50
Bexley	32,721	44	0·13	93	0·28
Brent	39,392	128	0·33	234	0·60
Bromley	44,376	82	0·19	301	0·68
Croydon	50,374	89	0·18	276	0·55
Ealing	42,345	185	0·44	292	0·69
Enfield	38,343	79	0·21	204	0·53
Haringey	35,193	47	0·13	188	0·53
Harrow	27,706	39	0·14	86	0·31
Havering	43,262	34	0·08	306	0·85
Hillingdon	36,842	72	0·19	211	0·57
Hounslow	29,685	88	0·30	174	0·59
Kingston upon Thames	18,898	30	0·16	110	0·58
Merton	24,599	80	0·32	168	0·68
Newham	39,611	18	0·04	312	0·79
Redbridge	32,833	40	0·12	74	0·23
Richmond upon Thames	20,305	73	0·36	153	0·75
Sutton	20,361	41	0·20	125	0·61
Waltham Forest	30,659	59	0·19	229	0·75

As may be seen in Table 6.6. there is a considerable variation in the proportion of schoolchildren who receive special education as a consequence of suffering from the two commonest handicaps–

maladjustment and educational subnormality. While some of the variation may be due to differences in social composition and circumstances of the l.e.a.s it is unlikely to explain it all. It is probable that some of the variation is explicable in terms of inadequate ascertainment. Perhaps too, shortage of places in a borough's special schools is a factor; there may be some reluctance to ascertain a child as needing special education when it will be difficult to place him in a special school near his home. It is noteworthy that the ILEA come out best by a significant margin in both cases.

Relationships with the schools. As always, generalizations are fraught with exceptions. Nevertheless most special school head teachers were not brimming with enthusiasm at the prospect of the reorganization and the former Middlesex ones were particularly regretful at the demise of the county. In those boroughs in which a particular interest seems to have been taken in special education head teachers are, however, by no means unhappy at the outcome. They sometimes enjoy more personal attention than previously and feel that there is often a quicker response to their problems.

On the other hand, some teachers complain of lack of contact with senior administrators and elected members compared to the old system. This seems particularly to be true of out-county boarding schools. The Middlesex County Special Services Sub-committee had visited each of these schools regularly, and the staff had looked forward to the visits. Not all the new boroughs have equivalent arrangements. Moreover, each head teacher used to present an annual report to the Middlesex Sub-committee and to go to Guildhall to answer any points raised. Most boroughs have discontinued the practice, thus reducing contact with members.

In general one can say that there is opportunity for more flexibility and more direct contact under the outer boroughs if they choose to use it. On the other hand they are able to provide less specialist staff than the counties or the ILEA, and their head teachers sometimes feel this lack.

School health and welfare. L.e.a.s are obliged to arrange free medical and dental examinations of their schoolchildren at appropriate intervals; where necessary free treatment must be provided either at a local authority's own clinic or elsewhere.[58] The school health service is not, however, confined to purely medical activities. Increasingly it is concerned with health education work in schools. Moreover, since health problems interact with social problems, most school health services now employ social workers. Throughout Britain the l.e.a. is also the local health authority (except in the case of the ILEA); consequently the most convenient and therefore the most common arrangement is for the medical officer of health to be appointed professional head of the school health service with the designation 'principal school medical officer'.

The school psychological service, although a distinct unit, is obviously closely allied to the school health and welfare services. As has been mentioned, educational psychologists are important for special education, especially for ascertaining mentally handicapped children. They can also, however, play an invaluable role in ordinary primary and secondary schools, for instance, in helping children who are only mildly disturbed or who have emotional problems. To be really effective this service needs to embrace three kinds of relevant staff-psychologists, psychiatrists and psychiatric social workers.

The school welfare service is a heterogeneous group of functions. The first historically was the duty of ensuring school attendance; indeed this is still a basic responsibility and truancy is in some areas again a problem. Certain other law enforcement work is carried out, for instance, that relating to the employment of children. In many l.e.a.s such regulatory activities have broadened into more of a social work approach. 'The purpose of an education welfare service is to supplement the work of the schools in educating their children in the widest sense so that not only do children attain the highest educational standard of which they are capable, but also learn to live and enjoy as full a life as possible, both in work and leisure'.[59] The LCC was a pioneer in this broader field. It established a school service under which care committees of voluntary workers, trained and advised by qualified social workers, were attached to each school. These were designed to promote the welfare of schoolchildren by maintaining liaison between school and home, and with statutory and voluntary welfare agencies.

Under the school welfare service l.e.a.s also provide clothing and footwear to children in impoverished circumstances. Finally, school meals and milk are usually considered part of school welfare, although the recent trend has been to curtail the subsidized provision of these services.

The situation in Greater London. The ubiquitous problem in almost all parts of the school health and welfare services in Greater London, as elsewhere, is scarcity of qualified personnel. Difficulties are particularly acute in respect of, amongst others, dentists and psychological staff. Government proposals for reforming the National Health Service and the possible effect on the local authority health services has caused some uncertainty about career prospects; this has had an adverse impact on recruitment during the last two or three years.

The London reorganization, in increasing the number of l.e.a.s worsened the staff position. Some of the outer boroughs have very thin staff resources in certain sectors of these services. While, therefore, medical services, and school meals and milk are adequately provided in Outer London, in some cases social work and welfare functions are less satisfactory. On the other hand the outer boroughs enjoy the considerable advantage of being both l.e.a.s and local health and welfare authorities. All have appointed their medical officers of health as principal school medical officers, and all have education welfare and child guidance provided by their new mandatory social service departments.[60] Co-ordination is greatly facilitated when all these functions are controlled by, and the principal officers employees of, the same local council. Moreover, compared to the counties the outer boroughs have shorter chains of command and closer links between the relevant departments. Their advantages are not indeed only administrative; it is of great *professional* benefit to have the social services for community, family *and school* provided by the same unitary authority.

The situation in Inner London has almost the directly opposite strengths and weaknesses. On the one hand the ILEA's size and resources enable it to deploy a corps of specialist staff which make it the envy of most l.e.a.s in Britain. Also its inheritance from the LCC gives it a massive advantage; for instance no outer borough can match the school care service bequeathed by the LCC. On the other hand, the great drawback is that while the ILEA is responsible for *school* health and welfare, the boroughs are the ordinary health and welfare authorities.

To run its school health service the ILEA uses the 13 inner borough and City medical officers of health. For this purpose they are *employees* of the ILEA. Even the Seebohm Committee could not grasp this arrangement. They thought that the school health service was provided *jointly* by the inner boroughs and the ILEA.[61] The

mistake was piously corrected by the ILEA.[62] While these arrangements are far from unworkable they are clearly most complex and it
is not surprising that they baffled the Seebohm Committee. The
ILEA's own medical adviser attempts to harmonize policy and
standards over the whole of Inner London but there are obvious
potential problems of friction and co-ordination when thirteen
separate and eminent medical officers of health have to be welded
into a common service.

The education welfare service in Inner London has, if anything,
greater administrative complexities. Until 1970 it was composed of
three separate units: school meals and milk were the responsibility of
the ILEA's education catering service, school attendance and related
matters that of the ILEA's school inquiry service, and general welfare
that of the ILEA's school care service. The catering service had its
own director and administrative structure; the school inquiry service
was divisionally based and came under the divisional officers; the
school care service was school based and had its own principal
school care organizer.

Moreover, the inner boroughs and the City as local welfare
authorities operated their own community welfare services. Thus a
family in difficulty might be visited by a social worker from their
borough welfare department, but if they had children they might also
be visited by someone from some part of the ILEA welfare service,
and even in certain circumstances[63] by a social worker from the
school *health* service.

The ILEA themselves recognized that these arrangements were
not ideal, and in November 1965 commissioned the social research
unit of the Department of Sociology of Bedford College to conduct
an inquiry into the welfare services for schoolchildren in Inner
London. The research team reported in November 1967.[64] They
concluded 'If education in Inner London were a borough-based
service, there would seem to us considerable advantages to be gained
from attaching social workers from a unified social work or family
service department to a school or schools to carry out the tasks at
present performed by school inquiry officers, school care organizers
and care committee workers'.[65] Given, however, the existence of the
ILEA, they recommended instead a unified education welfare
service.

The Seebohm Committee, which had studied the Bedford Report,
did not wholly accept these conclusions. They felt that it was wrong

to maintain the ILEA school social service separately from the borough's community-based social care services, because 'the child who presents problems in school (or which are detected there) is often in need of a wide range of help and, as we have already stressed, particularly help in his family and community setting'. They therefore recommended that joint arrangements between the ILEA, and the borough councils should be made.[66] These arrangements were basically that the borough councils, social service departments should provide the social workers for the schools, but that close liaison be maintained with the ILEA special services department; the school psychological service would remain with the ILEA.

The Seebohm proposals were rejected by the ILEA. Instead the Bedford unit's recommendation of an ILEA unified education welfare service was adopted.[67] The formerly separate school care and school inquiry services were amalgamated into a single, divisionally-based school welfare service, composed of both professional and voluntary staff. The school psychological service was, however, left outside as were the social workers attached to the school health service.[68] School meals and milk were also unaffected.

While the new unified service is undoubtedly an improvement on the pre-1970 arrangements, it perpetuates the fundamental irrationality which so troubled the Seebohm Committee. To have separate *school* and *family* welfare services, provided by separate authorities, makes no sense in social work terms. Some machinery exists which should reduce the dangers of duplication or even conflict. For instance in each borough there is a co-ordinating committee consisting of the representatives of the borough health and welfare departments of the ILEA, of the GLC housing department and of the probation service and the Department of Health and Social Security; there is also a Problem Cases Conference consisting only of the ILEA and borough personnel. Nevertheless such mechanics cannot eliminate all the difficulties inherent in the situation.

The school health service in Inner London also has peculiar problems because of its structure. While for example the borough health departments are responsible for ascertaining physical handicaps, they play no part in the ascertainment of mental handicaps (e.s.n., maladjustment, etc.); this is done by the ILEA's school psychological service. It is true that if a mentally handicapped child is to be placed in a boarding school the appropriate borough health social work service will usually be asked for a social report, but this is at a

very late stage in the case when the decision about boarding school
has already been taken. Indeed borough staff generally complain
that they are too little involved in the allocation of children to special
schools. The transportation of handicapped children is another point
of borough/ILEA friction. The ILEA is the transportation authority
and is more restrictive than the boroughs would like to be.

All this is not to say that the school health service in Inner London
is beset by ILEA/borough rifts. Far from it. On the whole, relationships are close and cordial. Also regular meetings of the inner
borough medical officers of health, and of principal social workers
help to harmonize policy and smooth out difficulties. Nevertheless,
there are illogicalities and tensions inherent in the situation.

Conclusions. In assessing the impact of the London reorganization
on the school health and welfare services one is faced with the problem that the two major considerations point in opposite directions.
On the one hand, the shortage of many kinds of professional staff
and the advantage of large resources (so that some can be spared
for minority aspects of the services) point to the case for large l.e.a.s.
In particular it supports ILEA. On the other hand the very close
links between the *education* health and welfare service and the
community health and welfare services indicate the desirability of
having one local authority responsible for both. Here the outer
boroughs clearly score over the ILEA.

Special education in London poses a similar dilemma. On the one
hand a large l.e.a. like the ILEA has very considerable advantages.
It can employ a corps of administrative and advisory staff devoted
exclusively to special education. Possessing a considerable number of
schools it can develop coherent special education policies. It has the
resources to devote to activities like designing furniture for special
schools. It has the size to organize events like a special schools
sports day.

On the other hand there are some potential drawbacks to size.
Perhaps more than for any other educational institution those
running special schools (especially the head teacher) need to feel
that they have the ear of the c.e.o. and the education committee and
that their unique problems are fully appreciated. Obviously there
can be less direct contact between top administrators and individual
special schools in the larger l.e.a.s. The ILEA get round this problem
fairly well by the establishment of an advisory special head teachers'

panel. Nevertheless in some of the outer boroughs where the c.e.o. has a particular interest in special education, head teachers greatly welcome the closer contact they enjoy. As was pointed out, however, by no means all outer boroughs *have* improved their contact with their special schools. A second danger of large authorities is of a certain administrative slowness and inflexibility. Havering, for instance, has been able to eliminate the two-year delay in allocating children to special schools which obtained under Essex. Even in Inner London, borough school health social workers, previously employed by the LCC stress the benefits of the shorter chains of command in their present authorities.

There are also great advantages for special education in close links between the l.e.a. and the local health authority. Ascertainment, for instance, is to a large extent a medical exercise. Again, therefore, the potentially optimum situation is where one authority is responsible for both. Nevertheless there are also very strong grounds for insisting that special education be administered by the same local authority as primary and secondary education. Special education is inextricably bound up with general schools provision and, if present trends continue, is likely to become more so. No administrative system should separate the two.

FURTHER EDUCATION AND TEACHER TRAINING

Further education is, broadly, education beyond the compulsory school age. This definition is, however, overlapped at both ends. At the lower end, education *in secondary schools* beyond the age of fifteen is not considered further education. At the upper end, education in *universities* was once distinguished by the superior appellation of higher education. Today, however, the term higher education is also used to describe the higher level courses in the institutions of further education.

Further education falls into two general categories – vocational and recreational. The former embraces full-time, part-time and sandwich courses which are of value to a person's career; the latter covers courses, often evening or weekend, which assist a person's recreational pursuits or broaden his general knowledge. An astonishing variety of recreational courses is offered in Greater London from continental cooking to Aztec architecture, from karate to bassoon playing. Most are provided in evening institutes which often occupy

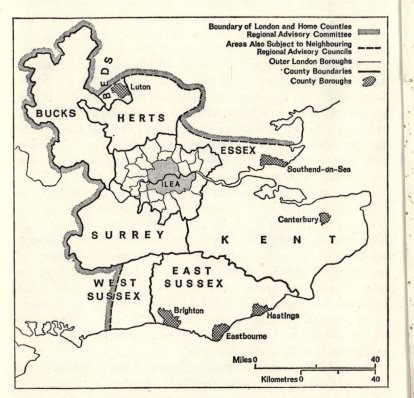

Figure 4. South East Regional Advisory Council for Technological Education.

a school building. Vocational further education also covers a very wide range from low level commercial courses and work of Ordinary GCE standard to advanced professional courses and degree work. This wide spectrum of provision was rationalized following the 1956 White Paper[69] into a four-tier system of colleges of advanced technology, regional colleges, area colleges and local colleges. The first were, however, transferred to the university sector on the recommendations of the Robbins Report of 1963.[70] The Labour Government decided that further expansion of degree level work should take place outside the universities. It proposed[71] therefore to develop a limited number of major centres in which a wide range of courses could be offered. These centres were to consist of several institutions of further education grouped together, or occasionally of a single college. They were to be given the title 'polytechnics', and would remain a local authority responsibility.

Teacher training for non-graduates is of course the responsibility of the colleges of education – some voluntary, some local authority controlled. The institutes of education offering teacher training to *graduates* are part of the university sector. Teacher training is a distinct activity and is not strictly part of further education. It is here considered together with further education, however, because it gives rise to similar problems.

The implications of the size of the new l.e.a.s. The catchment areas of colleges offering low level courses are usually quite small. As the courses offered become more advanced and more specialized the catchment areas become larger. At the top of the scale the polytechnics and the colleges of education have a national catchment area, although often a strong local connection too.

Given this, no l.e.a. can have a wholly satisfactory area for the administration of further education; even l.e.a.s the size of provinces would not solve every problem. Nevertheless, on the whole, geographically larger l.e.a.s have scope for more rational administration of further education. If only to avoid financial and administrative complications it is preferable for most of a college's catchment area to fall within one l.e.a. Moreover, the collection of data and the conducting of research into educational trends is more difficult when an area is divided between a number of l.e.a.s.

Apart from area, a related but distinct factor is the number of institutions administered by an l.e.a. Only when it controls a number of colleges can an l.e.a. have an effective further education policy, one which involves harmonizing and rationalizing provision. If an l.e.a. is responsible for only one or two colleges, such harmonization can only be achieved through inter-l.e.a. co-ordination or through the activities of the government or some ad hoc body.

Moreover, only when an l.e.a. administers a number of colleges can it develop real expertise in further education. If it has only one or two, its horizon must always be limited. Also, the small caseload precludes the employment of a wide range or large number of administrative staff for further education.

All this is directly applicable to the outer boroughs. Catchment areas for all but their lowest level institutions extend well beyond their boundaries. Since most have only one or two major colleges, their further education expertise can never be large. There is a danger on the one hand that colleges may be regarded as simply a kind of eccentric school and on the other hand that they may be treated with too much deference. Initially at least both kinds of attitudes could be found in the outer boroughs. The more long-lasting problem is of an l.e.a. striking the right measure of control, when it has little or no comparative experience with similar institutions.

No outer borough can draw up a very meaningful further education policy when each has such few institutions. Middlesex by contrast had a coherent county-wide policy; for instance the county graded colleges into three categories, related to departments not colleges as a whole. Thus one college might be in category A for engineering but category C for business studies. This ensured a reasonable geographical distribution yet retained considerable flexibility. No outer borough can do this sort of thing.

The ILEA not only enjoys the supreme advantage over the outer boroughs of continuity – having inherited the LCC's area, staff and institutions virtually intact – but because of its population, resources and number of colleges is in a different league in terms of size. The ILEA is indeed an outstanding further education authority with few peers either in Britain or internationally. It deploys an impressive range of further education staff – specialist inspectors, technical equipment assessors, experts in the complicated field of further education finance and so on. Its resources enable it to cushion colleges in times of economic stringency in a way that is quite beyond a small l.e.a. The large number and range of its colleges have enabled

it to develop great expertise and to apply coherent policies. No ou borough can hope to match this.

The disparity is if anything even greater in teacher training. T heavy concentration of colleges of education in the ILEA and t long experience of its staff in their control, bears no comparison the thin sprinkling of colleges of education in the outer boroughs a their paucity of staff experienced in this field.[72] Teacher traini indeed provides two examples of unusual administrative arrang ments, underlining perhaps the particular difficulty of fitting th part of the educational service into the new pattern of Londo government. First, Gypsy Hill College of Education in Kingston administered by Surrey. This was agreed initially for five years Surrey would like to continue the arrangement but Kingston is no keen to take over control. Secondly, Stockwell College of Educatio in Bromley is controlled by a joint committee of Kent County an Bromley borough. Kent were reluctant to have their relationshi with the College severed at the time of the reorganization.[73] The arrangement seems generally to be working well.

Finally, one of the great dangers of the new pattern of l.e.a.s in Outer London is of an increase in wasteful duplication of courses. With the disappearance of the county administrations and their restraining county-wide influence the possibility is that each borough will allow its one or two colleges greatly to increase their range of courses. Thus in neighbouring boroughs similar courses might be offered, both under-subscribed. This situation could arise from ignorance, boroughs being insufficiently informed of their neigh-bours' activities. It could, however, be more deliberate – a borough wishing to build up its college even if it means duplicating what is available in a neighbouring authority.

The body best suited to prevent these dangers is the Regional Advisory Council. Its activities must be outlined.

The Regional Advisory Council. The two main purposes of Regional Advisory Councils for Technological Education[74] as set out in the 1956 White Paper[75] are:

'(i) to bring education and industry together to find out the needs of young workers and advise on the provision required and (ii) to secure reasonable economy of provision.'

The second purpose is fulfilled through course allocation. Since

1957[76] no new advanced course can be established without Departmental approval and this approval is given or withheld in the light of the advice of the appropriate Regional Advisory Council.

The London and Home Counties Regional Advisory Council for Technological Education (henceforth referred to as the RAC), has an area of jurisdiction which extends well beyond Greater London.[77] In 1968 the RAC had a membership of 62; by far the largest block were elected representatives of the 35 l.e.a.s in its area.

The reorganization of London Government had a direct impact on the RAC. It meant a net increase of 16 l.e.a.s in its area,[78] all of whom had to be represented. Moreover, the crucial Distribution of Courses Committee which consists almost exclusively of l.e.a. representatives was nearly doubled in size.[79] In consequence, it is now compelled to operate much more through sub-committees. It has today 3 area sub-committees and a degree course sub-committee; 17 subject advisory panels, which are technically committees of the full RAC, also work on the Distribution of Courses Committee.

When an l.e.a. wishes to establish a new advanced course,[80] its proposals are first considered by the appropriate subject advisory panel and depending on the kind of course, either an area sub-committee or the degree courses sub-committee. The full Distribution of Courses Committee usually follows their advice. The avoidance of wasteful duplication is a major object of the exercise. The RAC is not itself, however, the final arbiter. The Regional Staff HMI has this role. In the great majority of cases, however, he follows the RAC's advice.

New non-advanced courses do not need formal Departmental approval, and do not have to be referred to the RAC. The London and Home Counties RAC now takes a far more active interest in non-advanced courses. According to its secretary, Mr R. D. Jamieson, this is largely a consequence of the London reorganization. Informal co-ordination by the outer boroughs proved insufficient – for instance they considerably overestimated the demand for OND full-time and sandwich courses and their provision was in consequence excessive. Because of such problems they have increasingly turned for help to the RAC. It acts as a clearing house of information on non-advanced courses not only for Greater London but throughout its area.[81]

The position is still by no means satisfactory, however. Wasteful duplication of non-advanced courses in neighbouring institutions undoubtedly occurs although it is very difficult to determine its extent. There are some significant pointers. For instance, the Institute of Legal Executives convened a meeting of all the colleges offering the first and second years of their certificate course because provision seemed to be too widely and thinly spread. In consequence of the meeting provision was rationalized in just three – Barking, East Ham and Tottenham technical colleges. Chief education officers themselves recognize the need for improvement – the Standing Conference of Education Officers of London and the Home Counties asked the RAC to make a start on the problem of rationalizing lower level courses. Perhaps the RAC should be given the same responsibilities in respect of non-advanced as it has for advanced courses.[82] This would, however, add enormously to its workload, which has already grown considerably. The increase in the number of course applications can be seen in Table 6.7. This increase is in large measure directly attributable to the London reorganization. Each of the new l.e.a.s tended to want its institutions to broaden their range of provision, whereas the former county administration achieved a measure of rationalization over their territories. There were, however, other factors – the expansion of demand for further education, the Industrial Training Act 1963,[83] new examination and changes in syllabuses,[84] and, finally, the implementation of the Pilkington recommendations on class sizes,[85] which led to re-applications for some courses which were dropped the previous year when they did not reach the requisite student numbers.

Financial Factors. Further education finance is a jungle. The basic complexity is that many students reside in one l.e.a. but attend colleges in another. At one time l.e.a.s recouped expenses on a bilateral basis. Because this was so cumbersome administratively, the Government introduced pooling arrangements. There are today an increasing number of pools – one for teacher training, one for 'no area' students[86] and one for advanced further education (AFE). L.e.a.s contribute to the pools on a formula basis. In the case of the last, l.e.a.s submit their expenditure on advanced further education, (excluding capital expenditure from revenue exceeding £10,000 in respect of any one college) for the previous financial year, and their likely expenditure for the present and ensuing financial years.

One half of this AFE expenditure is then allocated on the basis of an individual l.e.a.s total school population to the aggregate school population of all l.e.a.s; the other half of AFE expenditure is apportioned on the basis of each l.e.a.'s non-domestic rateable value to the aggregate non-domestic rateable value of all l.e.a.s. The needs element of the Rate Support Grant payable to each l.e.a. for any year, is increased or decreased by the amount by which the expenditure of the l.e.a. for that year either exceeds or falls short of the share attributable to the l.e.a. for that year.[87]

The other 'pools' are calculated in basically the same way but without reference to non-domestic rateable value, ie relying solely on school population. Settlement of these 'pools' between l.e.a.s amounts to a book-keeping arrangement in the Rate Support Grant and is unquestionably administratively simpler than recoupment between individual l.e.a.s.[88] The major problem so far as the London reorganization is concerned is the sheer complexity of the system. In many cases boroughs simply did not know how to operate it. The ILEA's experienced staff were inundated with requests for assistance in the transitional period.

Ignorance of the system is, one hopes, a transitory problem. The new pattern of l.e.a.s has more lasting repercussions. With more l.e.a.s there is more scope for friction. For instance the providing and paying l.e.a.s may disagree about the level of a course. There may be different practices about paying recoupment after the statutory time limit has expired.[89] Some outer boroughs are reluctant to introduce the ILEA's new simplified scheme of claiming for part-time students, perhaps because of fears of making mistakes. L.e.a.s are statutorily required to have regard to the fees charged in neighbouring l.e.a.s, but they need not charge exactly the same.

Pooling arrangements do not, moreover, cover the whole of the further education field; in particular they exclude non-advanced courses. Where there is considerable cross boundary movement in this field, l.e.a.s have to recoup on a bilateral basis. Hillingdon and Redbridge, for instance, have only one college (a new one) each providing non-advanced courses. Thus many students attend colleges in other boroughs, or the ILEA. This involves them in substantial recoupment payments.[90] The situation is much complicated by reorganization of London government. Hillingdon has today to make payments to, say, Ealing whereas previously since both were Middlesex-controlled such transactions were unnecessary.

In addition the finance of teacher training and student grants have also raised certain difficulties for the new authorities. It can be said, therefore that on the whole the London reorganization has had a most complicating effect on further education finance in the metropolis.

Relationships with the colleges. The ILEA's relationships with its colleges are generally excellent. While, of course, not everything is divine perfection – for instance, some ILEA inspectors are not thought to have anything to contribute to major colleges – most principals are more than content. Most would fight to the last ditch any suggestion that they be transferred to another l.e.a.

In Outer London relationships are satisfactory but there is not the same glowing confidence so noticeable in Inner London. As with head teachers of schools, it is unusual to find the principal of a further education institution who welcomed the reorganization of London Government. Most principals had few complaints about the previous administrations, and most have considerable reservations about the new. Indeed so appalled were the governors of Trent Park College of Education at the prospect of exchanging Middlesex for Enfield control, that they appealed to the government to be placed under the GLC or the ILEA. Their appeal was rejected.

The transitional problems seem often to have been particularly troublesome for further education. Few of the outer boroughs had previous experience of this branch of the service, and most had difficulty in recruiting staff with a further education background. Many of the tensions and frictions which developed between colleges and their l.e.a.s in the early years were the result of inexperience rather than incompetence or ill will. Harrow, for instance, caused great offence when it tried to give its lecturers the same conditions of service as its teachers. In some cases, however, overt hostility was displayed to further education; indeed it is still the official policy of the Conservative group in Richmond that further education should be taken over by the DES. As with the head teachers, the principals also regretted the disruption of long-established relationships with officers of the previous administrations.

There are more permanent repercussions. The c.e.o. and his senior colleagues generally enjoy much greater direct contact with the colleges. Principals generally welcome this increased contact. On the other hand, they resent the tighter controls which most boroughs impose. Principals tend to have less discretion in, for instance, financial matters and in staff appointments especially of a non-teaching character. Although tighter controls are naturally unpopular with principals they are not *necessarily* to be deplored.

A few principals complain of the increased politicization of their governing bodies. Sometimes elected members are given more prominence, for instance at Twickenham College of Technology the Chairman and Vice Chairman of the Governors are now councillors rather than independent businessmen as formerly. Sometimes although the role of the elected member has not changed, there is an increased tempo of party conflict as at East Ham College of Technology. There are exceptions; Harrow College of Technology has today only a minority of elected members on its governing body whereas under Middlesex they constituted 50 per cent.[91]

Finally, principals generally have less contact *with one another* than formerly. In Middlesex for instance the county convened regular meetings of its college principals. After 1965 these meetings ceased partly because there was no body to convene them and partly because the principals felt they no longer had as much in common. It is perhaps surprising, however, that meetings have not taken place on a Greater London basis. The LBA Education Committee could no doubt take this kind of initiative but perhaps it feels inhibited by the fact that ILEA is not a member.

Hornsey College of Art. The most dramatic episode in further education in London since the reorganization was the unrest at Hornsey College of Art which led to a six-week sit-in by its students in 1968. There were several causes of discontent but it is clear that the new administrative arrangements were a major contributory factor.[92] With the demise of Middlesex the College was taken over by Haringey; a large and experienced l.e.a. was replaced by a much smaller and, in the field of further education, inexperienced one. Haringey adopted rather restrictive attitudes to the College, especially in the field of finance. For instance Haringey insisted on paying the College strictly by the student hour as calculated from the registers; this meant the introduction of clocking-in arrangements for students and close supervision of their time. Also councillors would occasionally drop in to lectures and, if attendance were poor, suggest that the course be cancelled in the interests of economy. Moreover, the building programme was inadequate. All this fanned the flames of unrest.

Both for helping to provoke the student outburst and for his reactions to it, the College Principal was heavily criticized by two journalists but the borough too was considered a villain in the episode. 'It failed dismally to give the College proper accommodation. . . . In the field of human relations, they failed to notice that things were deteriorating in the College for a long time before the sit-in, and they handled the sit-in itself in a brutal and insensitive way.'[93] One cannot accept this assessment without qualification. Like most accounts Laurie and Law's is heavily biased in favour of the students. Nevertheless, there can be no doubt that Haringey's control of the College was far from satisfactory. It is bad logic to generalize from one extreme case and there have been student troubles elsewhere. Nevertheless it is additional evidence that the outer boroughs are not appropriate bodies to control major institutions of further education.

Conclusions. There are significant drawbacks to the outer boroughs administering vocational further education, especially institutions offering higher level courses. Catchment areas extend well beyond borough boundaries; each borough has too few institutions to be able to apply coherent planning or for its staff to develop great expertise; the range of staff compares badly with the counties' and with the ILEA's. The increased number of l.e.a.s and their small size has greatly exacerbated administrative and financial complications. The activities of the RAC have prevented wasteful duplication of advanced courses but not of non-advanced courses although its extent is arguable.

The new polytechnics also undermine the rationale of further education administration in Outer London. Three have been instituted in this area. Only one is fully contained within one borough – Kingston. The two others involve no less than six outer boroughs. It is likely that the problems of inter-borough co-operation and friction have helped to delay these projects. The London polytechnics are noticeably slower to get off the ground than those in the provinces.

Nevertheless the deleterious effects of the new arrangements are most in evidence when viewing the further education service *as a whole*. From the point of view of the individual colleges, the system is far from unworkable, partly because each enjoys a fair measure of autonomy (in professional fields at least). While therefore most principals would prefer to be controlled by a much larger l.e.a. most have settled down reasonably happily under the new regimes. Indeed they even recognize that there are some advantages, for instance often increased contact with the c.e.o. Moreover, the fact that in some cases the college is the borough's only pride and joy in further education has sometimes led to it being accorded greater priority and more deference than it enjoyed previously. And the London l.e.a.s have made some innovations in this sector of the education scene. For instance, Ealing took the initiative in establishing a day college of education for mature students, and the ILEA opened an annexe in East London to meet a similar demand.

Table 6.7

Applications to the Regional Advisory Council for their support in the Establishment of New Advanced Courses

Year	Applications Received	Applications Supported
1961–2	277	242
1962–3	339	289
1963–4	378	319
1964–5	448	382
1965–6	613	406
1966–7	700	415
1967–8	691	447
1968–9	819	559

Conclusions

The difficulties of assessing the effects of the reorganization upon the provision of education in Greater London are obvious. First, as has been stressed, objective criteria for measuring educational performance are rare, if they exist at all. Secondly it is extraordinarily difficult to isolate the particular impact of the structural reforms, because other factors have not remained constant. For instance, government policies have been amended in the field of nursery education and educational trends have accelerated the proportion of those staying on beyond the statutory school-leaving age. Above all, many changes seem to have flowed from the particular predilections of the newly elected members and staff.

Indeed, if one were to pick out the single greatest result of the reorganization it would be that it brought a large number of new personnel into the top posts and that they were given particular scope to introduce new practices and policies because the authorities too were new. A new c.e.o. appointed to an existing authority can have a considerable impact, but if he wishes to make changes always has to overcome its previous traditions. In the case of the outer boroughs the new c.e.o.s could exert their influence before traditions became entrenched. Of course there was always *some* continuity with the previous administrations but all boroughs used the opportunity to introduce some changes, and many for a fundamental review of administrative practice and educational policies.

These kinds of consequences would, however, have flowed from almost any reorganization – 10 l.e.a.s or 50 l.e.a.s. Indeed one could argue with more than a germ of validity that there should be a reorganization, say, every twenty-five years simply to provide a periodic administrative shake-up and thus to prevent l.e.a.s from getting fixed in a very deep rut.

What consequences, however, flowed from this particular reorganization? One of its most notable features is the preservation of the separate educational system in Inner London. The repercussions of this are examined below.

In Outer London the boroughs are able to provide a greater intensity of specialist advice than the counties could, but a smaller range. The education offices are on the whole more accessible to teachers and parents than the county headquarters but less than the divisional offices. Chief education officers and their senior colleagues are able if they choose to enjoy more direct contact with the educational institutions under their control than the county c.e.o.s, but less than the divisional officers. Consultation with teachers is far greater than formerly. It is difficult to generalize about the involvement of the public in the education service. Much depends on borough practice. Those which emphasize this, for instance in their appointment of managing and governing bodies, have made a noticeable difference. In others, however, it is difficult to detect a change. Public interest tends to be relatively high in education and since 1965 this interest has certainly been maintained if not increased. It is difficult, however, to attribute this to the new pattern of l.e.a.s. Rather is it the consequence of major educational controversies (such as comprehensive reorganization and 'banding' in Haringey,)[94] and the increased activities of certain pressure groups, eg the Comprehensive Schools Committee, the Confederation for the Advancement of State Education. There is, however, a tendency for more consultation with the public, for instance the open meetings held to explain and discuss comprehensive reorganization proposals. These would perhaps have been more difficult to organize on a county basis.

One of the advantages of the new system in Outer London is that it permits greater flexibility. The counties tended to apply certain uniform standards over the whole of their areas. The new boroughs are better able to adapt the service to local needs and wishes. On the whole there can be no doubt that the outer boroughs have proved themselves more than capable in the field of primary and secondary education.

In special education the counties had the great advantage of suffici- ent caseload and financial resources to provide a wide range of facilities and staff for minority needs. The outer boroughs obviously cannot each provide a similar range. Consequently they have been compelled to co-ordinate their facilities, so that, for instance, a specialist institution operated by one borough takes children from all the others. This co-operation seems, however, to work reasonably well. Indeed even the largest l.e.a.s cannot be wholly self-sufficient in special schools. Rather more cause for concern are the slim staff resources in many outer boroughs for special education and the allied fields of school health and welfare.

Further education does not fit happily into the l.e.a. structure in Outer London. Each borough has at most only two or three major institutions. In the first few years most boroughs were very un- certain as to how to handle them. The problems are more, however, than those of inexperience. There is a fundamental irrationality in l.e.a.s the size of the outer boroughs controlling powerful institutions with wide catchment areas; this applies especially to the polytechnics which are to all intents and purposes universities. There can be no meaningful co-ordination at borough level. Again therefore they are obliged to co-operate. And it works less well than in special education partly because the institutions are far more prestigious and power- ful, partly because the number of students involved is far higher and partly because so much more administrative effort and cost is involved. The RAC has had its difficulties increased by the reor- ganization but still achieves a satisfactory co-ordination of advanced course provision; nevertheless it has been unable to prevent much duplication in the non-advanced field. Moreover, the enormous administrative burden of interchange of students is very largely the fault of the structure. If, for the sake of argument further education were made a GLC responsibility a large percentage of these

H

difficulties would be eliminated at a stroke. There would still, of course, be some students residing outside the GLC and attending institutions within it and vice versa. Nevertheless, inter-l.e.a. movements within the GLC make up much of the administrative burden at present.

All this is not to imply that further education is in desperate straits in Outer London. Since each institution has a fair measure of autonomy, on the whole they operate effectively. The administrative structure is, however, a hindrance rather than a help.

To sum up, therefore, one can say that the new structure of l.e.a.s in Outer London seems to be working satisfactorily. It has certain marginal advantages over the previous system, such as greater flexibility, and certain disadvantages, such as in the field of further education. There is little evidence, however, that the old system was working extraordinarily badly. The main changes have stemmed from the impetus of new men in new authorities.

Inner London. The ILEA inheriting virtually intact the LCC's area, institutions and staff were far better placed than any of the outer boroughs to make the transition; indeed there was comparatively little transition to make. Moreover, given the LCC's excellence in many educational fields, the ILEA were indeed fortunate to succeed to such an inheritance. On the other hand continuity is not altogether an unmixed blessing; there might well have been some beneficial repercussions if education in Inner London had experienced the same kind of shake-up as education in Outer London. Some would claim that the ILEA is somewhat hidebound by its LCC tradition; its pioneering arteries have perhaps begun to harden.[95] Certainly the ILEA can point to few significant educational innovations since its establishment, and those few are mainly technical experiments (like the well-known Educational Television) rather than radical new policy ideas.

Of course the other side of the picture must be stressed. The ILEA undoubtedly remains one of the leading l.e.a.s in Britain. Its further education and special education services in particular are second to none. And in basic primary and secondary education it deploys a range and depth of specialist advice and administrative expertise, that only half a dozen other l.e.a.s come near to matching. In short it exploits to the full all the advantages of a large authority yet appears to escape most of the disadvantages.

The Redcliffe-Maud Commission recommended that l.e.a.s should not generally exceed one million in population.

'Once an authority's population goes much above 1,000,000 further gains in functional efficiency are unlikely to offset disadvantages associated with the management of such large units . . . Moreover, the bigger the unit, the more doubtful it becomes whether the individual citizen can have a real sense of belonging to it . . . The distance between the people and their authority, therefore must not be too great. This is particularly important for the personal services.'[96]

The Commission adduced virtually no evidence in support of this recommendation but they were clearly worried that very large units would on the one hand be remote and on the other suffer managerial problems – cumbersomeness, inflexibility, over-centralization and the like.

There is little evidence that the ILEA is remote. True neither its c.e.o. and other high-ranking officials nor its elected members are able to maintain much direct contact with its schools and colleges. On the other hand its divisional offices and its inspectors provide much of the personal link. Evidence of parents' feelings is scanty but the majority of teachers clearly find the ILEA a good authority to work for.[97]

The charge of inflexibility and overcentralization is not so easy to rebut. The ILEA would of course deny it and point to their divisional administration as evidence of delegation of responsibility down the line. Nevertheless teachers, social workers and other staff often complain that too many minor decisions have to be taken at County Hall, and that some rubrics are too rigidly enforced. One can only make an impressionistic assessment. To a limited extent this charge may apply to the ILEA but not so far as to threaten the general efficiency of its administration.

The greatest weakness of the Inner London educational system is school health and welfare. The administrative arrangements are exceedingly complex (with some parts of the services provided directly by ILEA staff, others by borough staff wearing an ILEA hat). Nevertheless on the whole the arrangements work surprisingly well. The relationship with the boroughs' family and community social services gives greater cause for concern. Despite great efforts to ensure co-ordination, friction and overlap are almost inevitable.

The Seebohm Committee predicted that the position would become increasingly anomalous in the future. The Redcliffe-Maud Commission, although their terms of reference did not extend to Inner London, endorsed the Seebohm Committee's view; '. . . it would be wrong for any new system of local government to divide responsibility for education and the personal social services between different authorities'.[98]

Possible Changes in the Administration of Education in Greater London

There is a great deal to be said for leaving the system as it is for a time, if only to give the administrators a period of comparative peace. A rapid increase in the incidence of coronaries among them would, one suspects, follow a further upheaval in the near future. One must, however, briefly consider the case for further changes. There are many possible changes but three appear most obvious and practicable: the breaking up of the ILEA among the Inner London boroughs, the allocation of higher further education to the GLC (perhaps accompanied by the abolition of ILEA), and the allocation of all educational responsibilities to the GLC.

Break Up the ILEA? There are some formidable arguments used in defence of the ILEA. First it is claimed that the unified system of educational administration in Inner London has a long history and is a response to the special needs of this area. The first point is not strong; administrative units with a far longer history have been swept away when circumstances so warranted. The special character of the area is a more substantial point. The density of population[99] and the excellent communications have meant a high measure of mobility of schoolchildren; borough boundaries, and the ILEA divisional boundaries, are constantly crossed. In 1969, 5,792 out of 30,992 primary pupils transferred to secondary schools in a different division from their primary schools. If primary and secondary education were transferred to the inner boroughs they would have to achieve a very high degree of co-operation. Once the comprehensives issue is settled, however, such co-operation would not be beyond the realms of possibility. Nevertheless, the risk of substantial disruption in primary and secondary education would be considerable.

Secondly the LCC had and the ILEA has a first-rate reputation as an l.e.a. To destroy it would be to impoverish education nationally.

All this is undoubtedly true and strongly influenced Ministry thinking at the time of the reorganization. Nevertheless, the ILEA must be looked at in the whole context of Greater London Government, and if to retain a good unit distorts and detracts from this whole pattern then there may still be a case for abolition. After all, other LCC services also had a high reputation yet were swept away. To put it another way, if in reorganizing local government nationally, the government were to retain intact as ad hoc bodies the six best l.e.a.s, the six best local highway authorities, the six best local housing authorities, etc., it would reduce the whole exercise to an absurdity. Thirdly, it is argued that abolition of the ILEA would be most unpopular among the majority of its teachers. Again this is true. It is likely, however, that most teachers were opposed to reorganization in Outer London too although they were not so vociferous or well organized. Yet they have generally settled down under the new regimes.

Fourthly, it is argued that the ILEA because of its size can allocate teachers flexibly over the whole of its area. If the inner boroughs were the l.e.a.s the less-favoured ones might have great difficulties in recruitment. This problem has indeed arisen in other services, eg the children's service. Some of the inner boroughs with high immigrant school populations and other social and professional disadvantages, would almost certainly find it far harder than the ILEA to recruit good teachers. Nevertheless, the supply of teachers is becoming a less acute problem and some of the l.e.a.s in Outer London with similar disadvantages, eg Haringey do not find the problem insuperable.

Fifthly, there is a high concentration of major institutions of further education, with very wide catchment areas in the ILEA, and the administrative staff concerned have long experience of handling them. It is thought inconceivable that these should be divided among the much smaller and quite inexperienced inner boroughs. It is difficult to quarrel with this judgment. The problems of the outer boroughs with further education would be multiplied many times in Inner London.

Finally, the ILEA has access to large highly professional GLC departments – Architects, Supplies, etc. This is undoubtedly valuable. Nevertheless the outer boroughs' architects and similar departments seem more than adequate for their educational needs. The ILEA's advantage here is only marginal.

There are three main arguments against the ILEA. First, it involves a disruption in responsibility for the personal social services. This has already been discussed. The arrangements are wasteful and inconvenient administratively and detract somewhat from the effectiveness of the services. Nevertheless they do not undermine them gravely.

Secondly, the ILEA can be attacked for producing an imbalance in the education service in Greater London. The ILEA regards itself and is regarded as a unique entity not as part of the general pattern of education in the metropolis. Thus, for instance, although the outer boroughs have bilateral arrangements with the ILEA, the latter body plays little part in the general joint arrangements. For instance, the ILEA is only represented by an observer on the Education Committee of the LBA. Yet it is clear that many of the items discussed could or should involve it. Yet it is so large that the outer boroughs are afraid it would dominate proceedings if admitted; indeed the ILEA itself is reluctant and would be embarrassed at being asked to participate.

Finally, there is the assertion that the ILEA is too big to be an effective education authority. Again this has been discussed. On the whole it has overcome the drawbacks of size.

In short, looked at in isolation the case for breaking it up is a thin one, although the dangers of so doing can be exaggerated (except perhaps in the case of further education). Looked at in the context of Greater London Government the case is stronger although still far from overwhelming.

The desirability of breaking it up has been argued entirely in administrative terms but there are now strong political overtones to the question. In the 1970 elections the Conservatives retained control of the GLC, but Labour captured the ILEA. Subsequently, it has become clear that the Conservatives find it irritating to have a virtually autonomous unit within the GLC controlled by their political opponents. Especially is it inconvenient when trying to restrict demand on the rates in an election year to have the ILEA making a substantial increase in expenditure. Conservative spokesmen have openly expressed their desire to see the ILEA abolished in its present form, and according to reports have set up a committee to consider ways in which this might be done.[100]

Allocate all education to the GLC? The Labour Government rejected the Redcliffe-Maud Proposals for education in the metropolitan areas and proposed that it should be administered by the top tier not the bottom. Applying the same model to London the GLC would become the sole l.e.a.

Such a change would greatly facilitate the administration of further education, and to some extent special education. It would also provide enormous financial and administrative resources. On the other hand it would cause an administrative split with the other personal social services. It is difficult to predict the effect on relationships with teachers and parents but there would obviously have to be a divisional administration. The managerial complexities could well be huge. Perhaps above all, another upheaval only a few years after the last one could well strain staff and elected members to breaking point. The case for such a massive change would have to be far more powerfully made out than it can be at present.

Allocate higher further education to the GLC? At present only the ILEA can claim to be an effective l.e.a. for higher further education, and there is a great deal to be said for its administration on a GLC basis. The main objections would come from those maintaining the 'seamless robe of education' argument; that is that education is a unity from nursery schools to further education and that there is no rational place at which it can be split between two tiers of administration. One could point out that it already *is* split because the universities are not part of the local authority education system. There is not space here to rehearse all the arguments for and against.[101] Suffice it to say that the present authors do not find convincing the objections to splitting off higher further education to an upper tier l.e.a.

If the GLC were to take over all higher further education, the ILEA's position would have to be re-examined. There would be a stronger case for its dissolution.

Summing Up. If any structural change in the education service in Greater London is to be made, the last appears the most justified. It must be emphasized, however, that the case for doing so is only moderately strong. Education is not going to break down in Greater London unless such a reform is undertaken. Moreover the consequences of another massive upheaval must be taken carefully into account.

Indeed one cannot help feeling that there is a danger of becoming over-pre-occupied with structural questions. It is difficult to imagine more horrendously complex administrative arrangements than those for school health and welfare in Inner London; yet the service is adequate, in some respects excellent. Given conscientious elected members and above all high-calibre officials and teachers almost any structure within reason can be made to work satisfactorily. At most the structure helps or hinders the education service in fairly marginal ways. Perhaps one should devote far more energy to trying to ensure an adequate flow of high standard recruits into educational adminis-tration than to devising the most rational structure of l.e.a.s.

Appendix

Unless exempted by the Minister, counties were required under the Education Act, 1944, to prepare schemes of divisional administration for their areas under which certain functions in primary and second-ary education together with local aspects of further education were delegated to divisional executives. These are usually composed of groupings of county districts to create a convenient unit of adminis-tration; in a few cases a single county district constitutes a divisional executive.

Boroughs and urban districts were empowered to apply to the Minister to be 'excepted' from their county's schemes of divisional administration and to prepare their own scheme of delegation as an excepted district. The Minister was obliged to grant the application where the county district's 1939 population exceeded 60,000 or its public elementary school population 7,000. Between 1958 and 1966 it was possible for new applications for excepted district status to be made from boroughs or urban districts with populations of 60,000 or other special circumstances.

Excepted districts usually enjoy considerably greater powers under the schemes which they themselves draw up, than counties allow their divisional executives. The latter are often not delegated power outright but are subject to county regulations and reserve controls – especially in the field of finance. The content of schemes of delegation varies, however, very considerably.

7
Housing*

Introduction

In the housing field, the post-war period might be characterized as one in which policies were designed to meet immediate needs. Fundamental thinking about the role of government in housing, and the relationship between housing and other government activities such as income maintenance, town planning and population dispersal policies were conspicuous by their absence. The administration of local authority housing departments has been seen primarily as a rather simple function of supplying houses below cost to homeless and ill-housed families. This view was strongly expressed by the Herbert Commission. 'Housing is essentially a local function', they claimed; and, again, 'housing is so closely connected with personal health and welfare that it must be essentially a borough service'.[1] This confident conclusion has been echoed by later inquirers[2] and reiterated with tedious frequency by local authorities jealously guarding their new responsibilities.[3]

In spite of these confident assertions most of the official reports have also recognized that there are close links between housing and planning. The Redcliffe-Maud Commission incorporated into their recommendations a housing/planning link in their ideas for the proposed metropolitan authorities which were allocated responsibility for the following major housing policies: the assessment of the need for additional building land; the formulation of policies for large-scale population movements and long-term urban redevelopment schemes and above all the determination and enforcement of rent policies over the whole metropolitan area. The Herbert Commission, in their discussion of the planning function, recognized that there were links between planning and housing policies but they nowhere discussed explicitly the relationship between planning and the local function of housing management. This is an extraordinary

* The authors wish to acknowledge the help given by Mr R. D. Callis of Kingston College of Technology in the early stages of research for this chapter.

gap in their coverage, because the major 'housing' problem of London has long been recognized as the way in which each local area treats its own housing problems in isolation from those of London as a whole.

The Milner Holland Report on 'Housing in Greater London' set out as one of the principal points to which it wished to draw special attention the fact that:

> 'Under the London Government Act 1963 the major housing function lies with the new London boroughs, but not all the boroughs will be able to solve their problems alone. In particular, land will be needed in the outer areas and outside London to receive persons displaced from inner areas. If the attack on the shortage and bad housing conditions is to be successful, it must be planned, applied and directed for London as a whole.'[4]

The Milner Holland Committee was a fact-finding committee and stretched its terms of reference to the limit by listing selected 'principles'. Had it been a committee asked to make 'recommendations' the quoted principle could clearly have been recorded as a recommendation that some authority should be given these housing powers. To obtain not only planning, but also 'application' and 'direction' it is clear that such powers would need to be extensive. The Milner Holland conclusions were by no means original and even the Herbert Commission had admitted that 'at present permitted densities there is little prospect that any county district in the Review Area can do more than keep pace with its own demands'.[5] From this knowledge, however, the Commission did not go on to consider whether (1) a single housing agent was required for the whole of the GLC area and/or (2) in what way a close association between housing and planning functions could be achieved.

By neglecting the examination of these possibilities they put on one side such controversial questions as policies on permitted densities and problems of building in the Green Belt. Instead the Commission took these policies as given datum and concluded that there was bound to be an overall short fall of supply within the GLC area which could only be met by building outside Greater London.[6] Hence they attached great importance to the building of New Towns and the Expanded Towns legislation. Having decided that Inner London population must be transferred beyond the Outer London boroughs they allowed this view to influence their proposals for the organization of the housing function.

The Herbert Commission's recommendations can only be understood with reference to their belief that housing was an adjunct to the personal health and welfare services and that its administration could be divorced from planning policies. In accordance with the first assertion that housing belonged with the welfare services, they assigned housing to their proposed Greater London boroughs; but to meet population movements beyond London they recommended that the new Council for Greater London should have certain housing powers. The Commission identified four problem areas which would need attention from a Greater London Council. These were:

i. problems within Greater London 'not susceptible of local solution';
ii. assessment of the need for building for Londoners outside Greater London;
iii. making and executing a long-term plan for housing outside Greater London;
iv. deciding and controlling policy for the allocation of tenancies in overspill housing.[7]

In spite of their recognition of the first point given above they stated that the ownership and management of *all* local authority houses in Greater London should be a function of the boroughs.[8] As a consequence of this decision the ambiguous proposal was made that 'it might be found convenient' to give the Council for Greater London the opportunity to nominate tenants to a certain percentage of the lettings transferred to the London boroughs. No explanation is given as to why 'nomination' might be convenient whereas ownership by the GLC would be considered inconvenient.

The recommendations of the Commission were in general adopted in the London Government Act, 1963, but some minor modifications were made. London boroughs were designated as the 'housing authorities' but the GLC was given some permanent housing powers. In addition to their permanent powers, the GLC was granted as a temporary measure certain of the LCC's Inner London housing powers. The immediate effect of the Act was not therefore very radical as far as the housing service was concerned. The transitional powers made it possible for the GLC to continue in much the same way as the old LCC, while the boroughs adjusted their management policies to the needs of the new larger boroughs. It was not until 1970 that the full impact of the reorganization became apparent when the GLC started the operation of transferring its stock of houses to the individual London boroughs. The long-term housing powers of the GLC are significantly smaller than those of the LCC. They are:

    i. to build within Greater London for the purpose of rehousing from a comprehensive development area or for rehousing people displaced by other activities of the GLC (eg provision of schools or roads);[9]

    ii. to conclude agreements under the Town Development Act, 1952, with authorities outside Greater London for the expansion of existing towns to house London families;[10]

    iii. to establish and maintain a record of need for housing accommodation;[11]

    iv. to establish and maintain facilities for the exchange of housing accommodation;[12]

    v. the GLC may also build houses within the GLC area for purposes other than (i) subject to the consent of the London borough in whose area it is proposed to build; or, if that consent is withheld, to the consent of the Minister.

These permanent powers fall generally under Part V of the Housing Act, 1957, and under that Act the GLC has the duty to 'consider the housing conditions in their district and the needs of the district with respect to the provision of further housing accommodation'. This is a general provision applying to all housing authorities and the Ministry of Housing and Local Government consider that this duty is associated with another housing duty 'to submit to the Minister proposals for the provision of new houses'.[13]

The principal transitional and temporary powers of the GLC are as follows:

    i. slum-clearance powers within the County of London; the Minister originally indicated that these might come to an end in 1970; but later this was extended to the mid-1970s;[14]

    ii. power to build within Greater London for all purposes without requiring borough consent: no date has yet been fixed for the termination of this power but it was indicated in 1965 that it was likely to be several years before this could be done.[15]

In addition, the GLC inherited on 1 April 1965 the whole of the LCC's stock of housing whether inside or outside Greater London. They were, however, required to submit to the Minister by 1 April 1970 a programme for the transfer of housing accommodation to the local authority in whose area the housing was situated, or to a housing association.[16] This did not imply that all the GLC's housing accommodation was to be transferred to the boroughs but that at least some of it would be.

The London boroughs do not have the power to make agreements under the Town Development Act, 1952, this power being limited to the GLC. They are also statutorily bound to provide the GLC with information about applications for housing accommodation within their areas to enable the GLC to carry out its function of maintaining a record of housing needs in Greater London.[17] Apart from these two provisions the powers of the London borough councils to deal with housing problems within their area are formally no different from those of county boroughs.

The housing provisions of the 1963 Act scarcely differed from those put forward by the Government in the original Bill of 1962. The proceedings on the Bill are nevertheless important for what they revealed about the Government's view of how the new system would work, and of the purpose of the transitional powers of the GLC.

When, for example, the Conservative Members of Parliament, Mr Robert Jenkins (Dulwich) and Mr Robert Allen (South Paddington) moved an amendment on behalf of the AMC to have definite dates inserted in the Bill for the termination of the GLC's inherited powers, Mr Corfield (Parliamentary Secretary, Ministry of Housing and Local Government) argued:

'it will inevitably take some time for the new boroughs to settle down ... it would be dangerous to try to forecast a definite date [but] the intention is that these powers will be brought to an end when it is clear that the new system is in full operation, that the new London boroughs are fully geared to carry out their slum clearance etc., powers and have the facilities to do so and the consent system embodied in the Bill is able to operate on the basis of a proper assessment of need after the necessary services have been undertaken.'[18]

On the other hand when the Labour members pressed strongly for the GLC to have permanent, concurrent slum-clearance powers Sir

Keith Joseph argued: 'I really do not think that the slum-clearance obligations falling on the London boroughs will be in any way beyond them.' In any case he saw 'no early likelihood' of ending the slum-clearance powers inherited by the GLC from the LCC.[19]

One of the few amendments which was accepted augured ill for the future co-operation of the boroughs; it was that over a limited period of years the GLC would charge any rate subsidy granted to the housing account mainly to the Inner London boroughs.[20] As any solution of London's housing problems requires a willingness to share the burden of costs this was a very poor start.

There are many crucial questions left unanswered in this view of how the new division of housing duties between the GLC and the boroughs would work. This can be illustrated from the important and urgent question of the estimation of housing 'need'. This function falls upon both the London boroughs and the GLC through the general provisions of the Housing Act 1957 which imposes on all local authorities a duty to assess the 'housing need' of their district. The London Government Act 1963 specifically placed this same duty on the GLC because it was the primary planning authority for the area. The generality of the statutory duty suggests that little thought was given to the meaning of the term 'housing need' when the 1963 Act was drafted and debated. There are, however, two quite distinct aspects of housing need which are not easily fused into one administrative operation.

The first meaning of the term relates to the current housing position. It involves an estimation of current shortages, vacancy rates and the identification of families in special need by size, income, mobility and length of time over which bad housing conditions have been endured. This may be regarded as the assessment of current housing needs and involves the intensive use of locally collected data. For example, the evidence of public health officials, a count of applicants on the current waiting lists, the knowledge of local estate agents and branch managers of building societies, and information obtained from social service departments and voluntary organizations are all highly relevant to the assessment of current needs. Much of this information is already in the hands of London boroughs and could most easily be co-ordinated by them.

A second, quite different meaning of the term 'housing need' relates to the long-term developments of an urban region. In assessing this need the current situation is of little value, since it must be

assumed that transport developments, long-term educational programmes, population and employment policies will all have major
consequences in creating or easing areas of housing stress.

Of considerable interest and relevance to the subsequent course of
events is Sir Keith Joseph's statement during the Committee proceedings of how the Government saw the new system working 'by co-
operation between the London boroughs and the Greater London
Council'. He said:

'We expect that the London boroughs will prepare a programme
of their housing needs ... They will then work out how many of
the needs they can satisfy by building inside their boroughs, how
much of their programme they can satisfy by building outside the
boroughs but within Greater London and how much of their
programme they can satisfy by overspill agreements through other
boroughs' building but within Greater London.

'Hon. Members may say that they will not get much co-operation
from other boroughs. That underestimates the powers of the
Greater London Council, because we must recognize that, if
London boroughs will not co-operate with each other to solve the
housing problems of the London boroughs with the greatest housing need, they are bound to find that the Greater London Council
is forced to seek building land inside their area. In other words,
co-operation between boroughs is the way to minimize, if London
boroughs wish to minimize, the work of the Greater London
Council within the Greater London area ... When these programmes have been made by the London boroughs, it will be for
the Greater London Council to judge from the survey of the whole
housing needs of Greater London how much will fall on it. It in
turn will decide how much that burden can be satisfied by overspill ... The Greater London Council will seek by consent arrangements to build the residue inside Greater London.'

Furthermore on the basis of the borough programmes and its own
surveys the GLC would make a five-year or ten-year programme for
which it would seek consents inside Greater London and overspill
agreements outside. But if it found that it could not get the necessary
consents and agreements it could then come to the Minister to seek
sanction to build inside Greater London. This would, however, be a
rare event: 'because London boroughs will operate their own building

programmes by themselves and will allow the Greater London Council to build through the consent procedure'.[21]

What is interesting about this statement is the degree of emphasis which it puts on the boroughs' capacity to deal with London housing problems, not only by building within their own areas but by making agreements with other boroughs. Sir Keith Joseph did not apparently see the GLC as taking any initiative at least in the earlier stages. It would act rather as a co-ordinator, although of course retaining the opportunity of seeking consents to build within the boroughs if necessary.

Unfortunately the division of duties in the London Government Act does not give any real encouragement to the GLC to develop the social planning techniques which would allow it to carry out this longer range planning function.

In the next two sections of this chapter we examine some of the other difficulties created by the Act and the development of housing policy within the GLC during the period 1965–70.

The working of the new system: provision of housing

In March 1965 the departmental committee which had been set up by the Minister of Housing and Local Government in August 1963 reported.[22] Their comprehensive report traced the development of London's housing problems during the period 1951 to 1964 and suggested that while the problems were complex they could be reduced in intensity if old prejudices could be buried and a new spirit of co-operation fostered. The immediate housing need was for an increase in the amount of rented accommodation throughout London, but as the shortages were concentrated in certain boroughs, other local authorities without such severe housing problems should be willing to co-operate with the hard-pressed areas of housing shortage.

In the post-war period the pattern of development in the GLC area had been confused by both outward and inward population movements. There were two major migration flows; the large outward flow of families with young children, counterbalanced by a substantial inward flow of single persons. These two primary movements did not however have an equal impact throughout the metropolis. There was a tendency for the inward flow to be concentrated on the boroughs lying to the north and north-west of the Thames but the outward family movements seem to have been more evenly spread throughout the suburban fringe.

The clarity of this picture was however much obscured by large 'planned' movements of population from the boroughs lying on the north-east and south banks of the river. These boroughs had been the centres of the worst nineteenth-century housing and they had suffered from fairly heavy bombing. Both the LCC and the metropolitan boroughs undertook massive urban renewal projects after the war. This meant that from these boroughs there was a greater outward movement of population than in the north-western boroughs, and that the inflow of population was curbed, the heavy concentration of small local authority dwellings inevitably reducing the chances of single people finding and retaining accommodation within the area. As each borough had some residential qualification for applicants wishing to enter their name on the waiting lists, migrants to boroughs with large local authority building programmes had less chance of remaining long enough to qualify for a council house than those going to other boroughs.

The outward flow from these boroughs was also rather more 'planned' than the private enterprise movement of families to owner occupation; although many East-end families were removed by the LCC to the outer boroughs, quite a large number were rehoused on inlying estates in such boroughs as Stepney and Bermondsey. Thus in the eastern boroughs local authority housing policy made it possible for many families to remain in the central area. For families with a very much higher level of income, private developers were also making it possible for parents to remain in central London. Some of the older districts of London such as Canonbury in Islington, Chelsea, small areas of Lambeth and Kensington were rehabilitated; houses which had long since been subdivided or allowed to deteriorate were modernized and let on long leases to 'owner occupiers'. Thus houses which might have been occupied by four or five householders with few, if any, children were converted back into single family occupation.

Both the private developer looking for Georgian or good quality Victorian houses, and local authority activity in Central London obscured the main demographic picture and caused the main impact of the inward flow of single people to be concentrated on a few boroughs or even parts of boroughs. If we measure population movements as the Milner Holland Committee did, rather crudely as 'adverse' if the changes produce a larger increase in households than in total population, Hampstead, now part of Camden, Kensington, Stoke Newington and Wandsworth suffered an adverse change in their population structure.[23] The change was less serious in Hampstead and Wandsworth as these more outlying boroughs do not appear to have attracted quite so many poor people who could only afford to rent one or two rooms, and the stock of houses was of a rather better quality. Kensington and Stoke Newington suffered acutely; the former because housing conditions in the northern half of the borough were already very bad and the latter because the size of the population increase was large relative to the smallness of its land area.

Population and Household changes, 1951–61

	Population %	Households %
Hampstead	+ 3·8	+ 7·6
Kensington	+ 1·8	+ 6·3
Stoke Newington	+ 6·2	+ 9·0
Wandsworth	+ 5·0	+ 7·9

These population movements were accentuated by the decontrol of rents during the period 1957–64. The provisions of the Rent Act, 1957, meant that more dwellings were decontrolled in the north and north-west where property was larger and therefore had higher rateable values than in the south and east. These were already the areas to which incoming single people gravitated and decontrol made it easier for them to obtain accommodation. Flats previously occupied by one family were turned into 'rooming houses' with one tenant for each room. The tenants were usually jointly responsible for the rent over a three-year period. Victorian houses which had been divided into two or three lettings for many years were further sub-divided into six, eight or even more lettings. At the same time the age of marriage was falling and many young people whose combined market pressures had created the great supply of one-room lettings and shared kitchens, found after marriage and the birth of their first child that they could not find suitable family accommodation.

While this was occurring in the private sector of the market the public sector was grappling with its own difficulties. By 1957 the supply of vacant, cheap bombed sites had come to an end, and in the areas of greatest housing stress new building could not take place until old buildings had been demolished. In the northern and western boroughs this meant the rehousing of all the new tenants from the densely packed Victorian terraced houses which had been released from rent control. As the boroughs could not, or would not rehouse all these tenants many were given notice to quit and had to fend for themselves. Inevitably this only worsened the situation in the remaining houses and the housing problems of Central London spread beyond the LCC boundaries into boroughs like Willesden and Southall.

Almost all new local authority building was used for the slum clearance programme and families on the general waiting list (however badly housed) had little chance of obtaining accommodation. Into this already very difficult situation there was injected the problem of racial tension. Before 1957 these difficulties had only been severe in the very worst housing areas of multi-occupied terraced houses. With the growing number of immigrants from overseas and increasing shortage of cheap family accommodation racial conflicts became widespread. The refusal of some of the indigenous population to let rooms to West Indians and others from abroad forced these people to buy houses and become landlords themselves. When this happened the new landlords treated English tenants much as they themselves had been treated, and refused to allow English tenants to stay in their houses.

Over this period there was a declining proportion of local authority building in the suburban authorities which was to some extent an indication that under the old system of local government, the outer boroughs did not regard the problems of their neighbours as ones which could or should affect their policies. The outward spread of housing stress was however beginning to affect the outer areas and the Milner Holland Committee laid great stress on the need for a 'London' solution to the problems. However before either the Committee's report was published or the London Government Act came into operation, a new Labour Government had taken office and they were committed to (1) the re-introduction of some form of rent control (2) a massive increase in house building (3) a shift from private enterprise to local authority building with increased subsidies (4) a vigorous campaign against bad landlord practices. Each of these election promises resulted in new legislation so that the new London boroughs inherited a mass of very grave problems, but at the same time seemed to have been given the economic and legal powers to deal with them.

The general background against which the new authorities had to draw up their policies reflected three main trends: i. the severe problems which they inherited; ii. the favourable circumstances provided by the availability of several large slum clearance sites ready for redevelopment, and by the election of a Labour Government which provided high Exchequer subsidies for new dwellings as part of a general policy to encourage local authority building; iii. unfavourable economic circumstances such as the high cost of renewing loans when they matured, high building costs and after 1968 restrictions on rent increases; to these economic factors might also be added the financial problems caused by the imposition of higher standards incorporating Parker Morris space standards and central heating.

In view of the very unfavourable economic circumstances prevailing from 1965 to 1970 the number of local authority dwellings completed in the period 1965 to 1969 is high. In both 1963 and 1964 about 14,400 dwellings were built by local authorities in the GLC area but by 1967–8 over 22,000 were being built each year. This high building programme was accompanied by a continuation of the slum clearance programmes and many of the new buildings were being used for those dishoused from clearance areas.

Table 7.1

Local Authority New Construction; demolished houses.
Greater London 1963–9*

Year	Dwellings built	Houses demolished	Estimated dwellings demolished
1963	14314	5241	10482
1964	14475	4789	9578
1965	17782	3279	6558
1966	19859	4641	9282
1967	22332	5864	11728
1968	22645	6883	13766
1969	23013	6892	13784

* Sources: GLC Annual Abstract 1967 T.715: Local Housing Statistics, (MHLG) and Housing Statistics: Great Britain (MHLG)

In 1966 the total stock of local authority houses in Greater London was about 600,000 which would have produced over 12,000 dwellings each year for re-letting. Between them therefore the local authorities had about 30–35,000 new and old dwellings to let each year. Many of these would have been allocated to families displaced by road improvement schemes, and other public and private developments which required the demolition of old houses. A few of the available lettings would, however, have been available for 'homeless' families and applicants from the London boroughs' ordinary waiting lists. The actual chance for any one family of being rehoused varied very greatly of course from one borough to another. The boroughs also varied greatly in the amount of housing stock which they inherited from their constituent authorities, and the distribution of local authority housing did not correspond to the needs of each borough. Thus, although two boroughs with the largest inherited stocks, Newham with over 24,000 dwellings and Southwark with almost 22,400 were also areas of great housing needs, Hillingdon with 16,273 dwellings and Enfield with 15,097 had relatively fewer problems and these were among only 8 boroughs which inherited more than 15,000 houses. On the other hand, Brent (8,786), Hammersmith (8,422) and Kensington and Chelsea (5,381) were among the boroughs with comparatively small stocks when set against their needs. However, it must be remembered that the last two, in common with all the boroughs formed from the old metropolitan boroughs, had access to the 210,000 houses owned by the LCC.

Table 7.2 shows the substantial increase in building which was achieved by most boroughs. The total built for the GLC area rose from 55,800 in the four-year period 1961–4 to 82,600 in the period 1965–8. The increased rate of building was heavily concentrated in the outer areas but two fairly central boroughs (Southwark and Camden) increased their building quite considerably. Greenwich, Hackney, the City of London and Kensington and Chelsea all reduced their building during the period although Greenwich and Hackney did maintain substantial programmes.

The figures have little meaning unless they are related to the size of the boroughs and for this reason the total stock of dwellings in each borough at the 1966 census has been added to the table. New building has not been given as a percentage of the total 1966 stock as it is so extremely low; but the reader will no doubt note that the rate of new local authority building is well below 1 per cent for most authorities. The figures also need to be related to the target programme of each authority. In September 1965 the Government published four-year building programmes for each of the London boroughs, the City of London and the GLC. Table 7.3 gives both the 'programme' targets and the number of dwellings started during the period. There is little correlation between the two and for the GLC area as a whole both completions and starts fell well behind target. Table 7.2 gives completions as 82,618 and Table 7.3 shows starts at 105,990 whereas the 'target' figure was 125,000.[24] This wide discrepancy is a not uncommon phenomenon in housing.

Table 7.2

Local authority dwellings completed by Greater London authorities and their predecessors (including the City of London).
Four-year average 1961–4; and 1965–8. Ranked in order of average number built 1965–8.

Authority	Number of dwellings in borough 1966	Dwellings built per year Average 1961–4	Average 1965–8
Newham	73,660	886	1151
Croydon	97,650	484	996
Enfield	84,900	384	924
Waltham Forest	74,830	410	902
Southwark	82,490	568	750
Ealing	86,280	265	660
Hillingdon	70,570	246	613
Camden	56,600	415	587
Havering	75,170	217	518
Islington	53,880	264	516
Wandsworth	87,970	*	489
Haringey	65,900	335	483
Greenwich	69,200	488	476
Tower Hamlets	56,920	411	476
Hounslow	62,070	410	472
Bromley†	92,700	265	466
Lewisham	79,870	315	429
Barking	52,890	254	425
Brent	76,200	206	421
Hackney	63,860	498	411
Barnet	93,420	248	324
Hammersmith	52,170	52	314
Lambeth	86,840	*	282
Redbridge	75,390	202	269

Authority	Number of dwellings in borough 1966	Dwellings built per year Average 1961–4	Average 1965–8
Richmond upon Thames	55,850	161	264
Westminster	67,660	190	258
Bexley†	68,490	214	252
Merton	58,810	130	212
Sutton	53,020	93	211
Harrow	64,870	90	144
Kingston upon Thames	45,450	81	140
City of London	1,120	178	93
Kensington and Chelsea	51,130	103	60
LCC/GLC		4098	5568
Total (average)		13965	20654
Total	(2,237,630)	(55861)	(82618)

* The average for Lambeth and Wandsworth combined 1961–4 was 790. It has not been possible to distinguish building in the different parts of Wandsworth Metropolitan Borough which was split between the two London boroughs in 1965.

† Bexley and Bromley figures estimated 1961–4 on the basis of including half the figures for Chislehurst and Sidcup in each.

Source: MHLG Appendix to Quarterly Housing Returns (to 1965): Local Housing Statistics Nos. 1, 3 and 9: Census 1966, Greater London, Housing Tables.

Some authorities in making the returns of 'programmes' simply stated the number which they planned to build when subsidies were lower, others the number which they would like to build if the level of subsidies from the Exchequer was beneficial while others might merely be indicating the number they 'wished' to build. The amalgamated figures scarcely represent a programme and it is no one's special responsibility to see that 'targets' are fulfilled.

It is evident that the GLC has played an important part in the local authority house-building programme in these years. Its relative importance is underlined by the fact that although the GLC only planned to build 22 per cent of the 125,000 target housing figures, it was responsible for 24 per cent of the 106,000 which were started and succeeded in completing almost 27 per cent of those actually built. The implication might be drawn that the GLC has been more

Table 7.3

London Housing Programme 1965–8 and Dwellings started
1965–8. Listed by size of programme

Authority	Programme 1965–8	Dwellings started 1965–8
Newham	6350	3708
Southwark	5700	5170
Waltham Forest	4900	3918
Hounslow	4875	3807
Camden	4275	2454
Brent	4100	3272
Tower Hamlets	4030	2792
Hackney	3825	3101
Greenwich	3775	2814
Wandsworth	3550	3317
Haringey	3425	3496
Islington	3400	3531
Ealing	3125	2790
Croydon	3025	3059
Westminster	3000	1535
Hillingdon	2950	2888
Enfield	2925	4200
Havering	2875	2308
Lambeth	2450	2083
Lewisham	2375	2096
Bromley	2375	1902
Barking	2300	2662
Barnet	2000	1757
Sutton	1900	1429
Bexley	1900	1274
Kensington and Chelsea	1700	516
Redbridge	1670	1679
Merton	1465	1563
Hammersmith	1460	924
Richmond upon Thames	1450	1127
Kingston upon Thames	1425	1125
Harrow	1300	743
City of London	775	965
GLC	28350	25985
Total	125,000	105,990

Sources: MHLG Press Release No. 212, 20/9/65: MHLG Local
　　　　　Housing Statistics.

willing or more able to see its plans through than the boroughs. Ultimately, however, the boroughs will, under the 1963 Act, be responsible for practically all new local authority building within Greater London. The relative roles of the GLC and the boroughs and how the relationship has worked in practice are therefore important in assessing both the effect of the 1963 Act, and Professor Cullingworth's[25] recommendation that the Act be implemented by the transfer of 70,000 GLC dwellings to the boroughs.

It was stated by the Ministry of Housing and Local Government in January 1965 that the main purpose of the transitional power given to the GLC for building within Greater London was to enable it to carry through the LCC programme which was already in train on 1 April 1965.[26] Out of nearly 41,000 families housed in GLC estates between 1 April 1965 and 31 December 1968 about 38,000 (92 per cent) came from Inner London boroughs; almost 60 per cent of the total was accounted for by 5 boroughs (Tower Hamlets, Southwark, Hackney, Lambeth, Islington).[27] The greater part of the new building by the GLC has also been in the Inner London boroughs, especially in Tower Hamlets, Southwark and Lewisham.

During the 1965–70 period, the GLC has simply been carrying on the LCC housing tradition; not only in devoting its main effort to rehousing people from the old LCC area, but also in concentrating its building there. (For the last fifteen years of its existence, the LCC had ceased to develop new 'out-county' estates and had concentrated either on town development schemes or on redevelopment within the county.) Many of the Inner London boroughs would have found it very difficult to make progress without this GLC help, and this is especially true of the rehousing of families on the waiting list; in Southwark, for example, 950 families were rehoused from the waiting list in 1967, 600 of them by the GLC.

The election of a Labour GLC in 1964, with Labour majorities in most of the boroughs, happened to coincide roughly with the election of a Labour Government. The coincidence produced two results: it put emphasis on an expansion of local authority building programmes, and, it led to a GLC initiative in attempting to foster co-operation between the boroughs. Mr Robert Mellish, the Chairman of the London Labour Party, was appointed Parliamentary Secretary of the Ministry of Housing and Local Government in October 1964 with special responsibility for London housing. He immediately began a well-publicized tour of the boroughs during which he used all his considerable powers of persuasion to get them to agree to increase their building programmes. The results of his efforts were embodied in the four-year programmes announced in September 1965.

Meanwhile, Mrs Denington, the Chairman of the GLC Housing Committee also set off on a tour of the boroughs; her mission was to get the outer boroughs to accept that they had a responsibility for solving London's housing problems by making available accommodation for people from Inner London.[28]

Table 7.4

Dwellings started and completed by the
GLC and London boroughs, 1966–9

Year	Started			Completed		
	GLC	Boroughs	Total	GLC	Boroughs	Total
1966	6884	15965	22849	5057	14802	19859
1967	6588	27038	33626	5398	16934	22332
1968	6766	21237	28003	6223	16422	22645
1969	4501	18001	22502	5279	17729	23008

Source: MHLG Local Housing Statistics

Thus, there was a clear strategy on which the Labour Government and the Labour GLC were agreed: encouragement of the boroughs to build as much as they could, with the GLC taking the initiative in trying to get co-operation between Inner and Outer London. By 1969, however, the situation had changed considerably and it was doubtful whether some of the earlier optimistic pronouncements about London housing could still be maintained.[29] The change was brought about partly by economic pressures, but partly by differing views about the GLC's role and about the extent to which local authority housing should be encouraged.

Two key events were the Government's economic measures of July 1966 and the Conservative victory in the GLC elections of April 1967 followed in May 1968 by the borough elections which left only four boroughs in Labour hands. The economic measures taken from 1966 to 1970 led to restrictions on public sector expenditure, and, more generally, to a slow down in the building industry; this with increasing land supply difficulties culminated in a sharp drop in the number of houses which were started in 1969, particularly in the public sector. In London, the momentum generated by the earlier building drive was sustained until 1969 so far as the number of houses completed is concerned. The number started, however, fell from a peak of well over 33,000 in 1967 to 22,500 in 1969. This was partly an indication that the Government was no longer prepared to encourage building at all costs as in the halcyon days of 1965. With the departure of Mr Mellish from the Ministry of Housing and Local Government in August 1967 no Minister was given special responsibility for London housing.[30]

The change in political control of the GLC led to a significantly different view of its role in London housing. Increasingly the Conservatives put forward policies to reduce the GLC's commitments, first, by a proposal to sell a considerable proportion of existing GLC housing to the tenants; secondly, by pushing on vigorously with plans to transfer roughly one-third of the existing stock of GLC housing to the boroughs; and thirdly, by giving greater emphasis to assisting housing associations as an alternative to council building. This Conservative philosophy was summed up by Mr Horace Cutler, chairman of the GLC Housing Committee, in a BBC television interview, when he said 'local authorities ought to get out of housing because they do not know how to run it';[31] although he claimed only to be expressing a personal view, many of the actions of the Conservative GLC were in keeping with this approach.*

This policy was the very reverse of that previously adopted by the Labour GLC supported by the Labour Government and not surprisingly, it led to a clash between the Government and the GLC. In July 1968, for example, the Minister of Housing and Local Government told the GLC that they would not be permitted to sell more than 600 council houses in the following twelve months, whereas the GLC had aimed at 6,000; the decision was described by Mr Desmond Plummer, Leader of the GLC as 'arbitrary and ill-considered'.[32]

Of equal importance was the change in the relationship between the GLC and the boroughs. The Labour GLC had emphasized its role as initiator and co-ordinator of a general strategy: 'a very close co-operation between the boroughs and the GLC with acceptance of a unified command', as Mrs Denington later described it.[33] But the crux of such an approach was the need to persuade the predominantly Conservative Outer London boroughs to make a real effort to help

* It is worth noting, however, that the GLC has continued to press ahead vigorously with the very large development at Thamesmead, designed ultimately to provide homes for 45,000 people.

the hard-pressed inner boroughs which, at least until 1968, were predominantly Labour. This approach met with some success in the early stages when the GLC negotiated agreements with the boroughs individually to provide nomination rights. These generally gave the GLC the right to nominate tenants to a proportion (usually 10 per cent) of the borough's new building.

Although the GLC emphasis at this early stage was on co-operation with the boroughs, there is no doubt that some of the outer boroughs were reluctant to have 'Londoners' foisted on them in this way. The GLC could, however, fall back on the threat of compulsory purchase of land to carry out their own building, knowing that they would get full backing from the Minister. Only in one case, however (Bromley), does this threat appear to have been used openly to persuade a borough to negotiate an agreement.[34] A number of other boroughs (eg Croydon, Hillingdon, Richmond upon Thames) have negotiated agreements on the basis that no building should be undertaken within the borough boundaries by the GLC.

The 1966–9 nomination agreements are given in Table 7.5. An average of approximately 850 families a year have been housed under these agreements, but significant contributions have been made by only a few boroughs which have either had large building programmes or made a more generous than average contribution, notably Croydon, Hillingdon, Ealing and Enfield. The Conservatives have not changed GLC policy on nomination agreements but it seems doubtful whether they will pursue it as vigorously as did Labour.

This is not, however, the end of the matter. Under the Labour GLC negotiations with the outer boroughs were, as has been seen, combined with an active GLC building role. The Conservative GLC policy is to shift more of the housing effort to the boroughs and voluntary housing associations. There is an obvious danger that with the decline in GLC initiative it will become harder to make progress with London's housing problems through a joint effort between Inner and Outer London. At the same time economic and political factors have combined to slow down the boroughs' own building programme. This has a direct effect on the number of houses made available under nomination agreements with the GLC because the latter are based generally on percentages of new houses. It is also affecting the boroughs' ability to deal with their own problems.

Earlier figures in Table 7.2 compared the four-year period 1961–4 with the four-year period 1965–8. This was not intended as a direct

Table 7.5

Nominations to Outer London boroughs

Borough	Agreed basis of Allocation	Housed between 1.4.66 and 30.4.69
Barnet	86 dwellings in 1967 and 1968.	47
Bexley	5 per cent of new dwellings in 1966: 10 per cent in 1967 and 1968.	53
Bromley	90 dwellings in 1967 (including 20 for 1966): 70 dwellings in each of the years 1968–70	169
Croydon	500 dwellings in the two years from 1.4.66. 100 for the period April–December 1968. Nil for 1969	617
Ealing	10 per cent of new dwellings + 10 per cent of relets during period 1.10.66–30.9.70.	251
Enfield	10 per cent of new dwellings for ten years from 1.9.66 + equivalent of 100 per cent of the product of certain industrial sites	243
Harrow	55 dwellings during the period 1.10.66–31.12.68.	45
Havering	15 per cent of new dwellings during the 4-year period 1967/1970	206
Hillingdon	25 per cent of new dwellings for 5 years from 1.4.66	456
Kingston upon Thames	20 dwellings during 1967. 12 per cent of new construction in 1968 and 1969	35
Merton	12 per cent of new dwellings during the 2 years from 1.4.66. Renewed for 2 years from April 1968 on the same terms	74
Redbridge	10 per cent of new dwellings during the financial years 1966–7 and 1967–8 + 30 dwellings. 10 per cent of new dwellings during the 3 years commencing 1.4.68	135
Richmond upon Thames	10 per cent of new construction + 10 per cent of relets for 2 years from 1.4.66. Renewed for 3 years from 1.4.68.	143
Sutton	15 per cent of new construction for 4 years from 1.12.66.	74
	Total	2548

Source: Communication from GLC Housing Department

comparison of the performance of the new and the old authorities. Clearly, houses completed in 1965 and 1966 and, to some extent, 1967 were planned by the old authorities. The figures were intended more as a guide to the developing housing situation in Greater London. What is important is to consider the effect of changes which will have their effect in the figures for 1969 and later years. It has already been pointed out that new houses started in the country as a whole fell sharply between 1968 and 1969 as a result mainly of the economic situation and the measures taken by the Government to deal with it from July 1966 onwards. Among these measures, high interest rates and increased Selective Employment Tax were singled out as particularly important in reducing the level of building activity. This period of economic restraint coincided with a period when practically every local authority of importance came under Conservative control, and both the economic and political factors combined to reduce the total of local authority building.

There were several indications that a number of borough councils, and especially those captured by the Conservatives from Labour in 1968, were not quite so anxious to put as much effort into council building. In some cases the new council proposed simply to sell off for private development, land which had previously been earmarked for council building. This happened with the 12-acre West Kensington goods yard site in Hammersmith. The proposal was rejected by the Minister who sent a stiff letter to the borough council telling them in effect to think again. The Chalkhill estate in Brent had an even more tortuous history: it was intended before 1965 by the Wembley Borough Council to be privately developed but the first Labour-controlled Brent council decided to develop it for municipal housing. When the Conservatives won Brent in 1968 they first tried to sell the estate to housing associations; this was prevented by the Minister and the council then proposed to charge much higher rents than would normally be found in municipal housing of this kind.[35]

Hackney in 1969 put forward a plan to halve its building programme, encourage private developers and housing associations, and sell off many of its existing stock of houses.[36] Haringey, like other authorities, ran into trouble with the Minister when it proposed to cut its building programme drastically and encourage housing associations as an alternative; the Minister argued that the borough should not shirk its responsibilities, and the borough retorted that the Minister's policy meant financial ruin.[37] Harrow, another Conservative council, not only reduced its housing programme to nil in 1969, but refused to discuss the housing situation with the Minister.[38] Although the GLC was prevented by the Minister from selling off more than a fraction of its housing it has gone ahead with plans to support housing associations.

The housing associations occupy a curious somewhat romantic position in English housing politics. They are on the one hand seen as a co-operative venture in the orthodox philosophy of socialist thought and on the other hand a private enterprise method of utilizing humanitarian motives and charitable funds for the provision of dwellings for the 'poor'. In the nineteenth century they were financed out of private funds which earned $2\frac{1}{2}$ to 5 per cent interest, but with the loss of semi-charitable funds they now rely almost entirely upon public funds and may be regarded as quasi-public institutions. Normally they do not have to pay income tax but they borrow from local authorities at the current long-term rates of interest and repay their loans over twenty-five to thirty years when purchasing and modernizing old property, and sixty years when building new flats or houses. The rents of new associations are inevitably above the average level of local authority rents as they have no pool of pre-1960 houses and they cannot therefore subsidize new dwellings by making 'profits' on old ones. They are eligible for Exchequer and Rate Fund Subsidies and in return for these they accept nominations from local authority waiting lists of housing applicants.

In 1969–70 the GLC planned to pay £180,000 from the General Rate Fund to housing associations in return for the right to nominate tenants; in addition to this rate fund subsidy an Exchequer subsidy is payable on the interest paid for loans for new building; on conversions the associations can obtain discretionary grants which were greatly increased by the Housing Act, 1969. The encouragement given to these associations by the GLC can be measured by the very rapid increase in loans made to associations since 1967. In 1967–8 £0·9 million of capital was loaned to the associations by the GLC; for the three years 1970–1 to 1972–3 no less than £25 million per year is being allocated for this purpose.

Since the associations depend for 90 per cent of their funds on local authorities, there is a strong tendency for them to be competitive with rather than complementary to local authority housing programmes. One of the major tactical reasons for the GLC and other Conservative authorities transferring part of their housing programmes to the associations was the desire to minimize the effects of the prices and incomes policy which had curtailed local authorities' freedom to raise rents. Housing associations are not subject to rent control or regulation legislation and may set rents at the market level. Thus by supporting these quasi-public associations the local authorities could introduce at least some degree of Conservative Party housing policy during the Labour Government's term of office.

A further difficulty in the already difficult London housing situation has arisen from the requirement in the 1963 Act that the Minister should make an order for the transfer of housing accommodation from the GLC to the boroughs if requested to do so by both the GLC and the boroughs; and the corresponding requirement on the GLC to submit a programme of transfers to the Minister by 1 April 1970.[39] This transfer of ownership was in accordance with the Herbert Commission's view that ownership and management of housing should be in the hand of the boroughs. Under plans agreed by the GLC in January 1969, over 70,000 houses, or approximately one-third of the existing stock, would have been transferred to the boroughs; the GLC was, however, to retain nomination rights to 65 per cent of the transferred accommodation. Some boroughs, such as Brent, Camden and Kensington and Chelsea, with by no means negligible housing problems, stood to gain very little in the way of transferred accommodation; at the same time, the GLC have stated that its loss of approximately 500 dwellings available for letting each year as a result of the transfers 'can only be at the expense of the quota which the Council gives to the borough councils on the basis of need'.[40] However, the proposals made in January 1969 were unlikely to have much immediate effect, though they further emphasized the difficulties of dealing with London's housing problems under the 1963 Act structure and soon ran into difficulties. Three of the boroughs which stood to gain most dwellings from the transfers rejected the proposals. They included the two Labour boroughs of Southwark and Tower Hamlets, and the Conservative borough of Hackney. The proposed allocations to these three boroughs represented nearly 19,000 houses or over one-quarter of the total. The primary reason for the rejection of these proposals was financial.

The majority of the boroughs were, however, not only ready but anxious to take over the houses offered by the GLC (now reduced to 46,060) seeing themselves, as the Herbert Commission had seen them, as the primary authorities for housing purposes. The Labour

Opposition on the GLC, however, saw in the transfer proposals not only a further step by the GLC away from its strategic role, but, more immediately, the prospect of considerable rent increases for its tenants transferred to the boroughs on 1 April 1970. Not surprisingly, therefore, in a GLC election year they were able to persuade the Minister to block the transfer proposals; the Minister laid the Order for the transfers before Parliament, as he was required to do, but a sufficient number of Labour members were mustered to object to it, and thus stop its progress.

Having blocked its passage through Parliament the Minister asked Professor Cullingworth to examine the question involved in the transfer of GLC houses to the boroughs. Cullingworth's inquiry was strictly limited by the exclusion of the financial aspects which were the main bone of contention. His report published in October 1970[41] recommended that the transfer proposed by the GLC should go ahead, but Cullingworth had so little faith in the administrative organization of London Government that he recommended the setting up of a powerful central committee which could enforce common management policies.

The evidence which the boroughs submitted to Cullingworth threw a good deal of light on their attitudes to their own role and that of the GLC in London housing. In particular, it illustrates how little common interest there was in 'London's' housing problems. Basically, the attitude of the boroughs was 'housing is our concern'; some acknowledged that there might be housing problems elsewhere in Greater London and conceded that the GLC might have some part to play in dealing with those problems, but this was essentially thought of as a very limited role.

Redbridge, for example, which had made available approximately 40 dwellings in the period 1966–9 through nomination agreements with the GLC regarded this as an adequate means of meeting Inner London housing needs without any need for the GLC to retain a separate pool of housing; similar arguments were used by many of the other Outer London boroughs, such as Barnet, Bromley and Richmond upon Thames. Hillingdon, more explicitly, conceded that such arrangements might not wholly meet the problem but complained that:

'The Council has never accepted the fact that the ultimate solution of the overall London problem will be found in this way; or if it is,

I

the effect on the Outer London areas will be disastrous. Fortunately the population of Inner London is falling far quicker than was estimated'.[42]

Even some of the boroughs which had benefited from its use of nomination agreements, like Brent, still thought that basically the GLC should not be a housing authority within Greater London. Lambeth too, although acknowledging the GLC's role in housing outside London, thought that housing should be as local a service as possible, and wanted the GLC to use its powers to build outside Greater London 'to take and absorb overspill from hard-pressed Inner London boroughs'.[43]

The difficulty with this approach to the problem is that there is no objective measure of the number of families who can be housed within a locality. In both Camden and Lambeth local policies will determine the amount of overspill which the GLC is expected 'to take and absorb'. The use of this very odd phrase to describe a movement of population may give some clue to the great contradictions we find in housing. On the one hand many authorities fervently believe that housing has a welfare aspect and must therefore be administered by the authority with responsibility for the social services. On the other hand there is a physical, almost mechanistic, concept of populations being moved and 'absorbed' into distant empty spaces. The GLC would find it very difficult to fulfil this second function and if it ceased to be a full housing authority the political incentive to perform this rather thankless task would be greatly reduced. The LCC did not build almost a quarter of a million houses as an aspect of its planning function, but because the people of London looked to it as a provider of homes. Now the people must look to their London boroughs and it is difficult to believe that the GLC will respond as quickly to Lambeth's call for 10,000 'overspill' houses as the LCC responded to 10,000 individual requests for accommodation.

Southwark was one of the few boroughs to reject the proposed transfer. It did so, as its evidence discloses, partly for financial reasons and partly because the transfer would have seriously reduced the number of GLC dwellings allocated to Southwark Council. This is the most serious mention of a reduction in the number of dwellings available for a borough which is undertaking a very large slum clearance programme. The expected shortfall seems to have arisen because the 'nomination' agreements granted by the GLC give a formal quota situation and the GLC has been able to supply many dwellings off-quota to the most hard-pressed boroughs. Thus Southwark obtained 1,432 dwellings through the GLC official quota arrangements and 2,127 through the off-quota arrangements. Southwark feared that the off-quota numbers would fall off considerably if the London boroughs controlled a large part of the GLC stock of dwellings. Each London borough would feel bound only by its quota figure and would use any excess for its own general waiting list. It is no doubt this factor, which was not brought out very clearly in other evidence, that has so greatly influenced most of the boroughs in their willingness to accept the GLC transfer offer in spite of the rather unfavourable financial arrangements which the GLC is imposing.

Wandsworth, unlike Southwark accepted the transfer of GLC property but like Southwark drew attention to the difference between the formal quota arrangements and the actual number of families rehoused. The quota number was 138 but 466 families were accepted; it is important to stress this difference between 'quota' and the actual number rehoused by the GLC because as more reliance is placed on a quota system there is likely to be a more rigid adherence to it. This is particularly true in the housing re-arrangements which are being made, because each local authority has its own non-priority waiting list and can absorb any lettings which fall outside an agreed quota. The present arrangement with the GLC does not give rise to this problem because the GLC keeps no general waiting list; when therefore it has lettings above the quota it has an interest in handing them over to the Inner London boroughs according to the priorities laid down by the formal quota arrangements.

The importance of this point was also stressed by Westminster which took a somewhat optimistic view of the way in which the new system might work. They stated that:

> 'the City Council has no objection to the G.L.C.'s taking the nominations of as many tenancies in the transferred estates as it wishes to have in order to discharge its statutory responsibilities. Even if the G.L.C. over-estimated its requirements it would doubt-less make the balance available to the City Council. In this connec-tion it is interesting to note that for the past three years the number of tenancy-nominations given by the G.L.C. to the City is such that in 1970 the total will be three or four times the quota, ie between 600 and 700 instead of 160.'[44]

The City goes on to stress its need for 'overspill' space and one assumes therefore that the City is expecting to obtain nominations on outlying estates after the new nominations agreements are introduced. But it is again necessary to question whether this would in fact be the case; it seems quite inevitable that the new owners of the old GLC estates will seek diligently amongst their own local population before making vacancies in excess of 'quota' arrangements available to relatively distant authorities. In this connection it is interesting to note the illusion of distance that seems to arise when the same distance is looked at from two different directions. Westminster would no doubt regard Hillingdon as a near-by site for its overspill; but Hillingdon looking in the opposite direction feels that it is too far away to be of much help to Westminster.

It has long been recognized that London's housing problems cannot be resolved by building only within London. The first generation of new towns were built after 1946 and now a second group are being created. The proposals for Milton Keynes, Peterborough and Northampton are rather outside the scope of this study, but of immediate relevance is the progress of town expansion under the Town Development Act, 1952. In origin largely a measure designed to help the LCC in its search for the provision of homes and employment for people living in the overcrowded areas of the county, it might be thought that schemes under the 1952 Act would assume greater importance under the new system of London Government in view of the GLC's diminished house-building role within Greater London.

By 30 June 1969 schemes of town expansion had been agreed providing for approximately 87,000 houses to be built and 37,000 had been completed. Many of these schemes had been initiated by the LCC which by 1 April 1964 had made arrangements for building approximately 70,000 houses; in some cases (eg Wellingborough) the GLC has greatly extended these original agreements but the largest new scheme so far concluded by the GLC has been one with Hastings providing for a total of approximately 3,500 houses.[45]

The GLC itself took the view in 1965 that the growth of expanding town schemes was a matter of urgency; this view was based mainly on the fact that proposals for new towns and other major schemes were bound to take some time to make an impact.[46] At that stage, Milton Keynes, Peterborough and Northampton had not been agreed. Within Greater London, the large-scale development of the riverside area on the borders of Greenwich and Bexley, later to be known as Thamesmead, was actively being planned but it would be some considerable time before the scheme reached fruition. Reinforcing the change to planning outer London developments was no doubt GLC awareness that in the long run its powers to provide housing outside Greater London would become its most effective means of ensuring its strategic role in London housing, much as the Herbert Commission had conceived it.

In practice, the GLC has followed the LCC in deciding as a matter of policy that the initial approach for a town expansion scheme must come from the potential receiving authority, thus limiting the choice of areas for expansion. Apart from this, the expanding towns programme needs to be looked at in relation to its effectiveness in meeting London's housing needs. An essential feature of town expansion schemes is that provision of employment and housing should go together. An Industrial Selection Scheme was operated by the LCC and taken over by the GLC on 1 April 1965 to assist in this aim. Londoners willing to move to new or expanding towns could register in the scheme and would be invited to move when suitable employment opportunities occurred in the new or expanding towns.

The GLC has declared that the aim of the Selection scheme is to give 'priority for houses and jobs in new and expanding towns . . . to London families in housing need, or to tenants of the Council or of the London borough councils whose move releases a dwelling in London which can then be used for an urgent housing case'.[47] In practice the great majority of those who register are on London borough waiting lists. Approximately 4,000 families a year are currently being housed from the Industrial Selection Scheme, three-quarters of them in expanding towns and the remainder in new towns. Between 80 and 90 per cent of London families going to the 'expanding' towns have gone under the Scheme.

The GLC is no doubt right to claim the success of the scheme in providing homes and jobs for families in housing need, but how far is it also true that it is 'making a notable and valuable contribution towards solving London's housing problem'?[48] In human terms, every family living in unsatisfactory conditions in London who can obtain good housing and employment in a new or expanding town, thereby gains. As however an employee is taken out of London by this selection process, the problems in London are only reduced in so far as the unsatisfactory accommodation vacated in London is not re-occupied by incoming families who may further sub-divide the accommodation and live at the same, or even more crowded densities, than the outgoing families.

The working of the new system: housing management

Management decisions.   The amalgamation of two or three existing housing authorities into one London borough created many difficult management problems. The immediate staffing problems were solved fairly easily by the inevitable introduction of more senior appointments and the absorption of junior staff in the new enlarged housing departments. This operation involved relatively few fairly simple problems of policy, but other questions such as those relating to rents, rebates and housing priorities became centres for political debate and prolonged negotiation.

Over the country as a whole there is little evidence that the political character of councils has a decisive effect upon the level of rents charged or the operation and scales of rent rebate schemes. Nor does it seem to affect the size of the rate fund contribution to the housing revenue account. In London, however, the exceptionally high cost of land and building tends to make the political attitudes of councillors of more importance in determining rent questions. Councils under Conservative rule tend to have fewer houses, to charge higher rents and to have a fairly comprehensive rent rebate scheme. Labour councils tend to have more houses (and therefore more older cheaper ones), to charge a lower rent, and administer only a small rent rebate scheme available for cases of extreme hardship. The comprehensive rent rebate schemes of Conservative councils combined with their policy of restricting local authority dwellings to fairly low income families means, however, that the rate fund contribution may not differ greatly from Conservative to Labour councils.

Even when the combining authorities were of a similar political complexion great difficulties arose in combining two different rent levels into one rational whole. During the 1950s local authorities introduced the habit of determining the rent of an individual house by reference to the gross value of each dwelling. These gross value assessments (set for each dwelling by the Inland Revenue) form the basis for the local tax paid by the occupier. Theoretically gross values are a reflection of the rent which can be obtained for each dwelling if it were let in the open market, but in the fifties the values were based upon 1939 rentals. In 1963 a new valuation list was introduced based upon current rentals but owing to rent controls and the increase in owner occupation and local authority dwellings there was little rental evidence upon which the Inland Revenue could base its assessments and there was some criticism of the variation found in different districts. By 1965 some housing authorities were still using the 'old' gross values as their base for setting rents while others used the current valuation lists. Thus one authority might charge 2·0 times 'old' GV (gross value) while another might charge 0·8 times current GV.

This complication was compounded by the fact that since gross values reflect the hypothetical open market rent, they varied from authority to authority to reflect differences in land prices, standards, types and sizes of dwellings. Thus, if a borough south of the river charged 0·9 times current GV, in 1964–5 average rents would have been about 31s (£1·55) for old flats and 49s (£2·45) for new ones. A borough north of the river charging 0·7 times current GV would be charging average rents for the same type of flat which varied from about 40s (£2·00) for old ones to 55s (£2·75) for new ones. Neighbouring boroughs which had been joined together suffered less severe differences caused by land value differences but they suffered even more acutely from age of stock differences. Thus for example in Chelsea where about half the stock was pre-war, 2·0 GV produced average rents of about 41s (£2·05) whereas the same multiple of GV produced higher rents in Kensington where only about 25 per cent of the stock was pre-war. If the unified borough decided to retain the GV rent setting method, the differential in the levels of rent would remain although Chelsea is normally regarded as a somewhat more desirable district to live than North Kensington where most of Kensington's dwellings were situated.

The reluctance of councillors and officials to conduct the debate in terms of the pounds, shillings and pence which were actually paid by the tenants, prolonged the debate and gave it a curiously unreal quality. However most boroughs terminated their debates by deciding the amount of rate fund subsidy they would make to the housing revenue account. The exchequer contribution was known and hence the total quantity of money to be collected in rents in order to balance the housing revenue account. Camden was exceptional in deciding to retain the three different rent bases inherited from St Pancras, Hampstead and Holborn until a borough inquiry had been made into the social and economic backgrounds of their tenants.[49]

The GLC were not immediately affected by this difficulty since they merely retained the LCC stock and under the first council continued the old LCC rent policies. They were, however, deeply concerned about the rent levels of the new boroughs since they had eventually to hand at least part of their stock of dwellings over to the boroughs. This meant that they had a special interest in a 'London' rent policy which would enable them to transfer tenants to any borough without causing the tenants undue financial hardship or creating real or imaginary grounds for complaint. Clearly if some tenants found that when their dwellings were handed over their rents were increased, while in other boroughs they were reduced, protests would be loud and possibly justified.

This difficult situation was somewhat complicated in 1967 by the election of a new Conservative council to the GLC. This political change introduced a desire for a different level of rents, over and above the desire for a uniform 'London' rent-setting formula. To meet both these needs the GLC proposed to charge their tenants rents which would equal the private landlords' regulated rents. These rents were set by rent officers under the provisions of the Rent Acts, 1965 and 1968, and were subject to appeal to a Rent Assessment Committee. This move was a return to earlier local authority housing legislation and its *raison d'être* was an attempt to avoid 'unfair' competition between local authorities as providers of accommodation and private landlords. In this politically oriented move it is, however, only fair to say that the rents proposed by the GLC fell far short of the regulated rents which were being determined by rent officers and appeared to be calculated solely with the desire of balancing the housing revenue account without rate fund subsidy.

However, before the GLC could put their policy into operation the Prices and Incomes Act, 1968, brought the central government into the controversy. The central government did not intervene in local government affairs to the extent of defining the term 'reasonable' rent; but under the Act it became obligatory for local authorities to submit proposed rent increases to the Ministry of Housing and Local Government for its approval. The GLC were unable to obtain this approval for the increases which they wished to introduce in 1969.

This controversy over rent levels has complicated the transfer of GLC dwellings to the London boroughs but by the end of 1969 the

GLC had almost completed negotiations with individual boroughs to transfer a third of their stock of houses to the boroughs. It is possible that the GLC's inability to pursue their adopted rent policy hastened these negotiations and made it rather more willing to off-load housing management responsibilities. It also seems clear from the legislation that central government intervention can only prevent increases in the overall levels of rent in one local authority. There is nothing to prevent an increase in rents being charged for the dwellings which are incorporated into a London borough's housing revenue account. Thus if the GLC handed over 500 of its dwellings to Westminster, the rents of these dwellings could be raised to equal the rents charged for other Westminster dwellings.[50] However, while such a move reduces the GLC problems it raises difficult political and management problems for the London boroughs which accept a part of the GLC stock.

The failure to reach a common rent policy is particularly serious because a real effort was made to examine and overcome the difficulties. In November 1966, the GLC agreed to a request made by the then Conservative opposition that there should be an investigation into 'the feasibility, the advantages and disadvantages' of a common rent structure with the borough councils. The report of this investigation, which was made jointly by the Housing and Finance and Supplies Committees, indicated that not only was there a 'strong logical case' for a common rent structure but also that it would facilitate transfers between authorities, and, in conjunction with a common rent rebate scheme, 'be an important step towards the recommendations in the Milner Holland Report', on the need for financial aid to poorer families. Yet, as the report acknowledges, such a scheme could only be achieved if the borough councils were willing to give up some of their freedom to set rents according to their interpretation of local needs. Further discussions were proposed and the report concluded, with remarkable optimism, 'We see no difficulty in arriving at a satisfactory basis for such a structure'.[51]

The boroughs' reaction was to undertake (with the help of the GLC's Director of Research and Intelligence) an investigation into methods of rent assessment, rent structures and rebate schemes operated by the GLC and the boroughs. Until this had been done it would be premature, they claimed, to consider the desirability or otherwise of discussions on a common rent structure with the GLC.[52] A year later, having completed this investigation, the LBA's advisory officers said they would be prepared to report on the advantages and disadvantages of a common rent structure but this would require many months of work. Moreover, there were 'considerable reservations' about the feasibility of such a scheme. The LBA therefore decided to ask the advisory officers to submit a brief report on the principles involved before deciding whether it would be worthwhile undertaking the detailed report.[53]

Effectively, therefore, the boroughs have made sure that there will be no progress towards a common rent policy or at least, that progress will be very slow. It is easy to understand why this should be so. Most borough housing managers had only just emerged from the struggle to devise common rent levels within their own areas. They were therefore reluctant to embark on a further and probably even more difficult period of negotiations on a common Greater London policy. Reinforcing their reluctance was borough unwillingness to forgo the traditional power of an authority to fix its own level of rents, although this problem might have been overcome by devising a scheme with a certain degree of flexibility.

The rent policy problems are reflected in the wide differences in the coverage of rent rebate schemes. In Westminster, for example, 85 per cent of tenants were granted rebates but in Bexley (another Conservative borough) the percentage was only 1·7 per cent. The same extreme differences are observed in the rate fund contribution to the Housing Revenue Account. Of the 34 housing authorities in Greater London, 4 in 1967–8 made no contribution and 2 others virtually none, whereas 6 relied on a rate subsidy to provide at least one-third of the income of their housing revenue accounts.

If the approach to a common rent structure has made disappointingly little progress, a slightly more optimistic view can be taken of the attempt to bring more uniformity into the assessment of priorities for housing need. Before 1965 each housing authority had its own method of determining which applicants on its waiting list should have priority for the allocation of housing. The usual means was a points scheme giving each applicant a basic number of points according to such matters as size of family in relation to size of existing accommodation, length of residence in the area with additional points for specific factors of hardship, eg for applicants living in overcrowded or defective or insanitary conditions. In most cases authorities required a qualifying period of residence in the area before putting applicants on the waiting list.

As in the case of rent structures, the introduction of the London Government Act, 1963, compelled the new boroughs to examine their methods of assessment of priorities since in most cases they had to assimilate two or more different priority points weighting systems inherited from the amalgamating authorities. However, a further incentive to look at the question from a Greater London angle was provided by two other aspects of the new system; first, the Act required London boroughs to accept any application made to them, whatever the length of residence in the borough of the person making the application;[54] secondly, the GLC was to keep records of the need for housing accommodation in Greater London. This did not make the GLC the depository of a gigantic Greater London waiting list; the boroughs were each to keep their own separate waiting lists and the LCC waiting list was divided among the Inner London boroughs. The GLC's function was rather to use the material supplied by applicants on borough waiting lists to assist them in assessing the size and incidence of London's housing needs. There is, however, no evidence that the GLC has used the data for this purpose.

The GLC thus had a very direct interest in seeking some measure of uniformity and at a very early stage this was considered in the Joint Working Party of Borough Advisory Officers and Officers of the GLC.[55] They proposed to the London Boroughs Committee (as it was then called) that the boroughs should adopt a common points scheme with a common residential qualification. The Ministry of Housing and Local Government, too, urged that boroughs should move towards uniformity over residential qualifications, even if they could not see their way to abolishing such qualifications altogether.[56] The LBC's handling of these requests is interesting not only for what was achieved but also for the manner in which it was done.

The Committee's Works Sub-Committee first considered that it was 'advantageous' to have such a scheme; the Committee in endorsing this view and recommending it to constituent councils was careful to point out that it recognized that each authority could come to its own decisions on special local conditions and matters of policy which might make variations from a common scheme desirable.[57] On the Ministry's suggestion over residential qualifications, the Works Sub-Committee took the view that it was essential to move towards a uniform policy, and the Committee therefore convened a conference of chairmen of borough housing committees and housing managers to discuss this question. The conference, on 26 April 1965, reached a fair measure of agreement on a basic residential qualification of one year in the borough and five years in Greater London as a pre-requisite to the offer of accommodation. The London Boroughs Committee strongly urged all boroughs to adopt this, at least for a trial period, 'in the interests of the London housing problem as a whole'; and regarded it as 'a very satisfactory degree of general agreement' on a controversial issue when 24 boroughs accepted the proposed residential qualifications and another 6 boroughs did so with the proviso that they wanted a longer residence in the borough.[58]

In contrast to the position over the proposal for a common rent structure, the LBC took the initiative in trying to get acceptance of a common points scheme with common residential qualifications, and largely succeeded. In both cases there is a potential conflict between what is desirable in terms of London as a whole and what an indivi-dual borough may regard as being in the best interests of its own inhabitants (or ratepayers). But this conflict is much less politically sensitive than in the case of rent policy since it has little immediate impact on the level of local rates and qualifications for being placed on a waiting list are of relatively little importance when few people are housed from that list. Nevertheless, it is interesting that the LBC were prepared, in the case of a common points scheme to go to some lengths by diplomatic persuasion to gain widespread agreement both on general principles and on specific controversial issues.

The common points scheme recommended by the LBC was based mainly on the extent of room deficiency and lack of (or sharing) basic facilities. Thus up to 20 points could be awarded to a family which lacked a room necessary to bring it up to a (defined) standard of accommodation, and 10 points if it had to share a living-room or kitchen. It is now operated by most of the boroughs, although many of them have incorporated relatively minor modifications or additions. Lewisham, for example, grants up to 20 additional points for certain medical conditions.

The success of these negotiations can largely be attributed to the initiative of a relatively few people who saw that the opportunity to make this change existed in 1965 and was unlikely to recur.[59] But although the boroughs have now achieved a fair measure of uni-formity, there must be reservations about the long-term effects of the scheme. The five years' residence qualifications in the GLC area is a seriously retrograde step as the LCC required only one year's resi-dence. Now the greater possibility of families moving within the

GLC area without losing the chance to go on a waiting list, has to be set against the fact that a stronger girdle has been put around the larger area of Greater London in an effort to stop new people coming in. It has also to be remembered that having one's name placed on the waiting list signifies very little if offers are only made to people after they have been duly registered for a certain length of time, which may be three or more years.

Housing departments. Almost all the new London boroughs set up separate housing departments under chief officers. In many cases this represented a change of practice as compared with the previous arrangements. Of the 3 metropolitan boroughs which amalgamated to form Southwark LB, for example, only Bermondsey had had a separate housing department. Only Ealing, whose housing manager heads a section of the borough architect's department, and Harrow, with a similar arrangement, did not initially take this step and they have both continued their original arrangements.[60] Some boroughs have regrouped departments since 1965 but this has generally left the housing manager in control of a separate, though subordinate department. At Sutton, for example, housing is now part of a Department of Health and Family Services under the direction of the former Medical Officer of Health, but the housing manager retains direct contact with the housing committee, though not with the council itself; at Haringey, departments have been divided into 5 groups, with individual chief officers retaining departmental authority; in this case, housing has been grouped with the architect's, engineers and valuation departments.

Unfortunately it is too early to test the important economic hypothesis that larger local authority units can and will introduce economies of scale. Nevertheless in view of the importance of the present review of local authority size and functions throughout Great Britain it seems worth reporting the little evidence available from London experience of increasing the size of management unit.

The available figures (Table 7.6) are taken from the IMTA housing returns and relate to supervision and management expenditure per dwelling per year. The years examined are those for 1963–4 for the smaller old authorities of the GLC area and 1968–9 for the new London boroughs. During these five years there were general price and wage increases and in looking at the percentage cost increase for the London boroughs it should be remembered that there was a 33 per cent increase in the supervision and management costs of all county boroughs and urban district councils. Thus if we accept a 33 per cent increase in costs as due to general inflation and changes in the quality of management, we have some indication of the existence of economies of scale if the price rise for the London boroughs is less than 33 per cent. Increases which exceed this amount suggest diseconomies of scale.

The available figures do not give any clear picture from which we could decide whether or not the larger boroughs have led to economies. Out of the 29 authorities examined, 18 showed increases which exceeded the national average and suggest diseconomies of scale. But of those 18, 2 (GLC and Harrow) had not changed the size or administrative functions of their housing departments during the period examined. They may, however, have been forced to respond to the changes generated by reorganized boroughs if they were to retain staff, or when they needed to recruit new officers to replace those who retired. If we deduct these 2 from the total we have 10 authorities where costs of management and supervision grew by less than the national average and 16 where it exceeded the average.

Further research over a longer time period would be needed to establish the reasons for the variations found. Our present knowledge suggests only that there is no evidence of very large economies of scale in housing management. It may however be that the period of three years, 1965–6 to 1968–9 is too short a time for administrative reforms to show any marked economic efforts.

Table 7.6

Average expenditure on supervision and management (general)
by (a) the old smaller authorities in 1963–4 and
*   (b) the new larger London boroughs 1968–9*

| London Borough | Average expenditure per dwelling | | |
| | 1963–4 | 1968–9 | % increase |
	£ s £p	£ s £p	
Richmond upon Thames	8 11 (8·55)	16 10 (16·50)	93·0
Barking	4 8 (4·40)	8 4 (8·20)	86·4
Ealing	7 8 (7·40)	13 4 (13·20)	78·3
Redbridge*	8 7 (8·35)	13 14 (13·70)	64·0

* Excluding, for 1963–4 the parts of Dagenham and Chigwell subsequently included in the London borough.

Harrow	7 12 (7·60)	12 0 (12·00)	57·9
		(no change in size)	
Brent	11 15 (11·75)	18 8 (18·40)	56·6
Hillingdon	5 1 (5·05)	7 18 (18·90)	56·4
Croydon	9 4 (9·20)	14 7 (14·35)	56·0
Tower Hamlets	7 11 (7·55)	11 12 (12·60)	53·6
Camden	9 14 (9·70)	14 15 (14·75)	52·0
Sutton	8 17 (8·85)	13 7 (13·35)	50·8
LCC/GLC	5 17 (5·85)	8 16 (8·80)	50·4
		(no change in size)	
Hammersmith	12 18 (12·90)	19 3 (19·15)	48·4
Newham†	5 6 (5·30)	7 13 (7·65)	44·3
Islington	8 17 (8·85)	12 15 (12·75)	44·0
Merton	8 11 (8·55)	12 0 (12·00)	40·4
Southwark	8 6 (8·30)	11 8 (11·40)	37·3
Havering	6 12 (6·60)	8 17 (8·85)	34·1
Haringey	7 18 (7·90)	10 9 (10·45)	32·3
Lewisham	13 0 (13·00)	16 17 (16·85)	29·6
Barnet	12 10 (12·50)	15 15 (15·75)	26·0
Westminster	19 6 (19·30)	23 1 (23·05)	19·4
Greenwich	10 4 (10·20)	12 3 (12·15)	19·1
Hounslow	7 10 (7·50)	8 18 (8·90)	18·7
Kensington and Chelsea	15 16 (15·80)	18 11 (18·55)	17·4
Enfield	5 15 (5·75)	6 11 (6·55)	13·9
Kingston upon Thames	10 6 (10·30)	11 11 (11·55)	12·1
Waltham Forest	7 15 (7·75)	8 5 (8·25)	6·5

Averages‡

London boroughs (Inner)			35·6
„ „ (Outer)			46·0
„ „ all			42·5

N.B. 1963–4 figures are averages of the combined expenditures of the authorities which were subsequently amalgamated to form London boroughs.

† Excluding, for 1963–4 the parts of Woolwich and Barking subsequently included in the London borough.

‡ Bexley, Bromley, Lambeth and Wandsworth are excluded from this table because it has not been possible to allocate the expenditure of Chislehurst and Sidcup U.D. and Wandsworth Met. B. for 1963–4 to the areas of the London boroughs. Figures for Hackney were not available for 1968–9.

Source: IMTA *Housing Statistics*.

A true measure of managerial efficiency cannot leave out of account the success or failure of the institution in achieving its aims. In the case of housing, the aim of the service is very difficult to define. Some housing authorities would take environmental improvement and slum clearance as the primary aim and attempt to measure economic efficiency in terms of improved general living conditions. Others would emphasize a simpler criterion such as the number of families who were on the waiting list and now have council houses. Others might go further and try to measure efficiency by taking into account all those who were seeking accommodation and have now found it; whether as owner occupiers, or tenants of an association, council or private landlord. This approach has been followed by Lambeth which has set up a Housing Advice Centre to help those who are seeking accommodation, those who have accommodation to let, sell or improve and those who have rent problems or difficulties in persuading their landlord to do essential repairs. If this very broad field of assistance is accepted as a measure of efficiency it must be firmly stated that most of the London boroughs fell very far short of the desired level of attainment.

During the 'sixties a rough and ready method of measuring the extent to which housing departments have achieved their housing goals has been the number of 'homeless' families. The voluntary housing organization called Shelter has extended the meaning of 'homelessness' to include families housed in sub-standard accommodation and their continuing publicity campaigns have suggested that housing shortage as measured by homelessness is getting more rather than less acute. In 1969 the Secretary of State for Health and Social Security (the Central Department responsible for homeless families) commissioned Professor Greve to report on the situation. His report has not been published at the time of writing but his findings have been 'leaked to the Press' and from these Press reports it would seem that between 1962 and 1968 the number of homeless families increased by 165 per cent in Inner London. In the Outer London boroughs 5·4 people were homeless per 10,000 of the population and in Inner London the figure was 22·6. In addition to these people who were members of a family the National Assistance Board reported that in 1966 11,000 single people were homeless in Inner London, and 6,000 in the Outer London boroughs.

There are three primary causes for this increase in the figures. First they are to some extent the outcome of more humane administration of the 'homeless' accommodation. After the scandals of the early 'sixties some of the worst accommodation was closed or improved and families were given accommodation of a type more closely allied to their long-term needs. The barrack-like arrangements were modified and in many boroughs husbands were permitted to live with their wives instead of being sent away to fend for themselves. The administrative revolution also included an attempt to provide the 'homeless' with 'temporary' accommodation which could be occupied for a longer period than 'short stay' accommodation. All these very real improvements in the method of treating homeless families has inevitably increased their number; not by making more people 'homeless' but by making the condition of homelessness a little more supportable and allowing people to stay longer so that families are not broken up, the children taken into local authority care or dispersed by parents amongst reluctant relatives.

However, two more sinister reasons for the increase in homelessness stem from inappropriate housing policies. The first is the continued unmonitored changes in the stock of dwellings in the central boroughs. Hammersmith, for example, suffered a decline in the number of dwellings in the borough of almost 15 per cent between 1961 and 1966. In Kensington the decline was 11 per cent. Some of this decline is due to slum clearance (eg the removal of one house containing 5 or 6 'dwellings') and some of it is due to private enterprise conversions. As these changes are not fully monitored the effects always seem to catch the housing officials unprepared and architects, engineers and builders still treat the land of London as if it was in plentiful supply. Land being London's scarcest factor it should never be left vacant for more than a month or two. But cleared sites are left idle for many months and closed houses are left vacant for so long that squatting has reached epidemic proportions.

The last problem which has greatly worsened the present London housing situation are the rents being charged in both the public and private sectors. In the private sector regulated rents are well above the means of families taking home less than about £20 per week, and although a rent rebate scheme may be of marginal help to these families, it must be remembered that almost all rent rebate schemes leave tenants with children at the official poverty level of living. This can be borne for short periods but if it is prolonged it almost inevitably leads to rent arrears followed by evictions. In the public sector the position is if anything worse because rents of new dwellings are much higher than the regulated rents for old dwellings. If the homeless are to be rehoused urgently they must be offered new accommodation since it takes time to wait for a suitable vacancy in old dwellings. But even as far away from Central London as Thamesmead gross rents are £8 per week and a rent rebate scheme that reduces the family to the 'poverty line' brings little comfort to the families housed on the estate. In this connection it must be remembered that new accommodation always means some new expense for curtains and furniture and there are of course the very large costs of moving in.

In the last five years or so many London boroughs appear to have neglected the poorest of their employed applicants. They have raised rents and reduced the landlords' duties such as internal repair and decoration; no one should be surprised if this has led to an increase in homelessness. Unhappily this tougher local authority attitude has coincided with a small increase in unemployment in the London region and much larger increases in other areas of the country. The countrywide increase in unemployment is bound to bring more people down to London in search of work and this must create some marginal increase in homelessness, although it is of course customary for the welfare departments in the London area to return 'homeless' families to the towns which they have left in their search for work.

An evaluation of the administrative structure, 1965–70

The administration of the housing service was seen by the members of the Herbert Commission and by the Conservative Government which was responsible for the London Government Act, 1963, as essentially a borough responsibility. The role of the GLC was a strictly limited one. It could, as Professor Cullingworth has pointed out, only collect and analyse information. It had 'no powers to guide or advise the primary housing authorities'. This meant that 'information was to be centralized [but] power to act was diffused'.[61]

The provisions of the Act were put into operation in 1965 and after only five years it is too soon to make a comprehensive evaluation of the administrative structure. But certain weaknesses have already become apparent and have led to a growing opinion that planning authorities should play a more active role in housing than that envisaged by the London Government Act. Criticisms of the Act rest on two main grounds. First, a rejection of the view that housing is simply a 'local' service which can be organized over small territorial areas. Secondly, a rejection of the view that housing is, or should be, a branch of the welfare services. Theoretically those who accept the 'local' and 'welfare' service view of housing could argue each point separately but they tend to fuse them together; however, the following two passages from the London borough of Camden's evidence to Professor Cullingworth distinguish between them and put the 'local' and 'welfare' view very clearly:

'An initial transfer of one-third of the G.L.C. estates to the borough would be acceptable, but the borough council is of the opinion that in due course, not unduly delayed, the whole of the G.L.C. estates should pass to the borough. The G.L.C. would then be in a position to frame proposals for coping with Camden's "overspill" and would not add to the borough's difficulties by importing tenants from other areas.'

'The London borough councils, following the reorganization of local government in Greater London, are the authorities responsible for health, welfare and children's services and the vital importance of a closer relationship between these and other social services with housing (including housing welfare) is generally recognized and advocated by all responsible bodies of opinion.'[62]

At the risk of being regarded as irresponsible the authors must question the opinions expressed above and point out that the view that welfare and child care services should be closely linked with housing has never been closely argued, it has only been asserted, much as Camden asserts it in the above paragraph.

In the London situation there are three possibilities. First to put almost all the housing powers with the London boroughs, secondly to place them with the GLC, and finally to give both tiers of local government equal housing powers, the GLC to exercise its powers with a bias towards large-scale planning projects, the London boroughs with a bias towards small-scale projects to meet the needs of minority groups. The arguments for or against one or the other solution to the housing problems of London can be summarized under the following headings:

(1) Exclusiveness
(2) Unitary Management
(3) Social Service emphasis
(4) Planning emphasis.

Before examining the arguments which fall under these four headings we would remind the reader that housing is a 'permissive' service and this differentiates it very clearly from 'compulsory' local services such as child care, education or public health. In most British cities local authority dwellings are only available to the minority of householders and in London only about 22 per cent of all dwellings are owned and let by a local authority. In such inner boroughs as Camden the percentage of local authority dwellings is only 20 per cent.

This fact is of great importance in considering the arguments put forward by the LBA for a clear-cut division of responsibility for housing in the London region. The Association wanted the housing function to be exclusively a responsibility of the borough in whose areas the dwellings were situated and said that the activities of the GLC led to 'misunderstandings and even suspicions between authorities'.[63] They regarded 'exclusivity' as all-important and overlapping of functions as most undesirable.

For a compulsory service such as the provision of primary schooling it is easy to visualize a situation in which overlapping of jurisdiction might lead to administrative muddle and a tendency to pass awkward children from one authority to another with the result that some children obtained no education. This position cannot however arise in quite the same way in the housing field because the local authorities only supply a small proportion of the total housing service of their area. In the London area, owner occupation, privately rented property and housing associations are all important competitors to local authority housing and people seeking accommodation do not regard the borough council as their only, or most significant source of housing. For readers who know little about the London housing situation it is perhaps worth stressing that only a minority of the applicants who wish to have a local authority dwelling, are ever likely to get one. This is the meaning of a permissive as opposed to an obligatory local authority service.

Exclusiveness (monopoly powers). The London boroughs demand to have a monopoly of local authority housing within their area rests partly on prestige and partly on the view that two authorities will compete for the available land. Land shortage in London gives strength to the latter argument until we examine the actual performance of authorities and their financial resources. Clearly the individual London boroughs have less resources for large-scale projects than has the GLC with its immense borrowing powers. As housing is a capital intensive industry it places a heavy financial burden on authorities during the land purchase and building period and this burden can force authorities to cut back on other capital expenditure. For a small authority a series of small- or medium-sized projects may be the only way in which they can build up their housing stock. As one project is finished and goes into letting a new one can be begun. The GLC can on the other hand undertake a series of large projects; the costs during the building period being spread over a larger number of ratepayers and a greater total debt pool does not create any severe local fluctuations in either rates to be levied, or debt pool management. The desire by London boroughs to have a monopoly of the service obstructs the possibility of any one authority undertaking a really large project.

To overcome this difficulty the defenders of 'exclusiveness' concede that the GLC should build, but not own and manage, houses within the borough areas. This solution seems, however, to be unworkable. It was tried in the early twenties by the central government (Office of Works) which built 'for' the boroughs and then handed over the houses for local ownership and management.[64] There were however endless arguments on the cost, design and quality of the building and the central government soon withdrew from the experiment. The same situation has already arisen in London in the 'sixties when the GLC thought they could build and transfer the completed houses to the boroughs, but the evidence submitted by Tower Hamlets to Professor Cullingworth suggests that this problem is no less real today than it was in the early 'twenties. Tower Hamlets stated that:

'1225 dwellings are included in Phase 1 (new dwellings built by the G.L.C. in the interim period) and, based on the estimated rate deficiency attributable to the Council's own construction of some £175 per dwelling per annum, the annual deficiency to be met from the rate is estimated to be of the order of £200,000/£225,000. This presupposes, of course, the GLC construction costs do not differ significantly from local costs.'[65]

In the GLC's own evidence they reported that they had wished to transfer 9,877 new dwellings to the London boroughs but only 4,780 were accepted. Thus rather more than half the offered dwellings were rejected – by the boroughs of Ealing, Hackney, Havering, Hounslow, Lewisham, Southwark and Tower Hamlets. It is hardly surprising that many authorities could not accept with equanimity the possibility of another authority imposing a rate burden upon them, and it is quite certain that all the boroughs will look most critically at every aspect of the GLC's designs and costs if they are to take over the burdens of ownership.

Two very important determinants of the total costs, whoever incurs them, are land prices and building densities. If land costs are high, density of dwellings per acre will be correspondingly high and from the point of view of London as a whole it might be better to build in an outlying borough rather than for example in Tower Hamlets. This was the great function performed by the LCC before 1965 and it was the reason why the LCC provided its dwellings at a lower cost than the old metropolitan boroughs. One of the consequences of the Herbert Commission's failure to consider the financial implications of its recommended reforms is that this supply of cheaper dwellings will not continue to be available when the GLC housing activities are terminated. Although the London boroughs have the right to build beyond their own boundaries, they have little incentive to do this on a large scale against the will of the receiving authorities. The administrative organization is such that each Inner London borough is hardly likely to proceed with plans to build beyond their boundaries without the active encouragement of the outer borough concerned. Since the outer authorities seem as convinced of the need for them to own and manage all local authority dwellings in their district as the inner authorities, we may be certain that few inner authorities will press the matter very hard against reluctant outer boroughs.

Finally the 'exclusiveness' argument has a strong element of borough patriotism. It is almost possible to speak of the 'Balkanization' of the London region and it is already possible to imagine the creation of a federation of semi-independent 'states'. Merton, in its evidence to Professor Cullingworth, put this position very clearly.

'If . . . this sense of community is in fact based upon that of a common landlord [the GLC], then it is suggested that it would be a healthier state of affairs for this to be broken down and for the tenants to have a greater sense of belonging to the Borough in which they live, along with the other residents be they owner occupiers or private tenants.'[66]

We have already quoted Camden's views on this subject of 'exclusiveness'. Lambeth's views reinforced Camden's and they asserted that 'housing should be as local a service as possible'.[67] Like Camden, Lambeth seemed to want the GLC to keep out of the borough; but on the other hand to help boroughs by providing accommodation for *surplus* population.

Unitary management. Arguments in favour of a single type of management and one rent, allocation and maintenance policy seem again to misunderstand or misrepresent the role and functions of local authorities as providers of accommodation. It must again be stressed that only 22 per cent of the population of London is in local authority housing and a great variety of management policies are encountered by Londoners. Since uniformity is neither attainable nor desirable over the whole housing market there seems no reason why two local authorities should not operate in one geographical district although they have different management policies.

Even those who maintain that there should be only *one* policy seek to encourage housing associations on the grounds that they create diversity and cater for needs neglected by the local authority. If this is accepted as a justification for the encouragement of housing associations, it should clearly also be accepted as a justification for allowing the GLC and the London boroughs to operate side by side. This comment points to the conflict of views held by the supporters of unitary management; the really important criticism of their position lies in an examination of the role of the central government.

From the earliest days of local authority housing the central government has sought to influence local authority management standards. At times the attempt has been slight but at other periods the central government has taken quite a strong stand. We are now in a period of strong central direction and even control of local authority housing functions. The Conservative Government is committed to a policy of radical reform which will drastically curtail local autonomy in the housing field. Rents are to be set throughout the nation on a uniform basis which leaves the individual local authority with little freedom; rent rebates are to be granted according to a national scale of income needs and will be administered according to central government rules. The introduction of these rules effectively destroys the London borough's case for unitary management as this will be attained, not by terminating the GLC's housing powers but by the introduction of nation-wide schemes.

Social service emphasis. During the 'sixties many reports have emphasized the need for 'co-operation' between housing and other local authority departments and the need for such 'co-operation' is clear. But it is by no means equally clear that the best, or only, way to obtain co-operation is to place housing with the social services, and there are several dangers inherent in such a move. First it must again be stressed that most of the people who use the social services (eg child care, welfare of the elderly, disabled or mentally ill people) are not local authority tenants. As 80 per cent of the people live outside the local authority sector of the housing market the officials of the social service departments have to learn to 'co-operate' with a great many independent landlords and owner occupiers. There seems, therefore, to be no reason why the activities of the GLC as another landlord should be attacked on the grounds that its existence as a housing authority makes things peculiarly difficult for the staff of the social service departments.

A second important consideration against placing housing too emphatically with child care and other welfare services is that there ought to be seen to be equality of treatment for all eligible people in the allocation of local authority dwellings. Already we have two conflicting situations in London which could be potentially dangerous if allowed to develop further. On the one hand we have a common allocation policy and common points scheme accepted; on the other hand the staff of the social services departments are allowed to allocate a small number of dwellings to their own 'clients'. These 'clients' are residents of the borough but they have a 'housing need' defined outside the public allocations and points scheme, and they obtain priority over other families who have not come to the notice of the social service departments. The danger of this type of 'back door' allocation of dwellings is that it must create doubts about (a) the fairness of the councils' allocation policy and (b) the efficiency of the 'points' scheme. If the social workers' clients have not sufficient points to obtain priority in the general pool of applicants, there may be something wrong with the points scheme and the weighting which is given to the various criteria of 'housing need'. On the other hand, it may simply mean that relative to other families' needs, the social services departments' applicants are not in a top priority bracket.

One of the reasons why the housing function was thought to be especially allied to the welfare services, is that by early post-war legislation care of 'homeless' families was made a 'welfare' responsibility. The 'welfare' authorities were the inheritors of the old Poor Law institutions and were regarded as the department most capable of meeting emergency needs. Housing departments were seen as providers of permanent accommodation and apart from the war emergency were not thought to be a suitable agency for short-stay hostel type accommodation. There are, however, no reasons of principle which would prevent the 'housing departments' from taking over the 'welfare' responsibilities for short-stay housing accommodation. Whether the period of stay is long or short the technical problems of land purchase, design and building, management and maintenance are the same and could be undertaken by one enlarged 'housing' department.

Thus we see that the main arguments for linking housing with welfare services are weak, and could each have been met by administrative changes which did not involve a single-tier housing authority. The disadvantages of having a monopoly provider of lower rented houses have long been recognized by the Conservative Party and it is interesting to note that while boroughs such as Islington have welcomed the termination of GLC housing powers, they have at the same time recommended the creation of housing associations within their areas, thus multiplying management, rent and allocation policies, while giving support to an opposed policy which is intended to introduce uniformity in these matters.

Planning emphasis. Those who are most concerned about the present state of London's housing, centre their anxieties around issues of land use, overspill and road and rail networks and employment opportunities. We have referred in the second section of this chapter to the reluctance of the outer boroughs to have inner borough overspill populations rehoused in their areas. In the chapter on planning, the weaknesses of the present system are stressed from the planners' viewpoint. Here we wish only to stress that with lower densities being imposed for planning reasons on the Central London areas the housing problems of London are a Greater London problem and by curtailing the GLC's housing powers the London Government Act ensured that there would be no 'London' housing authority to deal with the overall problem.

When Professor Cullingworth examined the proposed transfer of GLC housing to the separate boroughs he suggested that a powerful central committee of all the London boroughs should be able to impose a general London policy on recalcitrant authorities by majority vote. This proposal, made only five years after the reorganization of London suggests the extent of impatience and despair which the reorganization has induced. Some of the problems have no doubt been due to changes in government policy and the political beliefs of councillors and Ministers but the general feeling is one of such marked impatience with the parochialism of the London boroughs that their territorial monopoly is likely to be exchanged for an administrative loss of autonomy. The Conservative Government elected in 1970 was committed by its election programme to a radical reform of local authority housing subsidies, rents and rate rebate schemes. This will introduce some much-needed uniformity of management standards but it is doubtful whether it will create the conditions for greater co-operation between inner and outer London boroughs.

The London Government Act, 1963, in adopting the first of the three possibilities outlined earlier and giving almost all housing powers to the London boroughs, virtually ignored this link between planning and housing. In effect, Londoners have exchanged a small planning/housing authority and great diversity of management policies for one enlarged planning authority and 32 self-contained housing authorities with fairly uniform management policies. It is very difficult to see how this three-tier system (central government, GLC, London boroughs) will work in practice, but if the period 1965–70 is any indication of the future, we must expect a growth of central government intervention.

The alternative is to adopt the third of the possibilities mentioned above, and to give the GLC and the boroughs equal housing powers. This would amount to making permanent the GLC's present temporary powers to build within Greater London; it would require amendment of the 1963 Act. This is the only feasible way of achieving an effective approach to London's housing problems within the present local government structure. Housing is politically too important to be neglected by national politicians and if the planning/housing division makes it impossible for the redevelopment of Inner London slums to go ahead, it is difficult to see how the Secretary of State for the Environment will be able to stand aside. This issue is of great importance now as the reorganization of local authorities in England and Wales is following the same broad pattern as the London changes and housing is again being divorced from planning.

8
Highways, Traffic and Transport

Introduction

It is not easy to delimit precisely the subject matter of this chapter, but it is possible to distinguish three areas of inquiry. There is, first, the traditional highways function of local authorities – the responsibility for the construction, maintenance, cleansing and lighting of roads and streets to which has been added the increasingly important responsibility with the growth of motor traffic of making the best use of roads. Secondly, there is the responsibility for providing public transport. Thirdly, in contrast to these broadly executive functions, there are the responsibilities for planning, considered from two angles: the planning of communications and transport as a part of town and country planning; and transportation planning, that is, the planning of the transport network as a whole, so that, for example, policies for new roads are considered in relation to policies for the development of public transport and traffic and parking policies.

Although these three areas of inquiry are closely related to one another, together they constitute a very large subject for examination. Furthermore, in Greater London both before 1965 and to some extent after, the institutional arrangements for carrying out these functions were complex. In particular before 1965 the Ministry of Transport had greater responsibilities, especially for traffic measures, than was the case elsewhere in the country;[1] and at the same time, until 1970 responsibility for a large part of London's public transport was in the hands of a separate public corporation, the London Transport Board.[2] This was in addition to other arrangements peculiar to London, such as central government responsibility for the Metropolitan Police with their important duties in relation to traffic matters. Thus, analysis of the functions covered in this chapter involves consideration of the activities not only of the local authorities but also of these other bodies.

There is the further difficulty that the London reforms occurred at a time when policies and attitudes to transportation needs and how they should be met were changing rapidly. This is particularly evident in the road situation. Very briefly, in the immediate post-war period, to go back no further, little attention was paid to the need for new road-building. In the County of London, in particular, building of houses and schools was given high priority in comparison with expenditure on roads and road improvements. It was not until the later 1950s that the increased volume and use of motor traffic with their attendant problems, particularly of congestion, led to greater attention being paid to those areas. This took two forms: the application of measures designed mainly to ease the flow of traffic through existing streets, eg by the introduction of one-way systems or controlled street parking; and the production of plans for new and improved roads.

Both these approaches were, as will be described later, growing in importance in the period immediately before the London reforms took place, particularly from 1960 onwards, and they had implications for the institutional arrangements even before the reforms. At the same time, the possibility of transportation planning for Greater London, in the sense described above, was becoming a more active issue. Thus it is essential to see the London reforms within a context of change and, moreover, in comparison with the period 1945-60, of fairly rapid change.

This situation has implications for the way in which the effect of the reforms on the transportation function needs to be analysed. Not only does this function cover a very wide range which it would be impossible to deal with adequately within a single chapter;[3] but the very nature of the function, the number and variety of the bodies engaged in it, and the changes in views and attitudes to transportation questions all suggest a different approach from that in some of the other functional chapters. Direct comparison of indicators of performance by the pre-1965 authorities compared with the GLC and the boroughs is both more difficult and less meaningful, than it is in, say, the provision of welfare services which have simply been performed by a different set of local authorities since 1965. What is needed here is a broader approach which will take into account both the complexity of the institutional arrangements and the changing nature of the function. Accordingly, this chapter examines, first, the problems as seen by the Herbert Commission and the way in which the 1963 Act attempted to resolve them; secondly, what has been happening since 1965 in a number of crucial areas; and, finally, what impact the reforms have had on this group of functions.

The Herbert Commission's terms of reference did not include responsibility for public transport.[4] On the other hand, in their discussion of town and country planning they accepted the need for a comprehensive plan for Greater London which would inevitably involve: 'questions of housing, location of industry, highways, transport by rail and by tube and other matters'.[5] Indeed, the fact that in their view all these questions needed to be kept simultaneously in view, was a strong part of their argument for having a single Greater London authority.

In saying this the Commission were clearly influenced by the approach adopted by Abercrombie in his 1944 Plan for Greater London.[6] At the same time they thought that many of the specific assumptions on which the Abercrombie plan had been based had either not been realized in practice or had been overtaken by subsequent events. In particular, it had been assumed by Abercrombie that:

'the planning and construction of highways; improvement of facilities for London transport, particularly tubes; and suburban railway development . . . would fall within the general ambit of planning'.[7]

This had not happened and the consequent 'divorce between transport and planning' was, in the Commission's view, one of the reasons for a 'grave increase in congestion' both on the roads and railways; for example, although the siting of offices and housing had been part of the planning process, these had not been co-ordinated with the provision of the necessary communications between them.[8]

The Commission did not argue that the administrative arrangements for planning were the only or even the main reason for this divorce. The 'financial position of the British Transport Commission, which has been chronically in the red' was, they suggested, the main factor affecting the extent to which public transport had been brought into the planning process; whereas failure of the central government to provide the necessary funds was the main reason why highway planning had ceased.[9] Nevertheless, it was their view of the need for a single authority responsible for planning in Greater London which largely determined their attitude to the integration of land-use and transport planning.

In contrast to these firm and definite views, the Herbert Commission said very little about transportation planning, mainly because they were unable to discuss the place of public transport in their proposals; they recognized, however, that these proposals did not provide any means of comparing investment in new roads with that in other forms of transport, such as tubes; nor did they provide the means 'of correlating traffic regulation with the provision of transport services'. This consequence of the omission of public transport from their terms of reference left a problem which they thought 'worthy of study by some suitable body or by the Minister himself'.[10]

Thus although the Commission were quite clear about the need for planning to include and be closely concerned with transportation, they left somewhat in the air the question of whether and how transportation planning as such could or should be handled under any reorganized system of local government in Greater London. When they turned to consider what we have described as the executive functions connected with highways and traffic, they focused their attention on road traffic problems. From this point of view they examined the administration of traffic management ('the economical use of existing roads'), of highway construction, maintenance and lighting, and of traffic control. Their object was to see whether the existing administrative arrangements were adequate to provide for the proper measures to be taken to deal with the traffic problem. What their analysis mainly indicated was that there were obstacles to effective administration and that these obstacles were largely due to the fragmentation of powers and responsibilities between many different authorities.

Thus the Commission found that road improvements were often delayed because of disputes between authorities about who was to pay for them; that 'because each lighting authority is a law unto itself' and there were so many of them, standards of street lighting varied greatly from one part of Greater London to another; and that new road construction was also delayed by the multiplicity of authorities 'each with its own local view'. They concluded that the existing machinery was 'chaotic, inefficient and totally out of date' and that where changes had been made (eg the assumption of additional powers by the Minister of Transport) they had resulted in 'a jumble of ad hoc provisions, each of them designed from time to time to help paper over the cracks in the administrative structure'.[11]

The basic conclusion which the Commission drew from this situation was similar to that which had underlined their approach to planning: 'while improved administrative machinery will not of itself solve London's traffic problem, that problem is insoluble under the present machinery'.[12] In other words, their proposals were designed to remove the obstacles in the existing machinery rather than to suggest solutions to the problems which they had identified. These proposals were:

i. that a single authority should be responsible for traffic management in Greater London;

ii. that the same authority should be responsible for the construction, etc., of all main roads;

iii. that this authority should also be the planning authority for Greater London ie the proposed Council for Greater London;

iv. that responsibility for the construction, etc., of roads other than main roads should rest with the proposed Greater London boroughs.

Under the Commission's proposals, therefore, the Council for Greater London was to have a broad planning role and also a strong executive role in road-building and in dealing with road traffic problems. The boroughs were to have a much more restricted executive function for roads.

The new authorities and their powers

It was thus an essential part of the Commission's plan not only that there should be a Council for Greater London but that it should be a strong body. They were particularly insistent that so far as traffic management was concerned 'the only possible solution to the present administrative muddle is for the organization to be firmly in the hands of the Council for Greater London', with the Minister of Transport retaining only relatively limited powers.[13] The position was not quite so definite with regard to responsibility for roads. The Commission left it to the Minister to decide which roads were to constitute the main roads for which the GLC was to have responsibility. They also left for his decision the question whether trunk roads should be retained in Greater London.[14] On the answers to these questions depended the relative limits of responsibilities for roads of the Minister, the GLC and the boroughs.

These questions were still unresolved in the Government White

Paper of 1961, although it was accepted that 'ultimate responsibility for traffic management and main roads should be placed on the Greater London Council', leaving the complementary functions of the boroughs to be settled as a subsidiary question.[15]

The London Government Act, 1963, dealt broadly with highways and traffic in Greater London along the lines suggested by the Herbert Commission. In detail, however, there were some significant differences. No change was made in the position of trunk roads, but a new class of roads, known as metropolitan roads, was introduced for which the GLC was to be responsible. These roads were listed in the Act but not defined;[16] they were essentially roads considered by the Ministry to be the main through routes. All other roads were to be the responsibility of the London boroughs.

The position in the traffic field was more complicated. This was partly because the Act only went part of the way towards unravelling the complexity of existing provisions and partly because in spite of the White Paper's emphasis on the GLC's ultimate responsibility for traffic management, the Minister of Transport was still left with certain powers in this field. The position created by the 1963 Act can only be understood in the light of the historical background.

Before the 1963 Act was passed, regulation of traffic in London differed from that in other large cities. London for this purpose was the London Traffic Area which was created by the London Traffic Act of 1924. It was an area larger than Greater London for which, under the 1924 Act a London and Home Counties Traffic Advisory Committee was established, with representatives of the local authorities and road users in the area. For this area the Minister was the traffic regulating authority.

Outside the London Traffic area, the basic position was, under the Road Traffic Act, 1960, that the Minister made traffic regulations for trunk roads and the local authority (counties, county boroughs and the larger municipal boroughs and urban districts) for other roads. Orders made by local authorities were, however, subject to confirmation by the Minister, with certain exceptions (eg orders creating one-way streets did not require Ministerial confirmation whereas orders prohibiting overtaking or imposing speed limits did).[17]

Under the London Government Act, 1963, the Minister's powers to make traffic orders within Greater London were largely transferred to the Greater London Council and the London and Home Counties Traffic Advisory Committee was abolished.[18] The Minister retained power to make orders affecting trunk roads. He also retained default powers as well as some specific powers relating to roads other than trunk roads (eg concurrent powers, as elsewhere, to make orders relating to speed limits).[19] The net effect was to give the GLC rather greater powers than authorities elsewhere in the country (since its orders were not subject to Ministerial consent) but to give it a good deal less than 'ultimate responsibility'.

In other aspects of traffic management the boroughs were involved. Provision of off-street car parks, for example, was largely a borough function; it is true that the GLC was given a concurrent power to provide car parks but this was only exercisable with the borough's consent or, failing that, with the consent of the Minister.[20] The regulation of parking on the streets, eg by means of meters was, before the 1963 Act came into force, a Ministerial responsibility in the London Traffic Area. The Minister could make orders designating parking places both as a result of an application made by a local authority and on his own initiative.[21] The 1963 Act gave the GLC these powers concurrently with the Minister; applications by a London borough for an order were, however, to be made to the GLC except where they affected a trunk road.

The Act also carried out the Herbert Commission's intentions for the planning of roads. The GLC were required to draw up a Greater London development plan which 'shall lay down considerations of general policy with respect to the use of land in various parts of Greater London, including in particular guidance as to the future road system'.[22]

What is noteworthy is that only highway planning was specifically singled out in this way, the detailed contents of the Greater London Development Plan being left to be settled by regulations.

It seems clear in comparing the provisions of the 1963 Act with both the Herbert Commission's recommendations and with the previous situation that there was some ambiguity about the respective roles of the different authorities. This comes out most clearly in relation to the powers of the Minister on the one hand and the GLC on the other, in traffic management. One change in the situation here was that in 1960, when the Herbert Commission reported, the Minister had only just taken additional powers under the Road Traffic Act, 1960, and set up the London Traffic Management Unit.[23] The Herbert Commission's comments on the fragmentation of powers must refer largely, therefore, to the pre-1960 situation. The

K

establishment of the LTMU did at least ensure that there was a single body with oversight of traffic management in London.[24]

The success of the LTMU within its limited terms of reference of keeping the traffic moving may have contributed to the Ministry's unwillingness to surrender powers entirely to the GLC. This attitude came out most strikingly in the Committee proceedings on the London Government Bill. The provisions of the original Bill giving the Minister considerable power concurrently with the GLC to make traffic orders were justified by the Government spokesman on the grounds that someone ought to be in a position to intervene if the GLC were doing 'something that local people do not want and is quite wrong'.[25]

Thus the Ministry's unwillingness to trust local authorities too far was carried over into their view that the GLC 'will still be only a local authority'[26] in spite of its size and was evident in other provisions of the Act.[27] Nevertheless, it seems to have been accepted by the Ministry that there should be a single body within the new local government structure which, like the LTMU should have general control of traffic management.

There were, however, some difficulties over the division of powers between the GLC and the boroughs. In the first place, there was the question of responsibility for roads. The metropolitan roads listed in the London Government Act, 1963, constituted only just over 500 miles out of a total of over 7,000 miles of roads in Greater London; they were mainly Class I roads but even so 40 per cent of Class I roads remained in borough hands. Secondly, the division of powers in relation to parking was not entirely clear and logical. The Act limited the GLC to the power to regulate parking on the street, but left provision and control of off-street parking largely in borough hands.[28]

These examples illustrate some uncertainty about where the dividing-line should be drawn between GLC and borough powers. This was not due to a desire to boost borough powers as a counterweight to the GLC. The Government were committed to trying to create an effective highway and traffic authority for Greater London as a whole. But neither the Herbert Commission nor the Government seemed to have worked out the implications of splitting powers between the GLC and the boroughs in the novel situation created by the Greater London reforms. No doubt it was impossible in the circumstances under which the Bill was produced in 1962 to form a precise view of how the new system would work in practice. But there is at least a suspicion of doubt whether the Government were entirely clear about the proper relationship which should exist between the GLC and the boroughs.

Nevertheless, the broad lines on which the new system established by the 1963 Act differed from the previous system are clear; first, it created a single authority for Greater London responsible for strategic planning, for the construction, maintenance, etc., of the more important parts of the road network (excluding trunk roads), and for traffic measures; secondly, it created a much smaller number of authorities than before with responsibility for the remainder of the road system, and also with certain planning and traffic powers.

These were large changes in terms of the structure of local government, particularly the creation of the GLC. The remainder of this chapter will consider what effect they have had by examining the situation in transportation planning and traffic measures. It will also examine the reasons for further changes in the system which have been brought about by the Transport (London) Act, 1969.

Transportation planning

One of the difficulties of assessing the effect of changes in local government structure is that such changes may themselves take place in a context of significant changes in policies or attitudes towards the problems with which local government is dealing. It has indeed been argued strongly that the London reforms came into being at a time when new impetus was being given to problems of urban traffic and highway needs within the context of transportation planning.[29] This has a two-fold importance for this chapter; first, it stresses the need for seeing structural reform as only one of a number of possible causes of the changes which have taken place since 1965; secondly, it has a bearing on the particular form which developments have taken since 1965. The first question will be discussed further in the concluding section of this chapter; the second is particularly relevant to a discussion of transportation planning.

It was agreed by the Herbert Commission and by the Government in the London Government Act, 1963, that the GLC should be the authority for planning the main road system in Greater London. Yet in spite of the Commission's discussion of the relation between planning and transport, there was initially little recognition that road-planning formed only one part of planning transportation as a whole. When the draft Regulations for the Greater London Development Plan were considered by the GLC in December 1964 the Planning and Communications Committee complained that they concentrated on the GLC's responsibilities for metropolitan roads 'and give scant recognition of the Council's duties as the traffic authority for the whole of Greater London'. Moreover, they did not make any reference to other means of transportation, especially public transport.[30] In the Regulations, as they were finally published, the GLC were required to include in the written statement of the Greater London Development Plan:

> 'a statement indicating proposals with regard to special roads, trunk roads and metropolitan roads, having regard to all other methods of transportation, including public transport'.[31]

It will be noted that even in the final version the emphasis is still mainly on the road proposals. But before considering how the GLC has interpreted its duty it is necessary to indicate briefly the situation which the GLC inherited. From 1945 until about 1960 little money had been available for road construction in urban areas. Following the report of a committee under the chairmanship of Mr (now Lord) Nugent[32] the Government agreed to increase the allocation for London. Perhaps of greater significance in the long run was the decision, following another recommendation of the Nugent Committee, to set up a London Traffic Survey. This survey was put in hand in 1961, and the results of the first part (which dealt with the existing situation) published by the LCC in 1964; subsequently, the second part (dealing with forecasts of trends) was published by the GLC in 1966. The survey was then extended, as the London Transportation Study, to cover an examination of possible alternative plans for the future.[33]

As more money became available for road construction there was a shift in the policy of the LCC in the last two or three years of its existence. Instead of the £2½ million spent on road improvements in the county in 1959 £10 million annually was now being made available over a period of twenty years, and increasingly the LCC turned to a policy of developing an urban motorway network in place of the former policy of piecemeal improvement of existing roads.

The major and most controversial feature of the road proposals put forward by the LCC was a 'motorway box', a ring road encircling central London and passing through such places as Hampstead, Hackney, Blackheath, and Fulham. By 1963 the LCC Roads Committee was becoming increasingly attached to the motorway box proposal, although it did not become part of LCC official policy.[34]

By the time that the GLC assumed its responsibilities on 1 April 1965, therefore, planning for new roads in the County of London was under way. Moreover, as evidence from the London Traffic Survey began to accumulate, the GLC endorsed and extended this approach, first by suggesting that the motorway box might well be, 'a minimum requirement of a transportation plan for London', and, secondly, by stressing that further motorway ring routes might be needed in Outer London or beyond.[35]

Thus, the GLC started life not only with direct responsibility for over 500 miles of metropolitan roads but with at least the outlines of a plan of major new motorways. By contrast, not only did it have no direct responsibility for public transport but there did not exist in 1965 any comparable plans for the development of public transport.[36]

The creation of the GLC was an important factor in the increased attention given to the planning of a major new road system in Greater London in the later 1960s. It may be that, as indicated above, the early 1960s were a period when road planning came into prominence in contrast to the years 1945–60 when it had been virtually at a standstill.[37] Even without the creation of the GLC it can be assumed that there would have been greater emphasis on new road construction in the later 1960s particularly as more money became available for this purpose. But administratively this would have brought many problems. Even if the Ministry of Transport had taken a strong initiative, the formulation of a road plan for Greater London would have required some kind of joint organization including at least all the planning authorities in the area.

The GLC with its specific duty to include a road plan in the GLDP has gone ahead vigorously. A plan for a primary road network consisting of a series of three ring roads with connecting radial routes to be built as urban motorways, was first published in November 1967.[38] It formed a prominent feature of the Transport section of the GLDP where it was described as 'a major addition to the structure of London'.[39] Scarcely less important are the proposals for the improvement of secondary roads published in December 1969 as a basis for consultation with the London boroughs.[40]

It is a more difficult question to assess how far the GLC has related

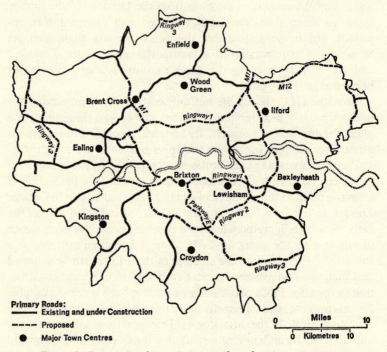

Figure 5. Greater London: primary roads and major town centres.

its road plans to other aspects of transport planning, and especially the planning of public transport. Certainly, in its published pronouncements the GLC has consistently stressed the need for both a road programme and the improvement of public transport.[41] Yet increasingly the GLC has come to be identified particularly with its road proposals. As its planners have themselves plaintively observed: 'The public seems to think that the plan which resulted from these surveys is really only one thing: a system of urban motorways. This, of course, is not at all true.'[42]

Whether true or not, it is not hard to see how this impression has formed. There is a sharp contrast between the precise road proposals in the GLDP and the much more generalized and indefinite proposals for improvements in British Rail and London Transport services. References are made to possible improvements such as the electrification of the King's Cross to Welwyn Garden City line or the building of the Fleet Line tube,[43] but these are nowhere near as clearly formulated as, for example, the stages of the primary road system; nor are they presented as a comprehensive set of proposals.

This emphasis on the road proposals in comparison with those for public transport could be justified on the grounds that the former would need to be carried out whatever was done about public transport.[44] Even if this were the case, it still seems difficult to reconcile with the insistence on 'five elements', each as important as the other,[45] as the constituents of 'the plan for movement'. And indeed critics of the GLDP's proposals for transport have particularly attacked them as being unbalanced and giving far too much weight to road-building as against other elements in the plan, the case for motorways being 'too much taken for granted'.[46]

What we must seek, therefore, are the reasons why the GLC came to accept that the case for motorways was so overwhelmingly strong in spite of their recognition that they could only form one element in any solution of London's transport problems. It is possible to identify four inter-related and important factors in the situation:

i.   the nature of the London Traffic Survey;
ii. the position of the public transport authorities;
iii. the initial structure and powers of the GLC;
iv. political and professional pressures within the GLC.

The first two phases of the London Traffic Survey were concerned, respectively, with a statistical survey of the existing travel situation and a forecast of travel patterns in 1971 and 1981 on the basis of existing trends. The third phase, called the London Transportation Study, aimed to test alternative plans for the future.[47] The second phase of the study was important in influencing the GLC's attitude since its results were becoming available in the early and formative years of its existence. It did indeed show that continuation of existing trends would result in a vast increase of traffic which would require a massive investment in new roads. This conclusion was immediately relevant to the Traffic and Highways Committee's assertion in 1965 that

'the preliminary results of the survey show that the feature which would be likely to provide of itself the greatest return to the community is a "box" of major roads around central London'.[48]

This assertion was directly related to the assumption in the London Traffic Survey of a large increase in both the number and use of private vehicles; from this it could easily be demonstrated that the existing road network together with improvements already planned would be inadequate to meet the demand for road space by 1971; and from this it could be argued that whatever improvements were made in the public transport system or through traffic management schemes, there would have to be a programme of new building.

The second factor which we have identified is the position of the public transport authorities. The 1963 Act made no change in the statutory powers and responsibilities of either the London Transport Board or the British Railways Board. Under the Transport Act, 1962, both were required, like other nationalized undertakings, to pay their way, a requirement which the railways had rarely been able to meet; London Transport, on the other hand, was able until the mid-1960s to meet this obligation. In addition, each Board had specific duties under the Act, the Railways Board to provide railway services and facilities with due regard 'to efficiency, economy and safety of operations'; the London Transport Board to provide (or secure the provision of) 'an adequate and properly co-ordinated system of passenger transport for the London Passenger Transport Area', again with due regard to efficiency, economy and safety of operation.[49]

These statutory requirements inevitably influenced the attitude of the Boards to their future plans. The railways, in particular, with their heavy annual deficits, were increasingly constrained by the need, as they saw it, to develop the most profitable activities for the future, such as high speed inter-city passenger trains and fast, regular, long-distance freight services. Improvement of London suburban services did not have the same priority. In any case, under the fragmented regional structure of administration which they had adopted, it was difficult for British Railways to concert a policy for such improvement. For London Transport it became increasingly obvious during the 1960s that their obligation to provide an adequate system of passenger transport was in conflict with their financial obligations, and this inherent conflict of obligations acted to inhibit their forward planning.

As against these limitations on public transport planning, the London Government Act, 1963, did introduce a new element in the situation in that the GLC could concern itself with public transport planning through its responsibilities in connection with the GLDP. Informal contacts between the GLC and BR and LT were put on a formal basis in 1966 when the Minister of Transport set up the Transport Co-ordinating Council for London.[50] Moreover, the London Transportation Study from the beginning involved both the public transport authorities with the GLC and the Ministry of Transport in direct participation in the planning process.[51]

The extent, however, to which these somewhat conflicting considerations enabled the GLC effectively to carry out comprehensive transportation planning turns crucially on the third and fourth factors. The GLC not only had planning powers but specific responsibility for main roads and for traffic management. It set up a powerful Highways and Transportation Department whose main expertise was in highway engineering and traffic management. There was thus an impetus towards dealing with matters where departmental expertise could be justified and used to advantage. This was a practical bias in the system. The GLC could not only plan roads; it could also build them; it had both the responsibility and the technical knowledge. By contrast, the improvement of public transport could not be secured directly by the GLC nor did it have the same degree of expert knowledge in this field. It could only operate by negotiation and persuasion of other bodies over whom it had no real control.

We may sum up the situation by saying that the circumstances under which the GLC was set up both contributed to the need for transportation planning and acted to limit its comprehensiveness.

Paradoxically, the GLC acting through the logic of its position as a strategic planning authority could see the necessity of transportation planning as an integral part of its planning responsibilities at the same time as the logic of its specific highway and traffic powers was influencing it in the direction of giving priority to a road solution.

Thus in spite of the GLC's insistence on the need for both urban motorways and improvements in public transport the emphasis in practice so far has been on the first rather than the second. Nevertheless, it has to be borne in mind that if the GLC has failed so far to achieve a comprehensive approach to transportation planning, the exclusion of public transport both from the consideration of the Herbert Commission and from the 1963 Act was a major reason for this lack of balance. Neither the Commission, who were not in a position to do so, nor the Government, who were, clearly saw it as an aim of the new system that the GLC should be the transportation planning authority. The GLC, which did see the possibility, was unable to transform it effectively into practice both because of its lack of powers over public transport and because of its inheritance of a road programme and road responsibilities carrying their own momentum. Nevertheless, the gulf between its planning powers and its executive powers in the field of transportation was one factor in the complex story which led to the GLC's assuming responsibility for London Transport on 1 January 1970.[52]

Traffic and parking

In this section, we are concerned, first, with traffic schemes, that is, with measures such as the introduction of one-way street systems, waiting restrictions and speed limits; and, secondly, with policies for controlling the parking of motor vehicles whether on or off the streets. Together these form an important part of what is often referred to as traffic management. Apart from considering how these activities have been carried out under the new system compared with the old, it is also important to see how far traffic measures have been co-ordinated with other highway and transport responsibilities, and, secondly, to see what the relationship has been between traffic measures and environmental considerations.

Under the pre-1965 system two phases can be distinguished in the administration of traffic management in London. During the period from the London Traffic Act, 1924, to 1960 the Minister of Transport was responsible for traffic measures in the London Traffic Area (which was much larger than Greater London) and acted with and through the London and Home Counties Traffic Advisory Committee. This was a large body; its 44 members included representatives of the central departments, local authorities, police, public transport authorities, and road users; the latter even in 1960 still included a representative of bicycle users and of users of horse-drawn vehicles. A large part of its work consisted of making recommendations on road improvements, waiting restrictions and proposals to impose speed limits or introduce one-way streets. During the 1950s it emerged as a strong advocate of the introduction of street parking meters, and generally of strict control over parking.

The LHCTAC was by no means a negligible body; indeed, as the Herbert Commission remarked, it had been 'a very useful body, and it is hard to see how any satisfactory results could have been achieved without it'.[53] But by its nature it was a very restricted body and could only advise on matters referred to it by the Minister. Furthermore, before specific measures could be introduced by the Minister, it was necessary to get the agreement of the individual authorities concerned. As the then Permanent Secretary of the Ministry put it:

'Thus, apart from the practical necessity for the Minister to carry the Advisory Committee with him, it is difficult for him to make improvements in London traffic control unless agreement is reached first with the police and the boroughs concerned . . . Thus, while the Minister is nominally the traffic authority, he is not in effective control and above all there are so many interests to be consulted that it is quite impossible to secure quick action, however great the need may be.'[54]

These difficulties led to the assumption of further powers by the Minister under the Road Traffic Act, 1960, and to the establishment within the Ministry of the London Traffic Management Unit. It was, in the Ministry's words, 'a special team of traffic-engineering experts, to study traffic conditions, which were steadily deteriorating there and to devise measures that would improve them'.[55] The creation of the Unit together with the Government's decision in 1961 to set up a Greater London Council led to the virtual extinction of the London and Home Counties Traffic Advisory Committee even before its formal abolition under the London Government Act, 1963; the

requirement that the Minister should consult the Committee before making traffic regulations was first suspended and then
abandoned.[56]

The most immediate effect of the creation of the LTMU was, as
had been intended, the introduction of a number of large traffic
schemes whose object was basically to get the traffic moving more
freely; the Ministry claimed that this had 'arrested the traffic paralysis
which was affecting Central London'.[57] Between 1960 and 1 April
1965 the LTMU initiated 75 major traffic schemes many of which,
such as the system of one-way working in the Tottenham Court Road
and Gower Street areas had a much greater impact on traffic in
Central London than the relatively smaller-scale schemes introduced
in the period before 1960. The LTMU also initiated a number of
experimental schemes, such as those for 'box junctions'.

The GLC has also been active in introducing schemes; in just
under four years from 1 April 1965 to 31 December 1968 it introduced 46 major schemes.[58] Many of these, especially in the earlier
years, were simply inherited from the LTMU and were already 'in
the pipeline' when the GLC assumed its responsibilities. Clearly, a
comparison of numbers of schemes introduced in itself means very
little. Not only does it become more difficult to devise major schemes
which are likely to justify the time and effort devoted to them once
the first flood of schemes has been introduced, but the creation of the
GLC itself raised larger questions than could be simply answered by
counting numbers of traffic schemes.

In a number of other ways, too, the GLC in its early years built
on Ministry experience in the introduction of traffic measures. For
example, the first 'urban clearway' in London was introduced in
1960 as one of the early measures of the London Traffic Management
Unit.[59] Since the GLC took over responsibility for London traffic
there has been a rapid extension of clearways; by the middle of
1968 240 miles of clearways had been introduced compared with
62 miles at 1 April 1965 and a further extension of the programme
was announced in October, 1970.[60] But perhaps the most important
of the complementary measures to traffic schemes which needs to
be considered here is the provision of parking facilities; and here
the position is complicated not only by the inheritance of the GLC
but also by the involvement of both the boroughs and the police in
the administration of parking measures under the 1963 Act.

The London and Home Counties Traffic Advisory Committee was

an early advocate of the use of parking meters to control street parking.[61] It was not until 1956,[62] however, that local councils were empowered to propose schemes for this purpose to be approved by the Minister of Transport. The City of Westminster was the first London authority to make use of these powers in 1958 but by 1963 a large area of Central London was a controlled zone for parking purposes and there were also smaller areas elsewhere in Greater London (eg Woolwich, Croydon, Kingston). In 1963 too, the Ministry of Transport published a booklet of practical advice with a foreword by the Minister, Mr Marples, specifically designed to encourage London local authorities in providing controlled parking.[63]

The London Government Act, 1963, transferred the Minister's powers to designate parking areas to the GLC, although the Minister retained concurrent powers in respect of trunk roads. Once again, therefore, the GLC came on the scene when, just as with other traffic measures, developments were relatively new and when, even if the GLC had not been created, there would almost certainly have been considerable extension of the use of parking meters. The 1963 Act did not, however, simply make the GLC the traffic king of London in place of the Minister of Transport. In the first place, there was still a good deal of division of responsibility for traffic matters between the Minister, the GLC, the boroughs and the police. This was one reason for further changes in the system under the Transport (London) Act, 1969.[64] From the point of view of the present analysis it emphasizes the difficulty of assessing the working of the system established by the 1963 Act since the period 1965–70 was in a sense a transitional phase between the old system and the more fully-developed new system which derives from the 1963 and 1969 Acts.

More immediately relevant here is the effect of transferring powers from a central government to a local government body. The Ministry of Transport, as is evident from *Roads in England and Wales*, was concerned largely with the immediate effects of measures on the traffic situation; this after all, was the *raison d'être* of the LTMU with its political antecedents in Mr Ernest Marples' desire to be seen to be doing something about London's traffic. But the GLC had wider responsibilities both as a highway authority and more broadly as a planning authority. The important question is, therefore, not only whether the GLC has been active in promoting traffic and parking schemes but even more how far it has brought these measures within a coherent policy for transportation, related also to its wider planning responsibilities.

Certainly, the GLC has asserted that this is its aim. At an early stage, Mr (now Lord) Fiske then Leader of the GLC told the House of Commons Select Committee on Nationalized Industries that 'we shall be much more concerned with the social consequences of traffic management'. On the same occasion Mr Stott, the GLC's Director of Highways and Transportation emphasized the new power which the GLC had of co-ordinating traffic management with the physical improvement of roads and he added 'we are trying to see if we can penetrate beyond the ad hoc aspect of traffic management to try to co-ordinate with London Transport connected schemes of traffic management'.[65]

Moreover, the Greater London Development Plan's section on Transport stresses the concept of balance between a road programme and public transport in drawing up 'a strategy for transport' in which 'a major programme of improvement of the road system' would be complemented by policies on parking and public transport and supplemented by 'carefully devised traffic and environmental management measures'.[66]

In practice, however, it is not clear that this aim is being realized. The question of parking policies illustrates the situation. An important statement of policy produced jointly by the GLC's Planning and Communications and Highways and Transportation Committees early in 1966 stressed that parking policy should not be considered in isolation; there must be a 'new initiative to improve public transport' and fresh attempts to deal with the staggering of working hours; and a number of more detailed matters, such as 'the amenity aspects of the proliferation of parking meters' would require special consideration. The main proposal, so far as on-street parking was concerned, was that the existing central zone of 8 square miles should be extended to an Inner London Parking Area of 40 square miles in all. The GLC claimed: 'we feel confident that the outline is sound, since our proposals are fully rooted in existing policies and in development flowing logically from them'.[67]

Here we see the main factors influencing the GLC. The impetus given to traffic and parking measures in the early 1960s was carried over into the GLC's early years. These measures were not part of an overall strategy when they were introduced under the LTMU because there was no overall strategy at that time. Nor did the strategy exist in 1966, so that in practice and in spite of the recognition that such measures should not be considered in isolation the GLC came down largely in favour of a continuation and extension of existing policies.

With the development of the GLC's programme for building new roads and with its assumption of responsibility for London Transport in 1970, it becomes more than ever urgent to ensure that plans for traffic management make sense as part of a general traffic and transport strategy. Already there has been criticism of the GLC's policy on traffic schemes on the grounds of 'the unforeseen and possibly unforeseeable effects of individual schemes on each other and on the network as a whole'.[68]

A new approach may emerge as a result in particular of the GLC's obligation under the Transport (London) Act, 1969, to prepare transport plans.[69] Elaborate machinery for consultation and preparation of such plans has been devised, including the creation of a new advisory group, the Greater London Transport Group, with representatives of the GLC, Ministry of Transport, London Transport and British Rail, and for the publication, after consultation, of 'green papers'.[70] However, it is still too early to say whether the GLC's potentialities as a co-ordinating and policy-making body will be realized in practice. It must, however, be pointed out that the possibility of co-ordination did not exist under the previous fragmented structure.

The last area of inquiry in this section is perhaps the most controversial of all. Major traffic schemes of the kind which the GLC has initiated have a direct effect on those who live and work in the immediate neighbourhood. A scheme which may benefit those who wish to drive through an area may bring considerable disadvantages to those who live in a street down which heavy traffic is diverted. The conflict is inescapable; the question is what effect the new system has had on the way in which such conflicts are resolved.

Before 1965 the Minister was required to consult the LHCTAC, before introducing traffic schemes, but in practice probably the most effective voices raised in defence of local amenities came from the various local amenity groups which came into existence under the threat of particular proposed schemes. Since 1 April 1965 the formal position has changed in two ways. First, the GLC is required, in carrying out its traffic functions to have regard to 'the effect on the amenities of any locality affected'[71] as well as to the desirability of securing and maintaining reasonable access to premises. Secondly, before making a traffic order the GLC is required to consult the London borough council (or councils) concerned.

At first sight, this seems to give adequate safeguard to amenity considerations. It is necessary, however, to examine further the GLC's obligations. The over-riding duty on the GLC is 'to secure the expeditious, convenient and safe movement of vehicular and other traffic (including foot passengers)'. Within this general obligation is the duty to have regard *so far as possible* to the effect on amenities.[72]

The wording of the Act suggests some confusion about how these powers should operate. The GLC is both a planning and a traffic authority, but the scale of these activities is different. As a strategic planning authority the GLC can and should relate traffic schemes both to other measures to deal with the traffic problem and to land-use planning for Greater London as a whole. But this is not the same as assessing the effect in environmental terms of a particular traffic scheme on the neighbourhood in which it is to operate. In the first case, the need is to ensure that the improvement of conditions for the movement of traffic is not achieved at the expense of deterioration in other living and working conditions over Greater London as a whole. In the second case, the need is to ensure that a particular scheme which may well be justifiable in terms of overall strategy is designed with due regard to local environment and amenity.

The distinction is not always an easy or obvious one in practice, but it arises precisely because of the way in which functions have been divided in Greater London. In a large provincial city such as Birmingham or Manchester, the conflict between traffic and environment is one which must be resolved within a single local authority responsible both for traffic measures and for planning at all levels. In London the planning function is split between the GLC and the boroughs in a way which makes it difficult for the former to relate particular traffic schemes to their immediate environmental and amenity effects without appearing to usurp the boroughs' local planning function.[73]

The Act does indeed expressly lay on the GLC the duty of seeing that traffic is kept moving. It is not to neglect other considerations but emphasis and priority are given to its traffic duties. It is, therefore, hardly surprising that there has been a tendency for the GLC to become identified with traffic interests and for the boroughs to have assumed the role of guardians of local amenity and environment. The boroughs have, it is true, complained of the 'tendency on the part of the Greater London Council to give greater weight to the solution of traffic management problems than to the preservation of a good standard of residential environment'.[74]

Such a criticism could only be sustained if it was argued in the context of the Greater London Development Plan, and if it could be shown that there the GLC was giving undue weight to its traffic management responsibilities. At the time when the criticism was made, however, the GLDP was only at the embryo stage. What the boroughs were really drawing attention to was the contrast between the GLC's explicit duty to solve traffic management problems and its broader and, indeed, longer-term duty to create the conditions for a reasonable environment for those who live and work in London.

Changes in the system

The first five years of the GLC's existence (excluding 1964–5, the year of transition) can be seen in retrospect as a period of adjustment to the new system of powers and responsibilities for traffic and highways. This was reflected partly in changes in the committee and departmental structure; but it also led to a redefinition of powers. Changes came about partly because of deficiencies in the original arrangements and partly because of external factors. Among the latter, particularly important was the commitment of the Labour Minister of Transport in 1966 and 1967 to a comprehensive approach to transport and the election of a Conservative GLC in 1967 pledged to 'get London moving'.[75]

So far as the GLC's internal organization is concerned, it seems clear that the committee and departmental structure which was devised in 1964 was to some degree experimental. The changes which have been made since then and the reasons for those changes are described more fully elsewhere.[76] Here it is important to note that the GLC began with a Highways and Traffic Committee and a Planning and Communications Committee; the corresponding departments were entitled Highways and Transportation, and Planning. In other words, the range of functions of each of these two major departments did not correspond to the scope of activities of the committees with which each was most closely linked; the Highways and Traffic Committee had narrower terms of reference than the corresponding department, whereas the Planning and Communications Committee, in theory at least, was concerned with all those aspects of transport planning which were directly relevant to land-use planning and the production of the GLDP. Since 1968 however, the GLC committee structure has been reorganized with the object of separating long-term planning from short-term planning and supervision of executive functions. The Strategic Planning Committee now deals with all aspects of the Greater London Development Plan, and the Environmental Planning Committee with such matters as the execution of the road programme, traffic management schemes and areas of comprehensive development. At departmental level the Planning and Highways department (known as the Department of Planning and Transportation) has as its joint heads the former Director of Planning (Mr B. J. Collins) and the former Traffic Commissioner and Director of Transportation (Mr P. F. Stott).

These bare facts about the major changes affecting GLC committee and departmental responsibilities underline the fact that the first five years of the GLC's existence have not been a static period. This is emphasized even more strongly by the way in which changes have come about in the GLC's responsibilities. At an early stage, the GLC drew attention to the limitations on its traffic management powers, claiming that:

'the arrangements envisaged in the Act enable the Council to co-ordinate traffic management with highways planning. There is still, however, no single authority responsible for traffic management, since the police retain separate but overlapping functions; the Minister of Transport has certain traffic management responsibilities; and the London borough councils have functions in respect of car-parking schemes.[77]

Although this statement carefully avoids any specific suggestion for change, its implications are clear – that there ought to be some strengthening of the GLC's powers at the expense of the other authorities.

At the same time, a number of factors were contributing to a change in attitude at the Ministry of Transport. At the time of preparation of the London Government Bill[78] and during its passage through Parliament the Ministry of Transport adopted a very cautious attitude to the question of giving the GLC wide powers. One result was that under the 1963 Act the Minister retained full control over trunk roads in Greater London. Apart from unwillingness to surrender powers, this caution perhaps also derived from a feeling that it would be necessary to see how the new system worked in practice before making such a change.[79]

At all events, the period since the GLC came into being has been marked by a considerable shift in national transport policies which in turn has affected the GLC's role in London. The election of a Labour Government in October 1964 and, more particularly, the appointment of Mrs Barbara Castle as Minister of Transport in December 1965, were significant factors in this situation, and led to the publication of a White Paper, *Transport Policy,* in July 1966.[80] The Government's policy was concerned broadly with the relation between transport and planning; it was also concerned with the relations between different forms of transport. Although 'the rapid development and mass production of the motor vehicle', was seen as the greatest problem, much stress was laid on improving and expanding public transport services. The White Paper showed the influence of current thinking;[81] its importance in the London context was that the Government's commitment to a new look at transport policy enabled London's problems to be re-examined.

Even before the White Paper, the Ministry had had a close involvement with the London situation. The London Traffic Survey and its continuation, the London Transportation Study, launched in 1965, is an example of joint working arrangements between the Ministry and, first, the LCC and later the GLC. Furthermore the programme for the primary road network was drawn up in joint discussions between the GLC and the Ministry. Formal acknowledgment that there were, nevertheless, gaps in the machinery established by the 1963 Act was provided by the Minister in March 1966 when she set up a Transport Co-ordinating Council for London under her chairmanship and with representatives of the GLC, British Rail, London Transport, the London boroughs and the trade unions. To some extent this move merely formalized existing informal contacts between the various bodies. Most of the work of the TCCL was done through five working groups, one of which was a Highways Planning Steering Group which in effect carried on under another name the GLC/Ministry joint discussions.

The 1966 White Paper stressed as one of its main themes the need to give greater priority to the problem of traffic conditions in towns, especially through 'integrated planning'; it also asserted that public transport 'must play a key role'.[82] In London it was claimed that the way in which problems were being tackled provided a guide to what was needed in other conurbations; at the same time

'these measures are a start towards the achievement of a properly co-ordinated transport system for London; as the longer-term studies are developed, wider measures involving legislation will be needed.[83]

The hint in this latter statement was taken up by the GLC who, in commenting on the White Paper, expressed the view that the Council's transportation policies could be better developed 'if it . . . were more directly concerned in the formulation of public transport policy'.[84] Since in practical terms association with British Railways involved far more complications than involvement with London Transport, the GLC confined itself to resolving

that, subject to the formulation of satisfactory administrative and financial arrangements, the Council be prepared to play a leading role in the establishment of a conurbation transport authority for London, in the realization that this would fully involve the Council in the finance and policies of the London Transport Board.[85]

Further consideration of the White Paper led the GLC to make explicit their claims for strengthened traffic and highway powers. It was made clear that the stumbling-block to GLC control of trunk roads was financial; the GLC had in mind control of the distribution of investment resources for all roads, including the local or borough roads; and it wanted not only to abolish the Ministry of Transport's residual controls over traffic measures,[86] but to explore the possibilities of determining policy in the operation of taxi services and of gaining some control over the revenues from parking. It was resolved to discuss all these questions with the London Boroughs Association.[87]

Thus by early 1967 it seemed that the GLC was moving towards a much more powerful role as the predominant traffic, transport and highway authority for Greater London than had been intended by the 1963 Act. There can be little doubt that both the Labour GLC and the Labour Minister of Transport were in basic agreement over the most important change, the need for the GLC to control the policy and finance of London Transport. Not only was this in tune with Labour philosophy, based on the need to strengthen public transport, but it also brought echoes from the past. When Mr Herbert Morrison (as he then was) as Minister of Transport in the Labour Government of 1929–31 was considering the problems of London transport he regretfully concluded that municipalization of public transport was not possible, and therefore proposed an independent public board. This was eventually established in 1933 as the London Passenger Transport Board. What could be more fitting therefore than that a Labour Minister nearly forty years later should revert to the original Morrisonian intention? Quite apart from this, Mrs Castle was anxious to win support for her controversial proposals for conurbation transport authorities. It would have been anomalous to have attempted to set these up on Merseyside and elsewhere if London had been left with a separate London Transport Board answerable only to the Minister.

The redistribution of powers which finally emerged, however, was somewhat modified by a further factor, the Conservative victory at the GLC election in April 1967. The Conservatives did not disagree with the principle of strengthening the GLC's powers but they put the emphasis rather differently. Just before the election during further discussion in the GLC of the 1966 White Paper *Transport Policy* on 14 March 1967, Mr Robert Vigars argued for the Conservative opposition that the ratepayers should not bear any deficits of the London Transport Board. In the Conservative manifesto for the GLC election published two days later and significantly called *Let's Get London Moving,* great emphasis was placed on the need for a Traffic Commissioner 'to cut through the tangle of overlapping committees and authorities' and to 'bring renewed impetus and urgency to bear on London's traffic and transport problems'. There was also emphasis on the road programme, but so far as public transport was concerned the Conservatives paid more attention to the financial problems of taking over London Transport and to the desirability of investigating new technical solutions (particularly a monorail) than to the possibility of improving the existing bus and underground services.[88] The Conservatives, however, accepted that the GLC would need to take over from the boroughs some of the latter's authority over roads and parking, as well as the Minister of Transport's traffic powers. They proposed a 'package deal' with the boroughs whereby the latter would gain additional housing and planning powers in exchange for the highways and traffic powers.[89]

It was in keeping with this approach that immediately after the election Mr Vigars, who became Chairman of the Highways and Traffic Committee, announced that the first task was to appoint London's Traffic Commissioner.[90] Although this idea, like that of the monorail, came to nothing,[91] the changed emphasis of the Conservative GLC soon became evident in the negotiations with the Ministry over the redistribution of powers and particularly over the terms on which the GLC would be prepared to accept responsibility for London Transport. With the GLC more immediately concerned over its traffic and highways powers, it was in a strong position to drive a hard bargain with the Government over the terms of transfer of London Transport, since the change of control in the 1967 election left the Government more anxious to carry through the transfer than the GLC. The latter's attitude now was that it would welcome control but only if the price was right. The details of the bargaining over this and over the parallel transfer of other powers to the GLC do not concern us here. But the GLC's new line helped to contribute to the view, evidenced in the GLDP that traffic and highways were at least at that stage of greater importance, in fact if not in theory, than public transport in overall transportation planning.

It was announced in December 1967[92] that agreement had been reached in principle on the new distribution of powers, a White Paper giving details was published the following summer,[93] and a Bill was passed in July 1969 to give effect to these proposals. The main changes made by the Transport (London) Act, 1969, were:

i. the GLC was to be the transport planning authority for Greater London with the specific duty of preparing and publishing transport plans; it was to aim at the provision of 'integrated efficient and economic transport facilities and services' (sections, 1, 2);

ii. the GLC was to appoint a London Transport Executive which would have responsibility for the operation of passenger transport services by bus within Greater London and by the underground system[94] (sections 4–6);

iii. the GLC was to control the general policies, budgets and fares policy of the London Transport Executive (sections 5–11);

iv. the GLC was to have power to make grants for unremunerative services (sections 25–6);

v. the GLC was to become the highway authority for all principal roads[95] (section 29);

vi. the GLC was to take over from the Minister residual powers
in relation to the regulating of traffic signs and pedestrian
crossings (sections 32, 34);

vii. the GLC was to have strengthened powers in relation to on-
street parking, eg by having a reserve power to require boroughs
to operate parking meter schemes; as regards off-street parking,
the GLC was to have the power to designate areas within
which car parks could only be provided by those who had been
licensed by the borough councils to do so (sections 35, 36).

A number of the increased powers which the GLC has now acquired
may be regarded as at last fulfilling the 1961 White Paper's intentions.[96] The transfer of responsibility for the policies and finance
of London Transport marks a much more significant change in the
system established by the 1963 Act. It still, however, leaves a number
of different bodies with powers and responsibilities in the field of
transportation. Apart from the GLC and the boroughs, perhaps the
most notable of these bodies are the police and the British Railways
Board.[97] Although it is clearly too early to say how the new situation
will in practice affect the problems discussed earlier, notably transportation planning, the concluding section of this chapter will
attempt to assess the significance of the changes which have been
made.

Conclusion

This study is concerned with the effects of the reform of the structure
of London Government which took place on 1 April 1965. In relation
to the group of functions examined in this chapter, the effects of the
reform have been treated rather more broadly than in some of the
previous chapters. The emphasis has now been on how the new
system has evolved and how it has operated against the background
not only of the ideas of the Herbert Commission and of the Conservative Government of the early 1960s but also of changing views
about the need for new roads and for transport planning. Correspondingly, there has been less emphasis on direct comparison and
in particular, statistical comparison of the performance of authorities
under the old and the new system. At the same time, the role and the

functioning of the GLC has received far more attention than that of the boroughs.

The reasons for this somewhat different approach are, first that the object of the reforms was not simply to transfer a given set of local government functions from one group of authorities to another, as happened essentially in the welfare or library services, for example. This, it is true, was part of what happened in that, for example, responsibility for the maintenance and lighting of roads was transferred from the authorities which existed before 1965 to those created by the 1963 Act. But this was only a part, and not the most important part, of the aim of the reforms which were also designed to create within local government a structure for dealing with matters which hitherto had been mainly dealt with outside local government, if at all.

Secondly, even to try to make direct comparisons is difficult since there are fewer reliable measures of performance in these functions than in other local government functions. Measures such as the amount spent on road maintenance per mile of road,[98] or even the amount spent on road construction and improvement are difficult to assess on a comparable basis over time. But even if they could be, they would throw only a limited light on the question of performance unless they could be related to other factors, eg in the one case to standards of maintenance and in the other the availability of central government funds.

Thirdly, the most difficult and important questions about the new system arise over the position of the GLC. The London boroughs, particularly since the Transport (London) Act, 1969, have a much smaller role in these functions. Whereas in housing, for example, it can plausibly be argued that the reforms were intended to give the boroughs the primary role, this was not so in the functions examined in this chapter.

One of the main reasons for this concentration on the GLC is thus the very strong executive role which it possesses, particularly since the 1969 Act came into effect,[99] and particularly in its responsibilities for major roads and for traffic measures. Transportation functions indeed loom very large within the whole range of the GLC's executive powers. In addition, as the body responsible for strategic planning in Greater London and, under the 1969 Act, for the production of transport plans and for policy and financial control of London Transport, the GLC is clearly at the focus of any discussion of the effects of the reforms and an assessment of how well they are working.

This situation is very much in accordance with the Herbert Commission's view of the need for a strong centralized and, above all, *local Government* body to be responsible for highway planning, main roads and traffic measures. Even under the London Government Act, 1963, with the continuing fragmentation of powers among other authorities including the Ministry of Transport, the London boroughs and the police, the GLC was in a strong position, as is evident from the evolving road programme and from the introduction of major traffic schemes. To that extent it is clear that one object of the reforms has been largely achieved.

Nevertheless, treating these functions in this way may seem to neglect one of the most important criteria suggested by the Herbert Commission. They argued, it will be recalled, that under the previous system the diffusion of responsibilities among a large number of different authorities created administrative confusion and delays. Should one not therefore examine the new system to see how far these defects have been remedied? But, again, the question really has broader implications. The burden of the Herbert Commission's argument was that necessary measures, such as road improvements and traffic schemes were held up by arguments over whose responsibility it was to carry them out and by delaying tactics. There is now less doubt about where responsibility lies, and to that extent the new system is an improvement on the old, but it is now clearer than perhaps it was at the time of the Commission that the system cannot be judged simply by whether it enables measures to be taken more speedily. Just as important (some would say more important) is the way in which measures are carried out; in particular, the extent to which they are related to a wider strategic view, and the extent to which environmental considerations are borne in mind.[100]

But in addition to what it has achieved in its executive capacity the GLC has also put forward extensive transportation, and especially road-building proposals in the Greater London Development Plan as part of its strategic planning responsibilities. Here two related sets of difficulties arise in considering the effects of the reforms; first, there is the question whether a further object of the reforms which was particularly important in the Herbert Commission's eyes, that is, the inclusion within the planning process of transportation, has been satisfactorily achieved. Secondly, there is the question which the

Commission did not really consider, but which has become of greater importance since then, of the effectiveness of the new system for securing the planning of the transportation network as a whole.

Both these questions, it is important to note, concern functions which were either not being carried out at all under the pre-1965 system, or, to the extent that they were, depended on central rather than local government initiative. The first relates to the way in which the GLC has carried out its planning function; there is no doubt that the GLC has in the GLDP brought its transportation responsibilities into the planning process; indeed, as was indicated above, criticism has been directed at the GLDP on the grounds that it is too much influenced by motorway proposals. The planning chapter of the present study discusses further how the GLC has carried out its planning role, but the wider issues raised by this question are discussed in the concluding chapter of this study.

The second question, concerning transportation planning, is complicated, as has been shown above, by the position of the public transport authorities. Failure even to consider what the effect would be of creating a powerful Greater London authority with extensive planning, highway and traffic functions alongside independent public transport authorities seems in retrospect to be the most glaring omission on the part of the Government. This is especially so if one contrasts it with the Government's intention (unfulfilled as it turns out) to transfer the powers of the Metropolitan Water Board to the GLC even though water, like public transport, was excluded from the Herbert Commission's terms of reference.[101]

The two cases are not, however, parallel. The transfer of water supply to the GLC would have been a relatively simple task compared with the issues involved in examining the public transport situation in London. But even more than this, there was little appreciation, whether inside or outside the Government, of the potential importance of the public transport issue. A recent inquiry commissioned by the Government has stressed how new is the concept of transportation planning, and how 'the idea of comprehensive planning of and for all forms of transport, private and public has only recently begun to be accepted'.[102] Moreover, even in those local authorities which had responsibility for bus services in their area, public transport tended to be isolated from the more traditional traffic and highways functions, for example, by being the responsibility of a separate committee.[103]

Thus, it could be argued that in 1962 when the London Government legislation was being prepared, there was no reason to suppose that it was necessary to go beyond a consideration of the distribution of powers in the traditional local government functions, which the Herbert Commission had examined, especially traffic powers and the planning, construction and maintenance of roads. Such a view shows a failure to appreciate that an entirely new situation was being created in Greater London. Nowhere else in the country did there exist a local authority responsible for an entire conurbation. Part of the reason for the failure of transportation planning to emerge in other conurbations was the fragmentation of planning and transport powers so that local authority areas were not meaningful in planning terms. The opportunity for looking forward existed in London in 1962, but it was not grasped.

However, even within the framework adopted by the Government, the London Government Act, 1963, had certain limitations from the point of view of the most effective working of the new system. There was, first, the retention of powers by the Minister of Transport; secondly, there was the imprecision of the line drawn between the powers of the GLC and the boroughs; thirdly, there was the greater power of the boroughs, particularly in planning matters, compared with the Herbert Commission's proposals, deriving in part from the reduction in the number of the boroughs for reasons unconnected with the functions considered in this chapter.

Much of what has happened since 1 April 1965 can be seen to have stemmed from the omissions and ambiguities of the 1963 Act. Certainly the GLC appears to have been pulled in two directions – on the one hand seeking almost from the start to strengthen its powers, on the other hand carrying forward powers which it inherited; on the one hand asserting its faith in comprehensive and balanced planning, on the other increasingly committing itself to a large programme of new motorway building. In many ways the situation is paradoxical. The *raison d'être* of the GLC is to play a strong strategic role combined with extensive executive powers, yet the limitations on its powers in the former capacity have left the impression that it stands for one kind of solution to London's traffic problems, the one which until 1970 was the only one which it had any real chance of carrying forward.

This is really the most important factor in the situation following the 1963 Act and before the Transport (London) Act, 1969, came into force. In other words, the period 1965–70 can, so far as these functions are concerned, be regarded as an intermediate stage in the development of London Government. This is not to deny the importance of the 1963 Act which, in structural terms, introduced fundamental changes; but rather to call attention to the fact that the 1963 Act was unsatisfactory viewed both from the point of view of fulfilling the Herbert Commission's intentions and from the point of view of coping with developing ideas of how to deal with problems of transportation in a metropolitan area.[104]

Even now the Transport (London) Act, 1969, although it puts the duty on the GLC of formulating transport plans, still leaves it unclear how far effective planning will extend to British Railways' suburban network. This is one of the big question marks hanging over the future. British Railways carry approximately 40 per cent of the commuters who work in Central London. A realistic transport plan for Greater London must bring fully into the reckoning the needs and possibilities of rail travel in the future. Yet essentially British Railways thinking is conditioned by its overall financial obligations. It is true that these were somewhat modified by the Transport Act, 1968, under which, for example, grants could be made to the Board for continuing to run unremunerative services. But the railways have for so long now run at a loss that it would not be surprising if a narrow view of profitability still tended to inhibit developments which might be desirable on wider grounds.

The danger, so far as London was concerned, was recognized in the 1968 White Paper. Not only was it proposed that the GLC should be given power to make financial contributions to British Railways for running unremunerative services,[105] but 'a new financial relationship' was to be established between the Ministry, the GLC and BR 'by the end of 1970 at the latest' with an immediate objective of achieving financial viability for the London commuter network by the end of 1972.[106]

At the formal level of authorities and powers, therefore, it is only now with the transfer of responsibility for London Transport to the GLC and with the strengthening of other GLC powers that a system is emerging which begins to make sense in terms of the aims of the London reforms; but this is only a part of the story. Limitation of powers is obviously important, but the working of the new system has been governed at least as much by the views which the authorities have taken of their own roles. The policies of the new authorities, and particularly of the GLC on which the attention of this chapter has been focused, are a matter for political decision. What this chapter has been concerned with is not so much the decisions and whether they were good or bad as with the structural and functional conditions within which those decisions have had to be taken.

It is therefore certainly no part of the argument presented here to attempt to show that had the Government acted differently in 1962 and given the GLC much stronger powers in the 1963 Act, the transport policies presented in the Greater London Development Plan would necessarily have been very different. No more could one argue that because the GLC now has power to make financial contributions to London Transport or British Rail it will do so. What we have been concerned with are obstacles and potentialities in much the same way as were the Herbert Commission. They wished to remove the obstacles to planning, to traffic management and to road-building so far as these obstacles were attributable to the administrative structure. By so doing they believed that they would create the conditions for dealing effectively with London's traffic problems. In a similar way we have argued that these conditions did not fully exist under the 1963 Act and still do not under the 1969 Act even though some of the obstacles have been removed.

The position may be summed up in this way. So far as powers and responsibilities for highways, traffic and transport are concerned the London reforms, taking into account both the 1963 and the 1969 Acts, represent both an improvement on the previous system and a reasonable attempt to meet the need for both a Greater London authority and local authorities, in the form of the London boroughs. This is not to suggest that the institutional arrangements for handling these functions are entirely satisfactory.[107] But it is to suggest that considering these functions in isolation, the more important questions now concern the policies which the GLC has adopted or will adopt, rather than whether it has adequate powers to carry out its metropolitan role.

Yet, paradoxically, this very success of the new system in one direction has raised doubts in others. Is the GLC too much of a transportation authority and too little of a planning authority? To be more precise, has the considerable range of powers at its disposal in these transportation functions served to mask the limitations on its powers in carrying out its strategic planning role? Answers to these questions are critical to an assessment of the new system since planning and traffic were at the heart of the Herbert Commission's case for a Greater London authority and, hence, for a fundamental change in the system of local government in Greater London. It is, therefore, of vital importance before attempting to answer these questions to consider the evidence not only of this chapter but of the planning chapter which now follows.

9
Planning[1]

Introduction: the comparison of planning systems

Prominent among the reasons for creating the new system of government in Greater London was the need for more effective metropolitan planning. The Herbert Commission saw the case for a strategic planning authority, covering the whole of the London conurbation, as a principal argument for the creation of the GLC. The boroughs were also given an important and novel role in the new planning system.

This chapter aims to give an account and a critique of how a new system of metropolitan planning was instituted, how it is now working, and how it might be improved. This is done by tracing the evolution of the new system and then analysing its performance in terms of plans, development controls, action areas, and organization. The sixth section turns to the crucial regional framework of London planning, and the seventh section sets up standards for evaluating the meaning and the quality of metropolitan planning.

One caveat has to be made. Both the system of London Government and concepts of the planning process itself, are in a state of considerable flux, and this chapter does not attempt to predict what further changes will occur in the near future. But there is enough experience to draw on for a first report to be worthwhile, while there is also enough confusion about the scope and aims of metropolitan planning for some elucidation to be desirable. This task is tackled in the conclusions, but by way of preface a brief review of the meanings of urban planning seems desirable.

The conventional meaning of urban planning in the British local government context is the regulation and development of land use under powers provided primarily by the Town and Country Planning Acts. An alternative and much broader view of urban planning, supported by some American theory and a smaller measure of American practice, is that it covers *all* long-range questions about the character and development of a city, and the welfare and prosperity

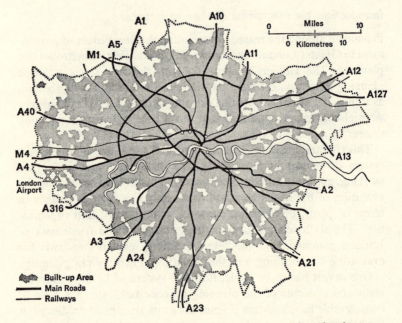

Figure 6. Greater London: built-up area and main roads and railways.

of its inhabitants, which are of concern to the city government. This view is gaining ground in Britain with a stress on PPBS and local government co-ordination, but is best stated by Robert Walker: *The Planning Process in Urban Government* (University of Chicago Press 1951). On the first view the planning agency is an ordinary department conventionally organized, which is British practice, while on the second view it is a 'staff arm' of the city government as a whole, and particularly of its political leadership.

This theoretical differentiation between two concepts of urban planning is blurred everywhere in practice. Local planning agencies in the USA rarely range as broadly as this theory might suggest, and in practice put particular stress upon physical planning. In Britain, the planning committee and department may look on a chart like any other branch of local government, but in practice they exercise some co-ordinating role in relation to other departments. Put another way, physical planning is always a major and crucial focus for the planning process in urban government – whether one approaches the subject first from the regulation of land use or from the integration of public services.

Nevertheless, there remains plenty of argument as to how the planning function in local government should be defined and allocated. The creation of a new metropolitan system of government for Greater London also offered the occasion and the opportunity for new treatments of the planning function. In particular, the scale, character and limitations of the 'strategic' planning role loosely assigned to the GLC represents a complete innovation in British local government, with a consequent need to reconsider what planning can or should mean in such a context. Although this debate is still far from finished – curiously, indeed, has hardly begun – it may be helpful here to indicate three planning issues which have special relevance and interest in the London context.

First, there is the increasing concern of physical planners with socio-economic objectives and options. In a sense this is nothing new – British planning has always been linked with definite social policies for environmental improvement. But the new approach is more open-ended about objectives, and predicates that physical plans should be drawn up not as ends in themselves but as the instruments for achieving politically agreed aims and priorities. Planning, it is urged, should also occur over areas broad enough to facilitate solutions for such social problems as urban obsolescence and

L

overcrowding, inadequacies of transportation, and conflicts between mobility and environment.[2]

These ideas have found some statutory expression in the gradual replacement of local development plans by sub-regional 'structure' plans under the terms of the 1968 Town and Country Planning Act, although full implementation of this Act depends upon a general reorganization of local government. The Greater London Council might be thought a particularly appropriate body to carry the flag for new modes of planning, because of its great population size and financial resources, and of its special 'strategic' relationship to the London boroughs, although the adequacy of its physical area for structure planning is much more dubious. At all events, it seems clear that the GLC should offer a broader interpretation of physical planning than is customary with British local government.

Secondly comes the question as to how the planning and development functions of a local authority ought to be related. One familiar view of physical planning requires aloofness from any sectional pressures for development, whether coming from private firms or public departments and agencies. In this way the planning department can act as an 'umpire', laying down general policies and standards which all developers must follow. The alternative approach is for the planners to co-ordinate their aims closely with those of the departments responsible for transportation, housing, schools and other forms of development; in this way, the joint impact of all public agencies will have a more positive effect upon the development of the city. The difficulty with the first theory is that it assigns to planning a negative and possibly ineffectual role, while the second approach carries the danger that planning will itself be subservient to the sectional aims of other departments.

The concept of planning as the 'staff arm' of the whole city government removes the planners (in theory) from entanglement with any particular interest. However, in practice the planners seem often to be too weak to influence significantly the actions of the major departments and may be driven into organizational tactics – for example, giving support to the claims of the weaker departments – as a way of achieving any results at all. Moreover, the generalized scope of the planning agency reduces its capacity to produce any positive strategy for physical development, thus again weakening its capacity vis-à-vis those departments which have a major stake in development.

In Britain it is more natural for physical planning to be directly linked with major public functions such as housing or highways. This linkage is expressed in the historic location of the planning office under the control of either the chief architect or engineer. This subordination of planning still continues in many county boroughs, although in counties it is now usual for a separate planning department to exist. Moreover, traditional planning policies have been primarily concerned with the improvement of the residential environment, and a close policy linkage has existed between planning and public housing. Recently, the growing problems of traffic in towns have caused a switch in the interests of planners towards transportation issues, and Ministry directives have laid increasing stress upon the need to link planning very closely with the provision of highways, public transport, and traffic management.

These developments have a special significance for the GLC because the most important of this body's limited executive duties concerns transportation. Thus Greater London has become the obvious area for the application of 'Buchanan-style planning', based upon a close integration of the transportation and physical planning functions. This has in fact come about in a purely organizational sense (see later). This development, despite its obvious advantages, has the further effect that the linkage between planning and other types of development such as housing becomes weakened, which is hardly consistent with the idea of broadening the social aims of planning.

Finally, there is the question (already raised) of the relationship between physical planning and general-purpose planning of the whole work of the local authority. The latter type of planning is gaining in importance, with an increased stress upon central management functions, resource planning, and variants of PPBS (planning – programming – budgeting systems). As the scope of both these types of planning broadens, conflict and confusion over their application seems inevitable – particularly, again, in the case of such a large yet executively restricted agency as the GLC.

One attempted solution, the American 'staff agency' concept, would in effect seek to pool these two types of planning. This approach now has some support in Britain, from physical as well as economic and organizational planners. Yet the difficulty remains that the perspectives, time-scales, and techniques of the two types of planning are substantially different – reflecting the different requirements of long-term physical development on the one hand, and of organization and financial management on the other. Pooling of experience and methods is obviously desirable, but complete fusion would probably entail dominance of one or other type of planning. Even physical planners who want to broaden their social objectives do not as a rule want to abandon the physical basis of their plans for a primarily financial or resource basis, or subordinate their vision of the 'city beautiful' to taxation strategy.

The alternative approach is to recognize two distinct types of planning at the urban level – physical and financial-managerial. This is in fact the solution which the GLC has so far adopted, but it leaves many unsolved questions. How shall the two spheres of planning be defined in detail, where shall each be located, how shall conflicts be resolved? There is also the interesting question of how the supply of information should be organized both for physical and general-purpose planning. The Herbert proposal for a high-level intelligence unit within the GLC gave special point to this question.

In talking about planning in this chapter, I shall normally be referring to physical planning. But enough has been said to show that the planning function, as thus understood, is broadening its scope and is colliding at some points with the new managerial-financial concept of planning. The relations between physical, economic, and organization planning are especially complex in such a vast and curious authority as the GLC.

The evolution of the new planning system

The 1947 Town and Country Planning Act transferred basic town planning powers from county districts to counties which, with county boroughs, became responsible for the preparation of local development plans. The area of Greater London became the concern of 5 counties and 3 county boroughs, and the relevant development plans of these authorities were approved as required by the Minister of Housing and Local Government, sometimes after a considerable delay, at dates between June 1954 and December 1958.

One of the counties (London) lay wholly within the metropolitan area as subsequently determined by the London Government Act, and one (Middlesex) almost wholly so. Three of the counties (Essex, Kent and Surrey) were each responsible for a large wedge of suburban London as well as for an extensive area outside the capital. One county (Hertfordshire) covered only a fringe of the capital, lying mainly beyond it. The 3 county boroughs (Croydon, East and West Ham) dated from a time when these places were substantial towns separated from the rest of the capital.

This planning system received from the Herbert Commission the obvious criticism that it operated through 8 planning authorities and 9 development plans (2 for Kent) when 1 of each would be more appropriate. This criticism could almost have been delivered *a priori* without hearing evidence. If (as the Commission claimed) London was one great city surrounded by a protected Green Belt, it followed that the old county boundaries were anachronistic for local government in general and the physical planning function in particular.

That there was hardly any evidence of this kind was due to a peculiarity of the London situation, namely the influence of the Greater London (Abercrombie) Plan produced in 1944. This advisory plan, commissioned by the Minister of Town and Country Planning and prepared by Professor Abercrombie and a modest staff, closely guided and conditioned physical planning within the London region up to the appearance of the Herbert Report in 1960, since a revised version of it was adopted as official public policy by both the Central Government and by the county councils concerned.

The Abercrombie 'strategy' required the stabilization of the population of Inner London[3] at a figure one million below that of 1939, the permanent protection of a substantial Green Belt around the conurbation, and the creation of 8 new towns and various town expansions beyond the Green Belt to absorb both population and employment that would be dispersed from London.

Despite the fact that the old planning system was harmonized to a surprising degree by acceptance of the Abercrombie principles, it was also clear to many observers and to the Herbert Commission itself that Abercrombie would soon be completely out of date. Abercrombie had not foreseen the general growth of regional population; the rapid expansion of office and service employment within central London; the consequent growth in long-distance commuting, rendered lengthier by the existence of the Green Belt; the plight of homeless immigrants; or the pressures to move and expand parts of the London docks. The rise of car ownership had also created a new range of transportation and environmental issues since Abercrombie reported.

The Herbert Commission listed many of these problems as clear evidence that a new Abercrombie-type plan was needed for Greater London. It did not inquire closely as to how far these problems could in fact be solved by a physical planning agency, or what kind of area, powers and relations with the Central Government this agency would need to possess. In a sense the success of the Abercrombie Plan may have blinded the Commission, as most others, to the much greater difficulties of 'comprehensive planning' in the 1960s or 1970s when compared with the immediate post-war period. It was also doubtless more politic for the Commission to stick to the general case for a new master plan, rather than to enter upon an administrative and political analysis of the conditions for effective planning.

To the Herbert Commission, then, the question became simply one of whether this broad, strategic planning for London, which ought to be done afresh and done regularly, was the appropriate sphere of central or local government. The existing local planning authorities themselves were prepared largely to leave the matter to central government, except that the LCC had considered proposing a joint advisory committee in 1960, an idea taken up by Surrey county council. Surrey later initiated the 'death-bed' repentance of all the counties who belatedly accepted the case for a joint executive planning board. These efforts eventually led to the creation of an advisory consortium of local planning authorities, known as the Standing Conference on London and South-East Regional Planning.

In a sense, however, the Herbert Commission would probably have regarded these developments as irrelevant even if they had occurred earlier. There is plenty of evidence that a joint local government committee cannot produce plans of a positive, original, or dynamic kind. Neither would central government co-ordination of local plans produce the desired style of planning, unless the central departments took the initiative and were themselves co-ordinated. The Commission was perhaps right in thinking that only a virtual Ministry for London could essay the task. Certainly a strong regional planning organization of some type would have had to be established by the Central Government.

The Ministry of Housing and Local Government (now Department of the Environment) has been drawn cautiously into regional planning because of the lack of any other appropriate agency to do this task, and from the need to establish more realistic guidelines for local planning. The existence of the GLC has not eliminated the need for broader regional plans sponsored by the Ministry; even while the Commission was sitting, a Ministry team of planners was producing the South-East Study whose aim was to provide local planning authorities with firmer policy guidance (see later section).

Thus the Royal Commission might logically have reviewed the planning of Greater London as primarily a task for the Central Government. Moreover this approach would have avoided the restrictions of London boundaries, and facilitated the genuinely regional type of planning which some witnesses told the Commission was necessary and which to some extent they accepted. But this solution would offend the basic values about local government held by the Commission and endorsed by (among others) the Permanent Secretary of the Ministry of Housing and Local Government. However blinkered or acquiescent the existing local planning authorities might be the Commission was at one with Dame Evelyn (later Baroness) Sharp in thinking that physical planning *ought* to be a local government responsibility. There was also the point that the Commission was expected (and wished) to reform *local* government, not to strengthen the central departments. The role and rationale of the GLC were settled accordingly and the problems of reconciling that role with the regional planning duties of the Central Government were left for later resolution – or deadlock.

The Herbert Commission proposed a singularly straightforward division of planning functions in London. The Greater London Council would prepare the development plan, but the boroughs would administer development control. The chief exception to the latter principle concerned the central area of the capital, whose functioning appeared to the Commission to be crucial for the planning of London as a whole. For example, growth of employment in the central area strongly affected transportation and housing policies for the whole conurbation, and indeed beyond it, while the protection of the cultural and historical amenities of Central London must be a basic feature of any worthwhile plan for the capital. Moreover, both under the Herbert Plan and the Government plan eventually implemented, the central area of London was divided between the City of London and several new boroughs; some coordination of these bodies seemed essential. The Commission drew a tentative map of the central area, and suggested that Ministerial regulations should settle the classes of planning applications within this area to be decided by the Greater London Council. It also foresaw that some major developments within other parts of Greater London would require similar treatment, especially in view of the decentralization of offices and other activities to places like Croydon, but regarded the situation as too shifting and complex to make firm proposals.[4]

Undoubtedly the planning proposals of the Commission, apart from its treatment of the central area, derived from its 'philosophy' of local government rather than from close analysis of the problems of physical planning. The Commission believed that functions should be clearly divided between the two levels of local government in London, and that each level should be as autonomous as possible in respect of its distinctive powers. This viewpoint was certainly influenced by the dissatisfaction among county districts with delegation schemes which the Commission unearthed, especially as regards education, and especially in Middlesex. It therefore set its face against delegation arrangements for any service, linking this distaste with the less managerial view that efficiency would be increased if local authorities did not interfere with each other's decisions.

In the case of planning, the Commission thought that delegation schemes were wasteful because of extensive double handling of applications. It accepted that any application which conflicted with the development plan ought to be referred to the GLC for decision, but it proposed that the borough should decide whether or not referral was necessary. It argued that 'almost no difference in practice' would have occurred if counties operating delegation schemes had done likewise, but this was certainly not the view of the county councils. Thus Middlesex found over 100 cases where a district council had been stopped from contradicting the county's policy; a number of these cases were later the subject of an appeal to the Minister, who usually had supported the county. The Herbert view was that if a risk did exist of the boroughs departing from the development plan, eg by building in the Green Belt, it was a risk worth taking.[5]

It is interesting to speculate how the Herbert Commission's proposals for planning would have worked in practice. Undoubtedly there would have been much greater variations between boroughs over the treatment of planning applications than previously occurred between county districts within the same county, due not only to the boroughs' greater freedom but to their greater stake in development, and to the impracticality of preparing a detailed development plan for such a complex area as Greater London. However, the GLC's sole responsibilities for the development plan would have facilitated an integrated strategy for the capital, as would the special arrangements for the central area. If greater GLC control of development had been found necessary in order to achieve this strategy, then additional powers of supervision might have been introduced later without cluttering up the GLC's initial task, which was to plan. The Herbert proposals would certainly have become operational much more quickly than those actually adopted; despite their lack of sophistication they had many advantages.

The actual system of planning introduced by the London Government Act, 1963, and subsequent regulations, differed in important respects from the Herbert proposals. The most important change was that the GLC was not given exclusive responsibility for the development plan for London, which became instead a joint product of the GLC and the 32 boroughs plus the City Corporation. The GLC was first to prepare a general plan which would require Ministerial approval in the normal way. Then each borough and the City Corporation would prepare a more detailed plan for its own area which would be expected to accord with the policies already settled by the GLC. However the approval of these borough plans was again a matter for the Minister, who would arbitrate any differences of opinion between GLC and boroughs. These combined documents would constitute the Greater London Development Plan. In the interim stage, which was certain to be a long one, the approved plans of the previous local planning authorities would constitute an Interim Development Plan for the metropolitan area.[6]

Other changes were made in the system of development control which was largely settled by Ministerial regulations made under the Act. One effect of these regulations was to give the GLC somewhat stronger powers of supervision over the decisions made by boroughs than Herbert had considered necessary. Another effect, however, was to eliminate any special treatment for the central area of London.

In Parliamentary debates on the Bill the Minister, Sir Keith Joseph, was sympathetic to the case for special treatment of the central area, but once the Bill was passed it doubtless seemed easier to Ministers and officials to 'wait upon experience', rather than to affront the City Corporation and the new boroughs which had just been created.

A planning system based upon local government considerations of this kind will not necessarily work well, or work at all. The opposition's criticism of this part of the Bill was that there must be a 'major partner' for a system of concurrent planning powers to be workable, and this could only be the GLC. This result would only occur if (a) the GLC had full responsibility for the entire plan, though in consultation with the boroughs; and/or (b) the boroughs acted only as agents for the GLC over questions of development control. The opposition was not quite clear as to whether both of these conditions must be satisfied fully (the latter of course contradicted the Herbert Report, as the Minister pointed out), although they could show that the Government scheme wholly contradicted previous planning practice as operated in the counties of London, Middlesex, and elsewhere. Sir Keith Joseph's defence was that the new London system could not be compared with arrangements elsewhere; the GLC was intended to be not a 'higher authority' but a 'wider authority' than the London boroughs. His defence did not of course settle the question as to whether the GLC could plan effectively on a wider basis without also being 'higher'.[7]

There was no room for doubt on one point: the Government system was vastly more complex than the Herbert proposals. The long delay before a new plan for London could possibly be produced and approved implied a long period of dependence upon inappropriate previous plans and ad hoc development control.

The Government's wish to have the GLC and the boroughs work in an intricate double harness led to complex provisions for Ministerial arbitration and intervention. For example, if the GLC and a borough could not agree over an 'interpretation' of the composite development plan (when finally produced) the Minister would arbitrate; and the Minister would also settle questions as to whether the GLC was correctly interpreting its part of the plan, and do the same in respect of the boroughs. These arrangements, as the Government argued, might represent only the adaptation of existing planning practice to the special circumstances of the new London system. But a new system might have been thought to require a revised conception of central control, especially as a major reason for the reorganization was to *reduce* the need for central government intervention in the affairs of London. The regulations also reflected a lack of imagination on the part of civil servants, who did not appreciate the force of the reform proposals in whose name they were acting.

Some brief conclusions to this section may be stated. The Herbert Commission's proposals for a new planning system in London were based less upon criticism of the actual working of the existing system than upon theories about the advantages of comprehensive metropolitan planning. The system proposed was intended to achieve such planning, for London generally and the central area in particular, while giving the boroughs considerable freedom over development control matters. The Government, however, from certain general considerations, shifted the emphasis towards equal partnership between GLC and boroughs, plus considerable Ministerial supervision, thereby weakening the independent planning role prescribed for the GLC by Herbert. In the event the double-decked system of planning brought in by the 1963 Act did prove to be very slow and cumbersome, and was abandoned suddenly in 1970; but not before it had strongly influenced the preparation of the first Greater London Development Plan.

The system in action: The Greater London Development Plan

The Greater London Development Plan appeared in 1969. Its form was governed by the London Government Act which required the plan to 'lay down considerations of general policy with respect to the use of land in the various parts of London including, in particular, guidance as to the future road system', filled out by a series of Ministerial regulations about the contents of the plan. The statutory plan comprised a written statement of policies, a roads map, and a metropolitan structure map. It was accompanied by a bulky Report of studies, comprising background research, and by a general essay, 'Tomorrow's London', intended to present the plan for discussion by Londoners.[8]

The purpose of this section is not primarily to give a critique of the plans, policies and techniques – several such critiques are in fact available[9] – but to consider how the GLC interpreted its planning function, and to look at the institutional and organizational factors which influenced the plan and its reception. A special and lengthy public inquiry into the GLDP started in the autumn of 1970, and it cannot be known how the Minister will amend the plan. A full verdict upon the new system will have to wait for many years.

The GLDP can be regarded in a sense as the first 'structure plan' to be prepared according to the provisions of the 1968 Town and

Country Planning Act. The purpose of a structure plan is to concentrate upon general strategies or policies for such matters as the distribution of population and employment; the location of service centres; the reservation of open spaces and recreational areas; and systems of communication. The Ministry's intention or hope is that, once these principles have been accepted following a full public inquiry, the local planning authority can proceed with detailed local plans and action areas with a minimum of public objections and Ministerial supervision. The actual GLDP regulations were prepared before the 1968 Act, but they took account of the views of the Ministry's Planning Advisory Group which inspired the Act, as well as of various Ministerial promises given during the passage of the London Government Act.[10]

The GLC was not only expected to pioneer 'structure planning' but to do this under the unique circumstance of expecting the boroughs subsequently to produce structure plans of their own. Partly for this reason much the greater part of the initial GLDP was general and tentative. In the debate on the London Government Bill, the Minister, Sir Keith Joseph, said that there was nothing to stop the GLC from prescribing basic standards, in the same way as the LCC did.[11] For example, the LCC specified residential density zones which controlled population levels, and plot ratio zones which controlled employment levels. But the GLDP rested content with much looser guidelines.

It gave proposed upper and lower population figures for each borough but left to the borough plan the specification of actual residential densities. In the case of employment, the plan included estimates of the labour resources of each borough, based on the population targets, and gave general indications of the volumes of new industrial and commercial floorspace which should be permitted, based on the expected labour supply and other factors; but these floorspace figures went only up to 1971, were given for broad sectors of Greater London and not for individual boroughs, and seemed closer to forecasts than to targets. Plot ratios were to be settled by the boroughs, within a policy framework which was looser than that for residential densities. It is interesting, however, that the GLC hoped to establish some direct control over the location of employment, by suggesting that it should administer the licensing of new factory and office premises which is done by central departments according to regional policies for economic development.

This would have helped the GLC to exercise a strategic role over employment questions without treading on the boroughs' toes; but the Central Government did not like the proposal.

The same tentative approach coloured most other aspects of the plan. For example, the Ministry's regulations required the GLC to list precincts and amenity areas of metropolitan importance, and to state policies for these areas. The criteria for conservation areas were explored quite carefully in the Report of Studies and 23 such areas were listed in the plan, but it is not clear that this listing will have much effect.[12] It is mainly for the boroughs, not the GLC, to take measures to protect or enhance the quality of these areas, and to add other areas if they wish. It is only if, for example, Camden Borough Council were disposed gravely to despoil the surroundings of Hampstead Heath, that the GLDP listing might exercise some restraining force. In this case it was mainly the Regulations which prevented the GLC from being more positive.

Another requirement for the plan was the specification of major shopping and commercial centres. This gave to the GLC the opportunity to reshape the functioning of the capital – for example through accelerating the development of major suburban centres such as Croydon was becoming, which would reduce the load upon the central area and provide focal points for increased residential densities. Highway plans might have been related to the creation of such centres, instead of the reverse approach. This opportunity was not taken. The plan's list of 28 important centres and 6 major strategic centres was based primarily upon the *existing* figures of retail turnover. The plan's estimates of required increases of shopping floorspace to 1981 proposed no more than a cautious continuation of existing trends towards somewhat stronger suburban centres. The nearest to a policy statement is the proposition that those town centre improvements which do not contribute to major highway plans *may* be delayed; but there was seemingly little attempt to adjust highway plans and priorities to a positive view of the development of centres. On this matter the regulations would have allowed more definite planning – if the GLC had dared or known how to do it.[13]

Various explanations can be given for the very tentative character of the GLDP, save for its highway aspects. For the GLC to lay down 'strong policies', as some officials in the Ministry hoped, was intrinsically difficult given the attitudes of the boroughs, and the protection which the Act and the Ministry's own regulations gave to those attitudes. Most of the boroughs wanted to do as much of their own planning as possible, and at this stage they held the trump card that their own plans would eventually be arbitrated not by the GLC but by the Minister. The GLC could only impose its own policies upon the boroughs with consistent and firm ministerial support, so that it seemed necessary to the GLC to come to terms with borough viewpoints.

The political history of London is also highly relevant. Observers all agreed that initially the GLC tried to act over planning matters somewhat like the 'LCC writ large', or in other words to boss the boroughs. This attitude reflected the traditions of the initial Labour majority, many of whom had served on the LCC and were anyhow more committed than the Conservatives to the idea of a strong central authority. The advent of a Conservative majority in 1967 produced a more conciliatory or placatory attitude by the GLC towards the boroughs' role in planning.

Moreover, the Labour Party on the GLC had inherited from the LCC and also Middlesex a fairly coherent set of planning policies and measures. These included determined efforts to disperse more population and employment from London, strict maintenance of the Green Belt, spreading of the housing load more equitably between inner and outer boroughs, limitation of residential densities on social grounds, and strict control of employment in Central London. The Conservatives, however, were dubious about almost all these policies. Their stronger business sympathies made them more protective of the 'economic vitality' of the capital, and consequently less inclined towards dispersal policies. They were anxious that labour supply within the inner boroughs should be adequate for the service requirements of the central area, and not much disposed to unload the housing problems of Inner London upon the outer boroughs which were mainly Conservative. However, these attitudes were not homogeneous, since some leading Conservatives had always supported the social goals of dispersal and housing equalization policies.[14] The Conservative leadership also had to contend with well-established professional views in favour of the continuation of existing policies.

The preparation of the GLDP proved a difficult process. The submission date first suggested by the Ministry was March 1967 but the GLC obtained a two-year extension to complete survey work. The boroughs disliked this delay. As an interim measure the GLC

tried to explain its intentions, but a committee of borough planning
officers urged a much less ambitious approach, which would exclude
attempts to demonstrate the economic viability of the plan or to
utilize mathematical models. The boroughs also wanted the Ringway
One proposals deferred as a definite element of the plan until more
supporting evidence was available. The GLC did not yield, knowing
of course how difficult or impossible it would be to secure borough
agreement.

The boroughs pressed for speed and simplicity. The GLC leadership, however, were increasingly worried about the acceptability of
the plan to the boroughs and the public. They believed that they had
to 'sell' a new conception of strategic planning to a critical public, and
they were none too sure themselves what form this basic strategy
ought to take. Moreover they knew very well that the urban motorway proposals, already in outline, would encounter a storm of
opposition. Their policy was to tread warily and speak softly.

These political cross-currents affected the final form and presentation of the GLDP. The plan purported to be based upon a series of
technical studies prepared by the GLC's substantial technical staff.
These comprised:

i.   completion of *London Transportation Survey* started by the
     LCC in 1962;
ii. a *change of land use survey* in Inner London and a complete
     land use survey in Outer London, begun in late 1965;
iii. an *employment survey* of Greater London begun late 1966 as a
     sample survey of 30,000 employers;
iv. a *housing survey* begun early 1967 (sample of 100,000 houses);
v.   *a Thames-side uses survey.*

(All these surveys were completed by the end of 1968.)

However, except for the London Transportation Survey, these
technical studies were not really policy-oriented. The initial approach
to the GLDP preparation was narrow and old-fashioned. Very late in
the day leadership attempted to broaden the research done in the
planning department, and to conceive the plan itself in wider terms.
One result was a last-minute juggling of professional staff and the
production of much supplementary material at the public inquiry,
somewhat to the confusion of objectors. The GLC also held a
succession of public meetings in different parts of London to try

and explain its plan, but the attempts at grass-roots consultation, though genuine enough, came at the wrong end of the planning process.

Despite the GLC's placatory attitude, some features of the plan were distasteful to a number of boroughs, particularly in Inner London. These boroughs argued that the population figures for their areas were too low and proposed higher ones. This was a confused mixture of technical and policy disagreement. The borough figures differed from the GLC ones partly because occupancy rates, for example, were calculated in different ways. In policy terms, these boroughs *wanted* to retain higher populations, partly as a matter of pride but also as a political response to local housing demands; whereas the GLC considered that it was only being realistic in projecting further declines of population within the more crowded boroughs, although it was prepared to accept a higher density of new housing than had previously been deemed desirable.

The strongest opposition to the GLDP was naturally directed against its road proposals, particularly Ringway One, the motorway box which would encircle Central London. This motorway would entail the destruction of nearly 10,000 dwellings and would cut a swathe through a number of dense communities which surround Central London, inflicting upon the adjacent inhabitants much noise and nuisance from traffic, depreciating property values, and interfering with the local patterns of urban life. These road proposals attracted criticism from a number of planners and economists.[15]

The GLC's general case was, of course, that the urban environment would eventually be improved by measures which removed a large quantum of through traffic from residential and commercial streets. The critics suspected that the motorway system would itself stimulate much additional traffic that would outweigh this gain, creating a case for more motorways and so on. Moreover the critics could point out that specific measures to protect the environment against the cumulative growth of traffic were not spelled out in the plan, except for an extension of parking controls. Only general aspirations were laid down, contrasting with the specific character of the highway proposals. Many critics also believed that an improvement of public transport was more desirable for a city such as London than grandiose road proposals.

In early 1970 the responsible Labour Minister, Mr Crosland, decided that the coming public inquiry into the GLDP should take an exceptionally elaborate form. Instead of the usual inquiry by a planning inspector, the Minister set up an expert panel comprising a legal chairman and four other members, who were given the further assistance of five specialized assessors. This inquiry was expected to last at least a year with an initial 88 days devoted to the 'principles' of the plan, followed by examination of the plan's local implications, and ending with a final review. The device of an expert panel was intended to ensure a skilled and wide-ranging examination of the plan. None the less the proceedings were fairly judicialized, with counsel for the GLC defending their plan against all comers in court-room style, and with the panel also legally represented. Discussion of the basic policy issues had to work through these legal forms.

The appointment of this special inquiry was partly due to the strength of the opposition against the Ringway proposals. This situation was rather ironic. It had been the specific aim of the GLC to integrate transportation and physical planning (see later section) but to little effect. Moreover in an operational sense the GLDP *was* very largely a highways plan, but road-building could proceed independently under its own legislation. The road proposals had already been discussed with the Ministry of Transport, and from 1970 the same Minister (Secretary of State for the Environment) became ultimately responsible for both highways and planning. Thus it would probably require a very persuasive report from the inquiry panel for any major road proposals, especially those that were well-advanced, to be greatly altered.

Mr Crosland's action in initiating a special inquiry was also influenced, without doubt, by a certain puzzlement as to what a plan like this should contain. The machinery which the Government had initiated for planning in London seemed to be working badly, and there was some hostility between GLC and boroughs over their respective roles despite the limited scope of the GLDP. 'Planning by inquiry' was a method of reassurance to public and critics which might ease as well as postpone the eventual political decisions. Mr Crosland had in fact adopted the same device in appointing the Roskill Commission on the third London airport.

However, in late 1970 the responsible Minister announced a major change in the system, whereby the boroughs would not prepare 'structure plans' but only 'local plans'. Although the Minister suggested that these local plans might cover large areas, possibly an entire borough, the practical point was that they would occupy a subordinate place to the plan produced by the GLC. This change could be easily justified, as it was by the Minister, by the obvious need to expedite the planning process in London; but it also had the effect of strengthening the strategic authority of the GLC, although the future relationship between the two levels of planning was still far from clear. (Outside London, the authority which prepared the structure plan would normally also prepare local plans.) An awkward implication was that the GLDP ought really to be reconsidered and redone, not just amended piecemeal through the work of the panel.

In concluding this section, it should be stressed that the production of a strategic plan for London is an intrinsically difficult task. The area of the GLC is inadequate for Abercrombie-type planning (see last section). Within Greater London, a strategic approach implies greater attention to economic and social needs than town planning as such can achieve.

The GLC realized this last point, but imperfectly and belatedly. If the GLDP has a basic objective, it is to safeguard the economic health and vitality of the capital. But the aim was poorly pursued. It was interpreted as requiring the retention within London of a sufficient labour force to man up 'essential' activities, without analysis of which economic functions *were* essential, how much they would expand, and how many less essential functions could conveniently be dispersed. Much stress was laid upon the importance of London's role as the dominant centre of finance, commerce and administration; upon the growing needs of tourism; and upon the need to retain some of the more productive and progressive industries located in London. But this analysis was entirely general. For example, about 30 per cent of Greater London's employment is still in the manufacturing sector, and something like another 30 per cent represents services needed by these industrial workers.[16] There is, therefore, a large scope for further dispersal of manufacturing employment without 'running down' the capital's economy, or affecting the unusually broad spectrum of employment opportunities in London which now exists. London is a very long way from becoming simply or mainly an office centre, or from polarizing into a city of rich executives and poor service workers, yet parts of the GLDP read as if this threat were justification enough for its policies.

Further, this economic goal was nowhere measured against the plan's second main objective of raising the social quality of the environment. The two goals were plainly in conflict, since efforts to retain the labour force implied also the retention of high population densities in Inner London, which would militate against improvements of housing, social facilities, open space, and traffic flows, and against the conservation of attractive areas. Models should have been developed for comparing variations of the London economy against variations in the pressures upon environmental capacity; and the presumed benefits of a larger economy compared with the higher public costs thereby entailed. The GLC proved too politically uncertain and too technically weak to essay these tasks – it assigned many of them to further research – but their absence rendered the 'strategic' character of the GLDP rather empty.

The system in action: development control and action areas

To appreciate the administrative and political problems of development control in London, one must remember the nature of the English town planning system. In theory, the control function follows upon and reflects the development plan prepared by the local planning authority. But in practice, there is greater flexibility and control powers can be fully exercised in the absence of a plan.

In the case of London, as we have seen, the first GLDP would not be fully operative until many years after the GLC's creation. In the meantime development control was guided by the patchwork quilt of the 9 plans of the previous planning authorities which together comprised the 'initial development plan'. These plans, however, became increasingly irrelevant to the new situation and did not bear any impress of GLC strategic guidance. In 1966 the GLC suggested that the policies contained in the 9 old plans might be made more consistent through an 'averaging' of densities and other standards over broader areas, so arranged as also to increase somewhat total population capacity. The boroughs successfully opposed this far from radical proposal.

One borough planning officer observed in an interview that the Herbert Commission failed to understand the importance of development control in the English context. This criticism seems to be true, but its significance had two opposite applications. The GLC could argue that some direct control powers were essential to its strategic role, whereas the boroughs could contend that since they were undeniably the primary control authorities, they ought also to shape their own plans – thrusting the GLC role to a still more generalized level. Development control represented an inevitable battleground between GLC and boroughs.

The system adopted gave to the GLC three types of control or influence over the handling of planning applications. A relatively small group of decisions became its direct responsibility. These included all cases relating to 9 comprehensive development areas which the GLC had inherited from the LCC together with one new such area (Covent Garden); cases relating to large transport centres, educational institutions, or public meeting-places; and mineral extractions covering more than 5 acres.

Secondly, a larger group of cases were referred to the GLC for advice, and were then subject to its formal direction from the GLC whenever the borough wished to grant permission. This group in-included shops of more than 250,000 square feet, factories over 5,000 and offices over 2,500; the erection of buildings over 150 feet high within central London or 125 feet elsewhere; development within 220 feet from the centre of an existing or proposed metropolitan (ie GLC) road and within 110 ft. of a railway station; car parks for more than 50 cars; development in the Green Belt; and the demolition or alteration of a listed building of architectural or historic interest.

Under a third type of relationship, boroughs had to notify the GLC of all applications within certain categories, but the latter had no powers of direction. This device was adopted to deal with a number of classes of application for which the GLC had wanted powers of direction, but the boroughs had resisted. These classes included a number of potential redevelopment areas; applications relating to certain precincts and amenity areas such as the South Kensington museum area, the area around St Paul's, and Hampstead–Highgate; and buildings on the *supplementary* list for architectural and historical interest. The device merely provided the GLC with a warning system enabling it at least to request the Minister to 'call in' the application for his own decision under special circumstances.[17]

A critical planning issue is the treatment of special areas which require some form of comprehensive planning and redevelopment. There are many such areas in London, where the need for comprehensive treatment arises for a variety of reasons, such as obsolescence and poor layout of existing properties; need for major transportation improvements; availability of surplus or reclaimed land offering new development opportunities; and areas of distinctive character and quality where substantial changes are unavoidable.

Until the 1968 Planning Act, the principal tool for tackling such areas was comprehensive development procedure, which enabled the local authority to acquire all land within a designated area and re-allocate it for private or public purposes according to an approved plan. Under the structure planning system introduced by that Act, CDAs will be replaced by action areas but the possible purposes of the operation remain broadly similar.

The new planning authorities in Greater London inherited 46 CDAs from their predecessors. The London County Council had been responsible for both the planning and implementation of CDAs with three exceptions, but in Outer London CDAs were designated in the county development plans but implemented by the boroughs or urban districts. An immediate question was how to handle these existing CDAs in many of which redevelopment was far from complete. (Action of this type is a slow process, sometimes lasting over 20 years.) It was decided that most of these CDAs should be transferred to the appropriate borough, but that the GLC should inherit 9 of the 13 previously managed by the LCC.

For the future, both boroughs and GLC were given powers to initiate and to manage 'action areas'. Borough action areas were primarily a local matter, although the fact that they had to be included within the GLDP gave to the GLC a limited authority (subject to the Minister's final arbitration) to review all borough proposals and decide upon priorities. GLC action areas would necessarily be a matter for close consultation with the affected borough, which also would play a variable part in the implementation of the action area plan. The first GLDP proposed 46 action areas, of which 10 would be the primary responsibility of the GLC and the remainder of the boroughs. These had proposed a larger number of areas, which the GLC had reduced.

Comprehensive development or action areas posed some difficult and delicate questions about the respective roles of GLC and boroughs which formal arrangements did not and could not iron out. So many types of areas required treatment and the stakes of the two 'partners' could be so variously interpreted. Which areas were so vital to the functioning of London that the GLC's role should be paramount? Covent Garden and Piccadilly Circus presumably – although the relevant boroughs were not wholly convinced – but what else? Moreover, besides planning principles there were operational stakes. The GLC was directly concerned with major highway projects. Both GLC and boroughs were concerned with the provision of new public housing and open space. The capacity of GLC or borough to negotiate effectively with large private developers was a major factor in some circumstances.

Moreover not only were the GLC, boroughs, and private developers closely concerned with action areas, but since this was London, the Central Government was also much concerned. The Government's intervention could take many forms – review of plans, call-in powers over planning applications, formal or informal arbitration between GLC and boroughs, direct action by central departments. Government took direct action over the redevelopment of its own office complex in Whitehall. The original consultant's plan for Piccadilly Circus commissioned by the LCC was not acceptable to the Ministry of Transport because its traffic capacity was deemed inadequate. Public disquiet over a proposed development in the Circus led to a public inquiry and to renewed attempts at producing a plan, first by the LCC and then through a Government-convened working party. Subsequently a further plan was commissioned jointly by the GLC and Westminster, but continuing conflicts of interest, relating to both traffic problems and potential developers, have continued to frustrate the planning of the Circus. The re-location of Covent Garden market at Nine Elms was carried out by a centrally appointed public agency utilizing special Parliamentary powers. The list could be continued. If one looks at the changing face of London, it becomes a tricky task indeed to disentangle the respective contributions of central departments, GLC and boroughs – not only because of the variety of formal powers and procedures which exist, but also the frequent informal accommodations.

The LCC had mainly concerned itself with the redevelopment of obsolescent residential and industrial areas. It had given less attention to the redevelopment of parts of Central London because of its strong concern with housing policy and the lower pressure (in its day) of transportation problems. The GLC changed these priorities. Urban redevelopment for housing purposes was relinquished to the boroughs, presumably in recognition of their political and technical strength. The GLC concerned itself with nodal points of communications, and 3 of its first 10 proposed action areas (Victoria, King's Cross–St Pancras, Piccadilly) were of this type.

Additionally the GLC (following the LCC) took the lead in securing the release of large blocks of surplus land from the services (Hendon and Croydon Airports, Erith Marshes), from British Railways (Marylebone Goodsyard, Somers Town), and later from the Port of London Authority (London, East India, and St Katherine Docks). Land scarcity made these acquisitions extremely important, and in the case of the GLC (unlike the LCC) it was an open question who should develop the land. Much of this released land was handed over to the new boroughs for housing, but for the larger schemes the GLC acted as co-ordinator and part developer and in one case the GLC itself was the developer of a new settlement for 45,000 people (Thamesmead, built on Erith Marshes).

Although the GLC took the initiative in new directions, its achievements were slow and its difficulties considerable. The redevelopment of nodal urban points is a very expensive and complicated business, requiring the co-ordination of numerous interests. GLC action easily got squashed between the layers of departmental supervision and borough independence or intransigence.

The GLC's concern with redevelopment or action areas became increasingly focused upon Central London. However, each problem area within Central London tended to become a special case, variously handled, and the GLC had no additional powers of planning or control for the central areas as a whole. The 7 second-tier authorities[18] particularly concerned with the central area set up a co-ordinating committee, but this device perhaps only emphasized the blurring of public responsibilities for this crucial area of London.

The system in action: planning organization

The design of a planning organization for the GLC posed a number of fascinating issues. The organization was required to carry out a novel kind of strategic planning, and to relate this to the limited set of operational functions entrusted to the GLC.

The most crucial organizational issue (as viewed by the GLC and relevant Ministries) was the relationship between physical planning and transportation. Additional questions included the location of the intelligence unit which the Act required the GLC to initiate, the internal organization of the planning staff, and the relationship of the planning staff to housing and redevelopment functions. There was also, of course, the practical difficulty of recruiting sufficient qualified planners.

The most obvious comparisons of the GLC's planning tasks were with those of the LCC and Middlesex, but the organization adopted followed neither of these models. The LCC had treated planning as a branch of the department concerned with architecture and civic design. This system reflected the strong operational concern of the LCC with a vast public housing programme, a number of massive redevelopment areas, and the expansion of many distant towns under the terms of the Town Development Act (1952). The LCC's eyes were fixed closely upon housing issues and the reconstruction of residential areas.

The LCC's Architect's Department had a high and justified international reputation. There was pressure from the Royal Institute of British Architects, among others, to create an architectural department in the GLC which could continue this nucleus of highly skilled staff. However the GLC's activity in residential construction was expected and proved to be less than that of the LCC, while its concern with highways and transportation was much greater. The type of planning to be practised was broader and less operational. Consequently the LCC linkage of planning and architecture was not followed.

Middlesex provided a quite different model since the county was not even a housing authority, nor did it operate overspill schemes. Its planning department comprised a headquarters division concerned with the development plan and technical services, and an areal organization dealing with development control and the joint area committees. This areal or field pattern of organization is highly important for most counties, since they must supervise the detailed work of development control. In the case of the GLC, however, development control is less important because of the position of the boroughs; work on the development plan, including basic information, is more important. Thus Middlesex or other county planning departments did not provide much of an organizational model either.

Until the major reorganization of October 1969 the GLC's Planning Department had three main branches. The Survey Branch conducted technical surveys and basic research; the Land Use Branch formulated policies and prepared the development plan; the Developments Branch dealt with redevelopment areas and development control and contained a small field organization for liaison with the boroughs. A separate team dealt with Covent Garden redevelopment. Overspill schemes under the Town Development Act were handled directly by the Architect's Department, and supervised by a separate GLC Committee for New and Expanding Towns. This organization roughly reflected the division between information, planning, and control which is sometimes recommended for local authority planning departments, but is less likely to be achieved by those which are immersed in details of control and have scant resources for information gathering.[19]

From the beginning the organizational integration of transportation and physical planning received close attention from the GLC. This followed from the Herbert Commission's contention that these were the two functions most needing integration at the metropolitan level but, in addition, the early years of the GLC coincided with the burst of administrative pressure for 'Buchanan-type planning' which followed upon the publication of 'Traffic In Towns'. Professor Buchanan dispensed the philosophy that transportation and urban structure ought to be jointly planned, so as to integrate the values of easy access and mobility with those of environment protection. The GLC provided an organization which might have been specially designed, and was in a sense designed, to attempt 'Buchanan-style' planning. Thus the early history of GLC planning was dominated by transportation issues which perhaps inevitably left less attention for other functional relationships.

Initially two separate departments were created for Highways and Transportation, and for Planning. However both were controlled by a Planning and Communications Committee, concerned with strategic issues, which had a higher status than the Highways and Traffic Committee covering operational matters. In practice, however, the Highways and Traffic Committee acquired more power than had been intended. Simultaneously, steps were taken to harmonize closely the methods of the two departments over development plan preparation. Surveys of existing conditions were followed by interim policies on land use and transportation, confrontation of these policies, separate projections based upon the interim policies, and final confrontation of policies. The research groups of the two departments were told to link their methods of data collection and their storage and retrieval techniques, while it was arranged that the London Transportation Study should incorporate three alternative urban form models within its investigation of transportation alternatives.[20]

In 1969, however, the surprising step was taken of formally combining the two departments into a single giant Department of Planning and Transportation. One reason was dissatisfaction with the degree of co-ordination achieved. Another topical reason was political pre-occupation with the GLDP and its controversial highway proposals which, in the view of many critics, had not been properly linked with planning proposals. A further related factor was the desire to strengthen the attention given to the strategic policy issues of both planning and transportation and to separate this work more clearly from the routine operations of the two services. A Strategic Planning Committee had been formed in 1968 to oversee the preparation of the development plan, which took over much of the policy work of the Planning & Communications Committee. (The latter committee continued its other tasks as the Planning and Transportation Committee, renamed after the 1970 election the Environmental Planning Committee.) Within the new department separate branches were created for Strategy, Plans, Traffic and Development and Construction. Strategy branch dealt with basic policy and research; Plans with plan preparation; Traffic and Development with traffic management, development control, and redevelopment areas; and Construction with engineering operations, primarily for roads. The new department was headed by Joint Directors of Planning and Transportation, initially Mr B. J. Collins and Mr P. F. Stott who were the previous heads of the two departments.

The imprint of these co-ordinative devices upon the GLDP was disappointingly small, although some of them appeared only late in the day. Detailed analysis of the inter-relations between transportation and planning policies was almost completely lacking, suggesting an abandonment of work done as being either inadequate or impolitic. In so far as a relation did emerge, transportation policy dominated planning policy. The measures needed to improve the mobility of Londoners were spelled out in detail, but measures to re-shape and protect the urban environment against traffic nuisance were indicated only broadly and left mainly to further action by GLC and boroughs.

Various reasons can be suggested for the difficulties of 'integrating' planning and transportation. One is that the techniques of the two

functions are traditionally very different, and that in the case of the
GLDP transportation techniques were considerably more sophisticated than those used by the physical planners. (Transport planning
had in fact reached its most advanced level in the London area
through a series of special studies, while physical planners brought
old techniques to new problems.) More broadly the *goals* of transportation planners tend to be more intelligible and operational than
the multiple aims of physical planners, particularly in the case of the
GLC where physical planning was subject to special constraints
and uncertainties. Moreover, the goal of 'getting the traffic moving'
was undeniably a task for an all-London authority, whereas environmental protection or enhancement plausibly seems a local matter
for borough action – even though the two types of action should
logically be related.

More broadly still it is necessary to note the intrinsic difficulties
of comprehensive Buchanan-style planning. One can state a general
policy but it is another matter to devise acceptable operational
goals and techniques for reconciling the conflicting claims of 'amenity' and 'access' or for anticipating and regulating the interplay of
transportation demands with locational choices over housing and
employment. Administrative integration on its own cannot compel
a harmonization of policies if the necessary political accommodation
and professional skills are lacking.

A separate question was the relationship between the planning
function and the 'intelligence unit' to whose creation the GLC was
committed. One view of such a unit is that it should concentrate
upon the objective collection of statistical and other information,
which would provide a useful common service to GLC departments,
other interested organizations (eg the London boroughs), and the
general public. A contrasting view is that the unit should undertake
basic economic and social research which would be closely geared to
policy formation. Both views could, of course, be combined and the
unit could seek to perform policy-oriented research in an objective
way as well as the collection of basic information. But the more the
Unit ventured into policy-geared research, the greater would be its
need for strong resources and a high status and prestige in order to
win the co-operation of major departments. Relations with a planning department would be especially difficult, because research is
basic to planning and on one possible view the Research and Intelligence Unit ought to be within planning.

The GLC initially inclined towards the first of these alternatives. A Research and Intelligence Unit was created within the office of the Director General, and a statistician appointed as its director. The director did not, however, have chief officer status and had to deal with the heads of powerful departments, such as Transportation and Planning, which preferred to control their own research programmes. The Unit included divisions for social studies, economic and population studies, and land use and transportation which conducted work that was highly relevant to physical planning, but the research viewpoints clashed. For example, the Unit's destruction of the myth that the population of the GLC could only be held down to 8 million through provision for 1 million overspill caused resentment in the planning department though it was later accepted. The Director wished to relate the Unit's research to policy formation but lack of co-operation and budgetery cuts negatived this aim.[21]

A further result was that the social and economic base studies required for physical planning were not initiated. These studies went beyond the technical surveys traditionally performed by physical planners, but also represented broader and less statistical types of research than the Intelligence Unit was expected or allowed to pursue. This was a most serious omission, because a strategic planning authority plainly needed to achieve as much insight as possible into such matters as the future economy and industrial structure of the capital, economic linkages, preferences for residential location and types of dwelling, and so on. It is paradoxical that London lacks the kind of long-term survey of economic trends which has been done for the New York region, but it is London which has the planning powers.[22]

The appointment of a Chief Planner (Strategy) in 1969 was the signal for more policy-orientated research to be undertaken in the Planning Department. In 1970 the Research and Intelligence Unit was transferred to the new Department of Planning and Transportation. It was soon decided, however, to shift the relevant divisions of the Unit to the new Strategy Branch of this department. The effect of these changes was to expand research in the Planning Department, but to reduce the role of the Intelligence Unit and to exclude it more definitely from policy formation in the major departments. Not surprisingly the first director of the Research and Intelligence Unit now resigned.

Within the GLC, physical planning is clearly differentiated from general organizational planning. Central management is provided by a Leader's Co-ordinating Committee comprising the chairmen of the important committees, and by a Chief Officers Board. Managerial services are provided by the Director General's staff. It is a consequence of this type of organization that the central management functions are rather narrowly interpreted, since many of the broader policy issues are primarily the concern of the Planning and Transportation Department; but conversely this department is not directly engaged in the central review of policy options. The introduction of PPBS served to complicate rather than change the basic pattern, although its application to an organization which has a large department concerned with 'planning' poses difficult issues.

The organization of planning within the London boroughs raised the question familiar to county boroughs, whether to establish a separate planning department or to place the planning staff under either the Borough Architect or the Borough Engineer and Surveyor. In 1969 11 boroughs had opted for the first alternative, 13 (including the City of London) for the second, and 7 for the third (the position in two boroughs was mixed or uncertain). The total planning staffs of the boroughs was 1,380 (with one borough excluded), compared with 479 at the GLC. Though individually much weaker than the GLC the total scale of borough staff was moderately impressive. There were considerable variations between boroughs. Hackney had only a small planning staff under the Architect, Camden more than twice as many organized as a separate department. Outer London boroughs were more inclined to establish separate departments, while in some boroughs with a rich architectural heritage (Westminster, City of London, Richmond) the Architect was in control.

But although apparently comparable to a county borough in terms of organization and staffing, the planning functions of a London borough were subtly different. The special problems of London (particularly Inner London) and the fact that initially the boroughs were intended to produce structure plans before this system was extended to other cities, were arguments for creating a *separate* planning department under a chief officer of high calibre. Conversely, the fact that boroughs only shared powers with the GLC was an argument for more modest arrangements, and the case for modesty was strengthened of course when the structure plan powers were withdrawn. In the initial period, however, the considerable demand

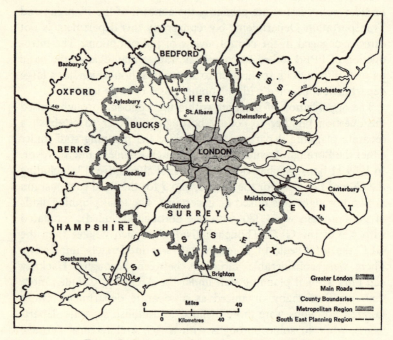

Figure 7. Greater London and South East England.

for planning staff in London, and the bidding for staff between boroughs and GLC created concern about staff shortages in London and elsewhere in the South East.

The regional framework[23]

Ironically the vacuum in strategic planning for the London region, which had existed since the Abercrombie Plan of 1944, was being filled at the very time London Government was reformed. Since that date three broad regional plans have appeared: *The South East Study* (1964). *Strategy for the South East* (1967), *Strategic Plan for the South East* (1970).[24] Each of these plans had a different inception, sponsorship, and character; their one common feature is that (like Abercrombie) they were all non-statutory and advisory plans, whose impact depended upon the willingness of central and local government to implement them. (Figure 7 shows the relationship between the GLC, the London Metropolitan Region and the S.E Economic Planning Region.)

This fresh burst of regional planning represented the real successor to the Abercrombie Plan for which the Herbert Commission had pleaded. The areas concerned were broader than the 1944 Plan – considerably so in the case of the South East Study – but this enlargement was a natural response to facts of population growth and spread. These plans attacked questions about regional growth and the relationship between Greater London and its region in ways that the GLDP could not compass. By virtue of their broader perspective and non-statutory character, these plans could afford to be and were a great deal more positive than any contrived by the GLC. A main aim of this section is to consider the relationship between the new planning system for Greater London and the changing character of regional planning.

Despite its title, *The South East Study* (1964) was in fact a plan for action devised by officials in the Ministry of Housing and Local Government. The bolder parts of the South East Study were its proposals for dealing with London's overspill, which was estimated at one million people, exactly the figure reached in 1944 by Abercrombie. This does not, of course, mean that planning had stood still. The million surplus population of Inner London which Abercrombie diagnosed had disappeared by 1964, and the new 'overspill' represented primarily the anticipated *future* creation of separate households for whom space could not be found in the capital plus relief to particular areas of overcrowding.

To absorb this million overspill the planners looked further afield than the new towns around London whose capacity was running out. Instead they advocated the substantial expansion of a number of existing towns or cities at distances 50 to 100 miles from London. Several arguments pointed to this solution. The basic one was that the 'counter-magnets' to the pull of London which the planners desired seemed more likely to materialize if the new developments were both large and distant. They could then become new centres for sub-regional growth, and offer a wide variety of services and facilities.

The Conservative Government in 1964 gave a broad endorsement to *The South East Study* proposals, and started to explore methods of implementation. The incoming Labour Government (1964) reviewed the Study, but in the end endorsed most of the proposals.

Only three of the seven counter-magnets proposed by *The South East Study* were, however, started by the Government. These were Milton Keynes, Northampton, and Peterborough, for each of which a new town development corporation was appointed on a basis agreed with the local authorities in these areas. Two projects (Ashford, Ipswich) were abandoned; one project (Newbury) was transformed into an extension of the GLC development of Swindon; and the largest project of all (Solent City) was remitted for prolonged study by a consortium of local planning authorities.

The next stage in regional planning centred around the activities of the South East Regional Economic Planning Council and the Standing Conference on London and South East Regional Planning. These two bodies were created after the London reform, and reflected the increased interest of both central and local government in regional planning and administrative co-ordination.

The SEEPC was appointed in 1965 by the Secretary of State for Economic Affairs and comprised prominent individuals drawn from industry, local government, public corporations, and academic life. They were appointed on a personal basis, though often on the advice of relevant organizations. The council's official role was to advise the Secretary of State on regional issues and to assist a regional planning board of departmental officials which was simultaneously created.

The Standing Conference was a joint advisory committee of local planning authorities, formed initially in 1962 to counter the weakness of regional planning stressed by the Herbert Report, but developing also as a local government counterpart to the work of the SEEPC. The SC enlarged and standardized its area to coincide with the economic planning region established by the Central Government. This region contained the GLC, 13 counties, and 10 county boroughs. It covered a similar area to that in *The South East Study* south and west of London, but a much smaller area to the north and east where a separate region was created for East Anglia. The Conference comprised a body of local councillors nominated by the constituent authorities plus a technical committee of planning officers, and had the aid of a small planning staff. It had of course no executive powers, and action depended upon the constituent authorities.

There was one other striking difference between the SC and the SEEPC. The SC produced excellent technical studies and papers, but was unable to prepare any very positive type of regional plan because of the different interests of its members. By contrast, the preparation of such a plan was seen by the SEEPC and particularly by its chairman as essential to the Council's role and rationale. The DEA encouraged all regional councils to produce plans or studies of a broad kind, appropriate to the conditions of the particular region. With a minimum of departmental guidance as to content, the SEEPC prepared *Strategy for the South East* (1967).

This plan was essentially a broad strategy for the physical development of the region. It proposed the establishment of 9 sectors of growth, which would extend along major communication lines from the outer edge of the approved Green Belt and culminate in at least four cases[25] at one of the proposed counter-magnets. This strategy recognized a need both for distant cities and also for considerable development closer to the capital where growth pressures were observed to be extremely powerful. The device of growth sectors also facilitated the designation of substantial green sectors where some degree of conservation could be practised.

With the publication of *A Strategy for the South East,* the initiative passed to the Secretary of State for Economic Affairs. A simple endorsement of the Council's strategy would be quite unacceptable to the other regional body in the field, the Standing Conference, partly because its thinking followed different and much more tentative lines, but mainly because the counties saw physical planning as *their* responsibility not that of a mere advisory body; while Government implementation of the strategy would fall largely to the

M

Ministry of Housing and Local Government, which also would naturally dislike physical planning being settled by another department.

The department accordingly adopted the device of persuading the SEEPC and the SC to join with the Central Government in commissioning a further plan. A joint planning team drawn from the staff of the departments and the SC was formed under the leadership of the chief planner at MHLG, Dr Wilfrid Burns, and produced *Strategic Plan for the South East 1970*. This regional plan investigated two strategies (labelled A and B), both of which assumed that future development in the South East should be concentrated to a considerable extent within substantial zones of development. Strategy B differed from Strategy A (which was the one based upon the SEEPC plan) in envisaging a faster rate of dispersal from London and in concentrating more new growth close to the capital. The preferred strategy selected by the planners was closer to B than to A. Since one major objective of the plan was the improvement of housing and environment within Inner London, the development of new concentrations fairly close to the capital seemed indicated, especially as other important factors (such as the functioning of transportation) appeared to be neutral. Of the five major growth zones proposed in the *Strategic Plan*, three were located within the metropolitan region relatively close to London,[26] and two represented a continuation of distant counter-magnets (South Hants, Milton Keynes) inherited from both earlier regional plans.

This brief history of regional planning can now be related to the analysis of planning within Greater London. One point is the effect of different perspectives upon the policies or strategies advocated. Thus from its standpoint, the GLC criticized some borough proposals as wanting a costly and undesirable intensity of development, just as the Burns Plan, from its much broader perspective, proposed measures of population and employment dispersal which had not commended themselves to the GLC. Similar points could be made about transportation planning. As the viewpoint moves from zones of congestion and population decline to zones of potential growth, the advantages of a more expansive and redistributive type of development strategy seem to increase.

The evolution of ideas about physical planning is clearly traceable in the three regional plans and in changing notions about the relationship of London to its region. *The South East Study*, coinciding roughly with the creation of the GLC, favoured a view of Greater

London as a fairly self-contained and stable entity, to be balanced eventually by distant, self-sufficient counter-magnets. The metropolitan region around London received less attention – it was recognized as an area of rapid growth but the assumption was that this growth could be diverted or strongly constrained. Events demonstrated that metropolitan growth was much more powerful than anticipated, and that any appreciable transfer of this growth to distant cities would be a very slow process. Indeed the most buoyant of the proposed 'new cities' (Southampton–Portsmouth) was expanding entirely through the impetus of its own economy and its attractions to migrants from all over Britain. Finally the planners' predictions were upset by the discovery in 1967–8 that the population of Greater London was not static but steadily falling although the Town and Country Planning Association had often forecast this development as inevitable.

In these circumstances the notion of London as a regional city extending well beyond its administrative boundaries became more current. These ideas were reflected tentatively in *Strategy for the South East,* and more markedly in *Strategic Plan.* The GLC for institutional and political reasons was on the defensive against this current of thought, but an increasing pre-occupation by GLC leaders and planners with regional questions, such as airport location, port developments, and recreation zones, demonstrated the growing reality of 'regional city'.

Despite their increasing intellectual persuasiveness, the operational effectiveness of the regional plans was as dubious as that of the GLDP, though for quite different reasons. The time-scales of the plans created absurdities. Thus the first GLDP was technically dated up to the year in which the Burns Plan was meant to take effect. This largely coincidental arrangement could be defended on the grounds that a long-term regional perspective would aid the examination of a current statutory plan; but the relationship was absurd inasmuch as the regional plan was as relevant for early changes in public policy as was the statutory plan.

The relation between Greater London and its region was not seen as a major issue either by the Herbert Commission or by the creators of the London Government Act. There was a certain tendency to equate the conurbation with the region, and to assume that the need for strategic physical planning could be satisfied by the GLC. Developments have refuted this assumption. The GLC has coincided with a much greater volume of regional planning activity, motivated in large part by the position of London as congested regional centre and generator of regional growth.

The extent to which London was becoming a regional city was demonstrated by the Roskill Commission on the third London airport which considered no site much nearer than 40 miles to the centre of the capital.[27] For physical planning factors relevant to the choice of site the Commission turned primarily to Dr Wilfrid Burns as head of the regional team. The two regional bodies also played an active part in the airport controversy, both strongly favouring the same site – Foulness (as on balance did the GLC). The apparent capacity of these weak regional bodies to express opinions upon this issue, in comparison with the powerful GLC, was illustrative of the growing significance of regional planning. It is also worth noting that the airport was the only subject on which the SC was able to speak with a strong united voice.

The GLC thus found it increasingly necessary and politic to participate in regional planning, but the influence of regional plans upon the GLC remained a more open question. The GLC is a member of the SC, but seemingly has not made much use of this body as a forum for the resolution of conflicts of interest; and as noted the SC is unable as a rule to show much policy initiative. The SEEPC could expect only to have a marginal influence upon the GLC, unless it were to be given a definite review function in relation to local plans. The chief role of these regional bodies was therefore as generators of ideas and possible support for such enterprises as the 'Burns Plan'. That plan itself was politic enough to avert direct conflict with the GLDP, preferring this implication of its strategy to be drawn by others.

In principle both plans accepted that it was desirable for London to retain the highest possible population consistent with good environmental standards; but whereas the Burns Plan saw these standards as requiring a rapid shift of industries and population out of the congested inner boroughs the GLDP threw its weight the other way. And while allegedly the objectives of the Burns Plan had emerged from study of regional problems, they actually reflected the current interpretation by professional planners of some long-established public policies – particularly the desirability of urban dispersal. This was being recharged at regional level while being quietly discarded by the GLC.

As the 1970s opened the planning of Greater London seemed headed for a conflict or deadlock of differing strategies. The value of the Burns Plan where London was concerned was that it gave some flesh and blood to the emerging concept of a regional city extending well beyond its administrative borders. Whether the regional city idea would be strong enough to overcome the GLC's itch for self-sufficiency has still to be seen.

The future of metropolitan planning

In this concluding section, the evolution of metropolitan planning in London will be considered from two viewpoints:

i.   How satisfactory is the reformed London system in meeting the general requirements of physical planning? In other words is there progress or regress in the planning of the capital, considering town and country planning as a service in its own right?

ii. How well is the GLC handling the planning function within its special block of allocated powers? In other words, how well has planning been fitted into the special organizational structure of the GLC?

A strong caveat must precede this section. The systems which we are discussing are in a condition of considerable flux and uncertainty. The arrangements now prevailing may be substantially altered over the next decade. This said, there need be no objection to some provisional conclusions which can be read both as interim verdicts upon the London reforms and as pointers to the desirable direction of further developments.

1. London Government must work the system of structure plans that is being gradually introduced throughout England and Wales under the 1968 Act. A structure plan is less detailed and more concerned with general policies or principles than was its predecessor, the local development plan; and it is supposed to cover a broader and better-balanced area.

These two functions are related, since adequate general policies *can* only be produced if the area around is satisfactory – or so the framers of the Act reasoned. The structure plan becomes the basis for local plans which govern the detailed pattern of land use and development, but these local plans do not require Ministerial approval and fall within the discretion of the local planning authority. Contentious issues are supposed to be resolved through the public inquiry into the structure plan and through Ministerial review of the inquiry's findings.

Clearly the whole system of structure planning faces formidable difficulties. Can the right areas be found for this purpose, where they do not at present exist, either through local government reform or the co-operation of local planning authorities? Assuming the areal basis is adequate, can planning at this rather rarefied level adequately guide the preparation of local plans? Even if it can do so in the eyes of planners, will the public be satisfied? There is the awkward fact that the hostility of objectors is typically directed towards particular and localized issues, rather than 'principles'; this situation creates the dilemma that either the implications of the structure may not be properly understood, causing subsequent wrath and discomfiture, or else that objectors – (fearing the implications) – may demand more specific structure plans and prolong the inquiry process. The numerous objections to the GLDP road proposals show how hard it is to formulate general plans without encountering objections to details – particularly no doubt with this function.

It is impossible to foretell whether or how these general problems of structure planning will eventually be resolved, but the application of this system to London poses additional issues. First comes the question of area. Greater London is both large and complex but is certainly not well-balanced, being mainly built-up and facing population decline, with major zones of growth lying beyond it. This is a serious point since the concept of the structure plan was particularly designed to tackle the problems of growing city regions or regional cities.

This situation accounts for the importance of regional planning within the South East of England, which as the last section showed has not been diminished by the GLC's arrival. A satisfactory region for broad planning purposes need not be as large as the South East economic planning region, but it would have to extend at least 30 miles from the centre of London. From the standpoint of physical planning, and probably also of transportation and the 'strategic' aspects of housing and urban development, a substantial extension of GLC boundaries would be very desirable. Admittedly such a vast and complex area could not easily be handled through a single structure plan, and there would need to be a series of substantial 'sub-regional' plans. However, political unification would ensure that planning policies were coherent and consistent throughout the city region, which would be desirable in itself and a necessary condition of successful structure planning.

The political difficulties of establishing so vast a public authority, only shortly after the creation of the GLC itself are fairly obvious and need not be discussed in detail. Failing such a measure, the problem for the 1970s will be to reconcile the very different perspectives of the GLDP and *The Strategy for the South East*. Some means exist for this purpose. The Standing Conference of Local Planning Authorities exists for the harmonization of the plans of the GLC and surrounding counties, but is unlikely to achieve more than minor adjustments. More to the point, the relevant Minister has some power to bring the GLDP into line with the regional plan, particularly if encouraged to do so by his expert panel of inquiry. However, it cannot be expected that the GLC will move far towards a genuinely regional viewpoint.

Next comes the treatment by the GLC of the various parts of Greater London. The structure planning approach requires positive guide lines for the redistribution of population within London itself, the location and development of major sub-centres, the intensification of residential development in certain areas, and the conservation of environmental values. For reasons already given, the first GLDP also did not provide adequate guide lines of this kind, although a principal cause of this failure, which was the independent powers of the boroughs, has now been much modified. It is possible, and certainly desirable, that the Minister will insist that the first GLDP should now be substantially rewritten and its time horizon considerably extended. The further delay would be worth accepting. If this should not occur, future structure planning in London should still be able to build on a sounder basis.

In a pamphlet published in 1962, I noted the limitations of the GLC area, but listed four valuable planning tasks which the GLC should accomplish within London itself. These were the planning of the central area; the location and development of major sub-centres; strategy for residential development, including stocktaking of the capacity of the various urban and suburban zones; and communications planning.[28] Progress with all these tasks has been disappointing or non-existent but all of them ought now clearly to fall within the GLC's grasp and capacity – except for the central area, which an energetic Minister could place there.

Of course the general concept of structure planning may be misconceived, particularly in relation to London. A manageable statutory plan of this kind may be sensibly produced for a small city-region or a well-balanced county, but the scale of a vast conurbation or city region is so enormous as to require detailed local plans of a very substantial kind. A broad general plan is still highly essential, but it would probably best take the form of a policy document for general discussion rather than of a statutory plan to which formal objections are made. Most of these objections relate to questions of detail or personal hardships and only obscure the intelligent review of general policies. But at a more localized level, detailed patterns of land use could be shown and formal objections sensibly considered.

The most satisfactory planning system for the London region would probably be a two-tier one. The Regional authority would produce a general plan (or rather plans), which would be formally adopted by the authority and confirmed by the Minister after the fullest possible public discussion, including mandatory conferences; but the regional plan would not go to public inquiry or be legally binding. This regional plan would then guide public developments throughout the region and provide a basis for the review by the regional authority of local plans, which would go through the inquiry procedure. The local plans would be prepared by the London boroughs and similar local authorities elsewhere in the region. Even if this system had been applied only to Greater London, the results would certainly have been an advance upon the pre-1970 situation, and possibly upon the post-1970 one. The verdict upon physical planning in the GLC area up to 1970 can only be that it has worked very badly. The results were no better, perhaps worse than under the previous system of planning authorities, despite the increase in planning staff, and falsified the hopes of the Herbert Commission – partly because of the Government's egregious alterations to that Commission's proposals. Moreover, a really satisfactory system of planning seems to require another considerable upheaval, involving the enlargement of the GLC and perhaps a reversal of the present treatment of structure and local plans.

But even without such changes, appreciable improvements in physical planning should be possible. There is now a fairly satisfactory outline plan for the London region, which might play something of an Abercrombie role over the guidance of local government planning. The GLDP could and should be brought more in line with it. Secondly, the ground is now better cleared for the GLC to prove that effective structure planning within London itself can after all be tolerably achieved.

2. Secondly, I turn to the position of planning within the organizational structure of the GLC. This is a perplexing subject, since (as in the last section) we have a tangle of general and special problems. The general problem is the location and functions of a planning agency in *any* large local authority, while the special problem is to relate planning to the unique powers possessed by the GLC. The general problem was discussed at the beginning of this chapter, and I shall approach the subject through consideration of the GLC's unique powers.

Environmental planning (broadly understood) is basic to the GLC's special set of responsibilities. Transportation and housing are two of its most important *functions,* and these are also the two public services which make much the heaviest impact upon the physical environment. Both these functions are very large users of land, and strongly influence the quality of the environment in many ways. Further, the physical functioning of a city is primarily settled by relationships between residential location, employment location, and transportation. Thus housing and transportation are vital elements of physical planning.

By contrast the GLC is not responsible for personal social services, save for the special case of the Inner London Education Service and for housing. Education, health, and welfare naturally have their connections with environmental planning, but very much less so than housing and transportation since their land needs are relatively modest and their locations are strongly influenced by housing. The co-ordination of these services, and of housing too from this angle, requires a type of welfare planning which is largely outside the powers of the GLC.

The GLC's other responsibilities, such as parks and smallholdings, sewerage and refuse disposal, fire services and ambulances, have a variable but sometimes considerable relationship to physical planning. To the extent that such relationships do not exist, the services in question tend to be somewhat technical and self-contained. The only types of co-ordination relevant to *all* GLC functions are universal organizational needs such as budgeting; but the GLC is unusual in having such a substantial block of its total work particularly concerned with environmental planning and development.

'Environmental planning' in this context should not be misunderstood. It does not, of course, imply a fixation upon the achievement of particular physical patterns and designs. Rather the planning of the environment is one important means for the realization of social policies such as greater equalization of housing conditions, or of economic policies such as improving the operation of the most essential activities located in Central London.

It can also be argued that the transportation and housing functions have their distinctive goals and methods that are different from – and usually clearer than – those of the physical planners. On a purist view, as was said earlier, planning might operate through the prescription of environmental policies and standards which would be as binding upon housing and transportation departments as upon private developers. In practice this is a recipe for frustration and ineffectiveness, in GLC planning because of the weight of these other functions. Moreover this relationship is not really desirable even in principle, because transportation and housing also regulate large areas of the environment which must be absorbed into a more general form of planning if such is to exist.

The GLC has changed its organization quite frequently during its limited life, but these shifts and expedients have so far failed to treat the planning function adequately. Originally, planning was conceived primarily as an ordinary department, although co-ordination with transportation was stressed. Later these two departments were formally merged. At the same time a variable set of co-ordinating committees and arrangements were introduced for the GLC as a whole, but the relationship between this overall co-ordination and environmental planning was never investigated or clarified. The Strategic Planning Committee was given a key co-ordinative role, but its energies were absorbed by the political problems of presenting and piloting the GLDP; it proved unable to tackle environmental planning on a broader front or to build a bridge between different levels of the organization. The Research and Intelligence Unit had initially an ineffective relation to planning, and was later shifted and then dismembered.

The GLC cannot be wholly blamed for this catalogue of mistakes or failures. Besides inevitable 'teething troubles' and universal problems over making planning effective, there is the point that physical planners are not well equipped to handle the sort of problems facing the GLC. Traditional town and country planning works largely through the use of physical standards (for housing, offices and factories, open space, parking, etc.), segregation of incompatible uses, conservation of attractive features, and ad hoc resolution of land use conflicts. These techniques have their uses, but they will not solve the social and economic conflicts relating to the physical functioning of the capital.

These inadequacies of the planners were revealed by the GLDP. This plan (perhaps reasonably) made the economic health and vitality of London its principal consideration, but the needs of the London economy were not analysed with any precision, nor was there any systematic attempt to match up possible models of economic needs and environmental capacity. Again the possible relationships which might hold between residences, employment and transportation were nowhere set out or analysed.

These professional or technical inadequacies of the physical planners went with ineffective relationships with both transportation and housing. That with transportation was palpably one-sided, with the physical planners failing to promote the control of traffic by environmental standards for which the Buchanan Report called, or to show how transportation policies might contribute to a better distribution of population and employment. The relationship with housing was too weak. There was no integrated attempt to look at the possibilities of redistributing the housing load within and beyond Greater London, and the social and economic effects of so doing.

This analysis suggests that a high-level agency within the GLC ought to take over the whole area of environmental planning, including strategic transportation and housing policies. Operating tasks would be handled by a series of separate agencies responsible for road construction and management, public transport, housing construction and management, land acquisition and negotiations with developers, parks and open spaces, and possibly development control. The exact structure of these agencies, and the question as to whether any should have a 'process' basis (eg as an architects or valuers department) cannot be discussed here. The basic point is that planning should be comprehensive and separated from operational tasks, and that the planning agency should be closely related to the political leadership so that integrated policies are adopted and implemented.

Admittedly, this proposal still poses many problems about keeping the planning function both manageable and operational. Clearly the new planning agency could not be filled with traditional town planners but would require a diversity of skills, foremost being the capacity to build bridges between different specialisms. But the alternatives seem a lot worse. A separate, traditional town planning department has proved a mistake, and complete fusion with transportation is palpably one-sided when compared with *partial* fusion with both housing and transportation.

Perhaps the most workable alternative to this model would be the opposite approach of strengthening and broadening the operating departments, and reducing the scale and ambitions of comprehensive planning. For example, a department of development could be established responsible for action areas and major development projects as well as housing, while all aspects of transportation would be handled by another department. Each department would broaden its staffing and approach so that environmental considerations were brought to bear on these major tasks. Comprehensive planning would then be dealt with in a smaller top-level unit, which would prepare and monitor the structure plan and advise the GLC leadership on general policies; but its authority over the major departments would be limited and would work largely through persuasion. This solution should commend itself to anyone sceptical about the possibilities of comprehensive planning, yet keen to make operating departments more environmentally conscious. It would also accord with the view (discussed earlier) that statutory plans might be better produced at borough level and vetted by the GLC against a general, non-statutory plan.

Finally the relation between environmental planning and overall organizational planning must be considered. The latter is necessary in any organization for controlling the flow of basic inputs of money, staff, and legal authority, to the various departments, and for co-ordinating policies. Such co-ordination is essential for political purposes, but it cannot cut very deep when resources are being allocated between disparate functions such as the fire service and housing. On the other hand, within the field of environmental planning, the uses of resources are frequently inter-related. For example, housing can be provided much more cheaply in the outer than the inner boroughs but costs of transportation may be thereby increased; or again the purchase of industrial sites is one way of reducing housing costs. The structure plan itself needs to be considered for its effects both upon the costs of public services and the capacity of rateable and tax resources to support these services.

Physical planning in Britain is too divorced from these fiscal and economic considerations. It does not even have the degree of responsibility for capital budgeting which is a feature of many US Planning Commissions. If, however, these issues are left to a planning unit covering the whole organization (as is being tried in the GLC), the treatment tends to be somewhat superficial and based upon formal lines of departments which disguise the key role of environmental planning. In the case of the GLC it seems desirable that the resource implications of environmental planning should be carefully covered by the agency responsible for this subject, although working in conjunction with the body responsible for overall budgeting.

Somewhat similar considerations apply to the research and intelligence function. In principle there is much to be said for a central agency, 'independently' located, which will undertake key aspects of this work for the principal departments; but this system will not work easily or at all where the principal departments are few but powerful, and particularly of course where one of them is labelled 'planning'. It is true that the GLC R and I Unit proved itself considerably more objective and accurate over population forecasts than the planning department, although this fact also reflected the limitations of traditional town planning. Conversely the argument that sufficiently objective research will almost *automatically* suggest the right possibilities is not generally true. The aim of keeping research as objective as possible has to be weighed against the need to relate it closely to ongoing problems and policies which a centralized agency can only partly appreciate.

Because environmental planning covers so large a chunk of the GLC's work, and because broad research is basic to such planning, it seems right in this case that the planning agency should be responsible for all the research relevant to its tasks. This would hold true, whether one accepts the model of a large planning agency controlling operating departments or of a small agency with more limited functions. (In the latter case, departments of transportation and development would also have research functions.) It may well be that there is also scope for a small central R and I Unit providing some statistical services of general interest, and assisting smaller departments and the London boroughs with relevant information. This can be treated as a separate issue.

In conclusion then, how has planning worked within the GLC area? Any direct comparison with the previous governmental system is not very profitable because costs and benefits are so very different. What can be said is that the new planning system has proved distinctly disappointing, when set against either the expectation of the Herbert Commission or against the high costs of staffing it or against the apparent potentialities of the London reforms.

At the same time the system *could* work very much better, even without entertaining the ideas of a radical enlargement of the GLC area or a revision of structure planning. Substantial improvements are possible if the GLC creates a better type of planning department or agency, and utilizes (with the Minister's support) its new opportunity for more effective structure planning. The role of the boroughs is also of course very important, and has been lightly sketched here only because of limitations of space and relevant data. London has a long way to go to achieve a worthwhile system of metropolitan planning, but it need not give up the chase.*

* The author wishes to acknowledge the help given by Mr John Shepherd in the preparation of this chapter.

10
Finance*

Introduction

The financial consequences of reorganizing local government in Greater London seldom entered the discussions which preceded the 1963 London Government Act. This was surprising, especially when the earlier emphasis placed on this aspect is taken into account. The Ullswater Commission's terms of reference, for example, implied that structural changes should only be recommended if they could be shown to produce 'greater efficiency and economy'.[1] Several factors lay behind this changing emphasis on the words 'efficiency' and 'economy'.

First, although the Herbert Commission's terms of reference did not specifically include these words, the Commission did discuss them, but in a different light:

> 'We have borne in mind throughout the need for economy. By economy we mean economical performance of functions undertaken either compulsorily or voluntarily by a local authority, and not necessarily the mere amount of money expended in total. An efficient and economical local authority may well be one which undertakes a wider range of activities than another which has a more limited expenditure per head of population'.[2]

In the Commission's view, it was sufficient for the new structure which they recommended to attract 'good councillors and officers' which they believed to be 'the best guarantee of efficiency and economy'.

However, when the Commission came to discuss the specific financial implications of their proposals they were to claim: 'our recommendations should not involve any material change in the aggregate rate expenditure in the Greater London area'.[3]

* The author would like to thank Mr T. W. Sowerby, Treasurer of the London Borough of Bromley, for his generous help and advice in the preparation of this chapter.

This suggests, as has been stated earlier,[4] that the Commission were not seriously concerned with the financial effect of their proposals; for not only did they fail to consider the foreseeable consequences of their recommendations involving on the one hand the expansion of the GLC's role in planning and traffic management, and on the other hand the fragmentation of the services they proposed giving to the London boroughs, but they also made no allowance for the additional costs that would arise in the period of transition and the possible stimulus to the demand for some services. It is true that the Commission did forecast an increased rate burden for Inner London ratepayers as a result of their proposals, but they expected a corresponding reduction in the rate burden in Outer London. These changes in rates levied would be a result of the amalgamation of areas, the re-distribution of functions, and the wealth of the central area becoming available to the Greater London area, and not a consequence of any increased expenditure by local authorities in Greater London.

Since both the Royal Commission and later the Government were putting forward proposals for a fundamental reorganization of local government, which in itself was sufficient to arouse heated controversy, it would be understandable if those who advocated reform were to refrain from fanning the flames further by introducing to the discussion the additional expenditure which might be incurred by a reorganization of local government.

Secondly, no similar reorganization on the scale of the London proposals had taken place before, and until the new authorities had begun work, no one could predict with any certainty either the nature or the scale of any additional expenditure that reorganization might engender.[5] Thus, if the Herbert Commission had had any strong desire to cost their proposals it is unlikely that they could have made any very precise predictions.

Thirdly, the opposition to the Government's proposals might have been expected to emphasize the costs to which reorganization would give rise, had not more fundamental political arguments been raised against the reform proposals.[6] This was in part due to the way in which the Government had set out their case. In the White Paper giving the Government's proposals for reorganization it had been emphasized that reorganization was essential both for the efficient performance of functions and for strengthening local government. It followed therefore that: 'In the Government's view the financial arrangements should follow consequently on changes which are necessary for other reasons.'[7]

This subordination of financial to other considerations was perhaps based on the assumption that local government reorganization would produce economies through the more effective use of staff and other resources. The Herbert Commission had merely hinted at possible savings through the removal of duplication and overlap.

However, financial savings of the type suggested are not likely to occur overnight. The additional administrative burden imposed by reorganization combined with the desire of the staff commission to avoid redundancies meant that initial establishments were not merely an amalgam of existing authorities departments but also included additional staff recruited in the transition; administrative savings could not therefore be expected in the short-term.[8]

Even if it is accepted that financial considerations as such are subsidiary to the major issue, that being the creation of a more effective and convenient local government system the cost of providing such a system cannot be ignored and must form part of any assessment of the effects of the 1963 reforms. Moreover, if the cost of providing local government services in Greater London today is higher than before 1965 this could be caused by a number of factors; increases due to inflation and rising prices, increases caused by greater provision in response to an increased demand on the services, or by an increase directly attributable to the reorganization. Thus, if financial considerations are to be included in a discussion of the effects of the 1963 Act, it is not enough to say by how much expenditure in Greater London has risen or fallen since 1965. The question is to what extent such increases or decreases can be attributed to the reorganization rather than to other factors.

However, as the preceding functional chapters have illustrated, it is by no means easy to separate the effects of reorganization from other factors which have contributed to change and development since 1965 in particular local government services. Perhaps the most difficult question of all is whether, if there are increased costs attributable to the reforms, they have been matched by corresponding benefits.

These questions will be discussed in the main part of this chapter. Before that, however, it is necessary to discuss changes in the financial situation of the London authorities directly resulting from the reforms, since these have a distinct bearing on the way in which developments have taken place since 1965.

Financial situation of the new authorities

The 1963 Act influenced both the resources and the needs of local authorities in Greater London. It affected resources first by the combination of local authorities so producing units of higher penny rate products which in the main were well above the average for the rest of England and Wales. Secondly, the Act included in Greater London some of the most wealthy parts of the counties of Middlesex, Essex, Kent, Surrey and Hertfordshire. The total rateable value of the LCC area before 1965 amounted to £301·4 millions. In 1966 the rateable resources of the GLC were £635 million giving a penny (d) rate product of over £2·5 million. This concentration of resources within the GLC area represented 30 per cent of the total rateable value of England and Wales. However, within this administrative area there were some wide differences in terms of the distribution of rateable value.

While the total rateable value of the GLC area was equal to £635 millions, more than one-third of this was concentrated within the four authorities of Westminster, Kensington and Chelsea, Camden and the City of London. The rateable wealth at the disposal of the various new London boroughs ranged from Westminster with a total rateable value of £104 million to Barking with only £9 million. The Herbert Commission had earlier argued that the amalgamation of authorities to form 52 London boroughs would even out some of the variation in resources and make for greater financial stability. The Government's decision to establish 32 London boroughs did in fact reduce this disparity still further, but as can be seen from Table 10.1 5 London boroughs had rateable resources of less than £50 per head after 1965 while two had rateable resources of over £300 per head.

Some indication of the effect of the amalgamation may be gathered from a comparison of the authorities in receipt of rate deficiency grant under the various proposals. This form of Exchequer aid was given under the 1958 Local Government Act to those authorities whose resources measured in terms of a weighted per capita rateable value fell below the national average. Before 1965, 17 authorities in the Greater London area qualified for this form of assistance; under the Herbert proposals for 52 boroughs 7 would have been eligible,[9] but of the 32 authorities as constituted in the 1963 Act only the borough of Lewisham had resources below the national average and thus ranked for the grant.

Table 10.1

Rateable Value

A comparison of rateable value per head of population for:
  (1) Constituent authorities of Greater London in 1964
  (2) The new London Boroughs in 1965
  (3) County Boroughs in England and Wales in 1964

Range of Rateable Value per head of population £	Constituent authorities in Greater London 1964		Greater London 1965		County Boroughs 1964
	Metropolitan Boroughs & City of London	Remainder of Greater London Area	Inner London Boroughs	Outer London Boroughs	
Above 300	4	—	2	—	—
200 – 299	1	—	—	—	—
100 – 199	4	2	2	—	—
80 – 99	2	2	—	—	—
70 – 79	4	5	2	4	—
60 – 69	3	11	2	3	4
50 – 59	3	22	2	10	6
40 – 49	7	13	2	3	27
Below 40	1	3	—	—	41
Total	29	58	13	20	78

The 1963 Act affected the 'needs' of local authorities in Greater London as a consequence of transferring to the London boroughs responsibilities for the health and welfare services, and in Outer London, responsibility for education. In total, these are amongst the most expensive local government services and accounted for 56 per cent of the total revenue expenditure by Greater London authorities in 1965–6. Before 1965, these services were the responsibility of the LCC and the surrounding county councils and county borough councils, and any difference between expenditure and income from charges and central government grants was met by a rate precept upon county districts and metropolitan boroughs. After 1965 this difference was met by the individual London boroughs by means of a rate levy on their own rateable values. Not only was there a disparity of rateable values amongst the new London boroughs as has already been shown, but also total populations varied. More important, as far as demand upon the education and personal social services were concerned, the age structure and social and economic composition of these populations varied widely.

Table 10.2

Inner London: Net Rate poundage 1966–7

London Borough	Total Rate for Borough purposes		
	s.	d.	p.
Camden	5	6·69	27·79
Greenwich	9	10·66	49·44
Hackney	10	2·24	50·93
Hammersmith	7	10·85	39·52
Islington	7	4·66	36·94
Kensington & Chelsea	3	3·60	16·50
Lambeth	7	4·84	37·02
Lewisham	11	5·31	57·21
Southwark	11	3·47	56·45
Tower Hamlets	10	2·96	51·23
Wandsworth	9	2·65	46·10
Westminster	1	11·87	9·95
City of London	2	2·72	11·13

Source: Annual Abstract of Greater London Statistics, with decimal equivalents inserted (GLC Volume 1 Table 241)

Table 10.2 gives some indication of the disparity of resources in relation to expenditure, which is especially marked in Inner London. Although it is not the case that those boroughs with the lowest rateable values necessarily have the heaviest demands upon their personal social services, it is clear that the cost of these services does impose a disproportionate strain on the finances of these authorities. As Professor A. R. Ilersic of Bedford College, London, later remarked

'it is hardly credible that these differences between the rate levies required to finance borough services could be accounted for solely by the individual council's attitudes to expenditure'.[10]

To summarize, therefore, the 1963 Act reduced the disparity of rateable resources amongst local authorities in an area which was already the wealthiest part of the country. Yet the combination on the one hand of boroughs in the central area with high commercial rateable values, and on the other hand in other parts of Greater London, boroughs with relatively higher 'needs' in terms of the demand upon the personal social services meant that the former boroughs would have to levy very much lower rates than the latter.

It was this concentration of rateable values in Central London combined with the disparity in rate poundages which their proposed distribution of functions would produce that led the Royal Commission to study the possibilities of continuing the existing metropolitan boroughs' rate equalization scheme in Greater London. In the short chapter dealing with the financial implication of their proposals, no less than three and a half pages are devoted to equalization. After looking at the history of equalization schemes in the administrative county going back to 1867, and examining the working of the most recent scheme which had been instituted in 1959, the Commission concluded that a comprehensive equalization scheme on the lines of the existing scheme was neither necessary or desirable but that a more selective scheme should be introduced.

Rate equalization, as practised amongst the metropolitan boroughs while the Herbert Commission were sitting, consisted of a pool equal to 70 per cent of the aggregate local expenditure of all the metropolitan boroughs, financed by rateable contributions from the boroughs. The effect of this scheme was basically for all the expenditure of the LCC and also 70 per cent of the metropolitan borough expenditure to be met rateably by precept of the LCC.[11] The Metropolitan Boroughs' Advisory Body of Treasurers did not, however, accept the form of rate equalization recommended by the Royal Commission but favoured a continuation of the principle of equalization established under the Scheme of 1959. Following these views the Government included in the 1963 Act a provision[12] enabling the Minister to make 'a scheme or schemes' for the purpose of reducing the disparities in the rates levied in Greater London. A further clause was inserted in the Committee stage of the Bill, enabling any such scheme in Inner London to re-allocate the general grant payable to the Inner London boroughs.

The 1963 London Government Act did not make any changes in the grant structure or the financial framework in which local government was operating in London. The major upheaval of the Local Government Act, 1958, when a general grant took the place of former specific education, child care, health and fire service grants was applied to the new structure of local government in London, but the London boroughs and not the GLC became the grant-receiving authorities. The general grant, which was assessed on a national basis was distributed among local authorities by means of a complex formula intended to take account of their need in relation to the population, number of children and old people in their areas. Since the two most important factors in the formula were total population and school population it was frequently argued[13] that the grant was primarily a specific education grant based on objective criteria.

The GLC received no part of the general grant and met its revenue expenditure by precept upon the boroughs. This in itself was a complex operation, for the Act stipulated that separate accounts[14] were to be kept in respect of those functions which the GLC discharged over the whole of Greater London, and those for which it had a responsibility in certain areas only. Thus the revenue expenditure on such functions as refuse disposal, fire and ambulance, was met by precept over the whole of Greater London, while other special accounts met the cost of education, housing and parks in Inner London, administration of justice in Outer London, sewerage in the sewerage area and land drainage in the land drainage area.

The LCC and the county and county borough councils which had previously been in receipt of general grant had also had responsibility for all the services which the grant was designed to replace. In 1965, the situation was complicated by the fact that the boroughs were now in receipt of general grant,[15] but in Inner London, education was the responsibility of the ILEA, and apart from a specific school meals and milk grant, the major proportion of the authority's expenditure would have to be met by precept upon the Inner London boroughs.

The anomalous position of education finance in Inner London, combined with the Royal Commission's assertion that Inner London boroughs might expect substantial rate increases upon reorganization,[16] led the newly constituted Treasurers Advisory Committee of the London Boroughs' Committee to state that there was not yet any 'evidence of a need to extend the principle of equalization of rates to the whole of the Greater London area' but that 'discontinuance of rate equalization in Inner London is not a practical proposition'.[17] The main justification for this view, being, that despite its faults the previous scheme had resulted in some reduction in the disparity between resources and needs of the various authorities. Moreover, in three cases only were the new Inner London boroughs a combination of 'receiving' authorities under the old scheme, while Kensington and Chelsea was a combination of two 'paying' authorities.

In March 1965, the London boroughs' committee agreed on the recommendation of the TAC, to an equalization scheme which would apply to Inner London only. In order to take account of the new borough services which had been transferred from the LCC 75 per cent of expenditure was to be pooled, this to be financed by a pooled general grant, and any difference to be met by a levy on rateable values. The purpose of pooling the general grant, it was argued, merely continued the practice of earlier years, when the LCC received all the general grant accruing to Inner London.

The scheme undoubtedly achieved its primary object of maintaining the pre-1965 pattern of rate poundages and in fact a small reduction in the range of rate poundages was achieved as can be seen below:

Table 10.3

Rate Poundages

	Metropolitan Boroughs 1964–5			Inner London 1965–6		
	s.	d.	p.	s.	d.	p.
Highest	9.	4.	46.7	10.	8.	53·3
Lowest	7.	2.	35·8	8.	8.	43·3
Range	2.	2.	10·8	2.	0.	10·0

However, the authorities who 'gained' and 'lost' under the scheme did not correspond with those authorities generally regarded to be the 'richer' and 'poorer' London boroughs. Professor Ilersic later remarked that the outcome did 'not accord with what might be expected from such a scheme'.[18] While the 'richer' central boroughs of Kensington and Chelsea and Westminster contributed to the scheme, so also did the 'poorest' Inner London borough, Lewisham. Amongst the recipients, some of the poorer boroughs such as Tower Hamlets and Southwark received payments but the wealthy City of London and Camden were also amongst the beneficiaries.[19] The reason for these anomalies was simply the fact that while 75 per cent of a London borough's expenditure was eligible for equalization purposes, so also was the whole of the general grant that a borough received, which was of course population-based.

This scheme was, therefore, a positive attempt to maintain as far as possible the pre-1965 pattern of rate poundages in Inner London and did not commit the new authorities either to the principle of equalization or to the inclusion of Outer London in the scheme which metropolitan boroughs had previously enjoyed.

As early as October 1964 a move had been made by one Outer borough to have the principle of equalization in Greater London discussed by the London Boroughs' Association. However, as has been shown, the Treasurers at this time believed that Inner London would experience the most substantial rate increases, and the effect upon outer boroughs of having to bear the full burden of education expenditure was not fully appreciated.

In the course of the next two years the 1965 equalization scheme was criticized on a variety of grounds. On the one hand it was criticized by some of the 'poorer' contributing authorities who considered their needs to be greater than those of Camden or the City and on the other hand, the fact that it was an expenditure-based scheme led to the traditional criticisms of these types of schemes which had been levelled at the 1959 scheme. One of the strongest criticisms came from Kensington and Chelsea which paid in 1967-8 nearly £2 million into the Inner London pool; their argument rested on the fact that because the resources of this borough were primarily domestic property and those of the City of London commercial, this latter class of property was in fact receiving a subsidy from the domestic ratepayer.[20]

Outer London boroughs also criticized the Inner London scheme on the grounds that some of the receiving Inner London authorities had rateable resources in excess of their own. Moreover, within a few months of the new authorities assuming responsibility for education in their area the full impact of this costly service became apparent. Professor Ilersic was later to describe Havering's education service as 'an expensive luxury'.[21] It was argued that the Herbert Commission had intended some of the less wealthy outer boroughs to receive some form of aid from the central authorities through the equalization mechanism.

It was against this background that the London Boroughs' Association invited Professor Ilersic in April 1967 to prepare 'an impartial and objective' investigation of rate equalization in Greater London. The report took six months to complete and was put before the LBA in December 1967.

The report[22] argued that there was a clear case on economic and social grounds for treating the inner and outer boroughs as a single entity for administrative, planning and financial purposes. Given the disparity between the resources and needs of the individual boroughs 'some form of equalization is essential'.[23]

Moreover in the past, said Ilersic, London equalization schemes had paid too much attention to reducing the disparity in rate poundages throughout the Inner London area, but the only effective way of achieving this object was to pool 100 per cent of all expenditure and charge this rateably over all London. Previous schemes had therefore the nature of percentage grants which reduced the incentive to economy with the criticisms outlined above.

Ilersic proposed a new scheme[24] which, although described as a rate equalization scheme, aimed more directly at a re-distribution of resources from the richer to the poorer boroughs. The scheme was based on an assessment of an authority's 'resources' and 'needs' according to objective criteria, thus overcoming the disadvantages associated with percentage-based schemes; any economy achieved by a London borough in its revenue expenditure would result in a positive saving rather than a reduction in the equalization grant for which the authority was eligible. The scheme was accepted by the LBA, and the Minister of Housing agreed in February 1968 to implement it under the powers given him in the 1963 London Government Act. However, in view of the strong opposition from a small number of authorities who would not benefit by the full implementation[25] of the scheme the Minister decided to establish a working party to 'consider whether any modifications should be made or some alternative scheme might be preferable'.[26]

The result of the scheme in the first year of full operation is set out in Table 10.4, one of the main results being for some £4·3 million to be transferred from the richer Inner boroughs to Outer London. In Inner London only four authorities now contribute to the scheme, but these correspond to those Central London boroughs with the highest concentrations of rateable value. In Outer London, three authorities contribute more than they receive from the pool as a result of higher commercial and industrial values within their boundaries. On the other hand, all the others receive net sums from the pool of which Lewisham and Wandsworth in Inner London, and Bexley and Havering in Outer London are the most significant.

The Ilersic scheme also included as an integral feature a transfer to the ILEA of 40 per cent of the needs element of the Rate Support Grant accruing to Inner London boroughs. This proposal was intended to counter the anomaly already mentioned whereby the ILEA did not receive Rate Support Grant. Although this was merely a bookkeeping operation from the ILEA point of view – the increased grant income making possible a reduction in the education precept upon Inner London – the effect upon Inner London boroughs was to modify the impact of the new equalization scheme. This occurred because those boroughs with relatively low education 'needs' were also those with higher rateable values, and thus gained more through the reduction in the ILEA precept, while the loss of 40 per cent of the Rate Support Grant needs element meant more in cash terms to those boroughs with higher education 'needs', who also tended to have lower rateable values per head.

Table 10.4

Greater London Rate Equalization
Payments and Receipts in respect of financial year 1971–2

		£	Gain d	Gain p	Loss d	Loss p
Inner London						
Paying Authorities						
City of London		4,709,681			25·0	10·4
Camden		1,122,976			8·4	3·5
Kensington & Chelsea		891,275			8·8	3·7
Westminster		8,699,670			21·2	8·8
	Total	15,423,602				
Receiving Authorities						
Greenwich		1,230,784	24·6	10·3		
Hackney		1,301,372	22·7	9·5		
Hammersmith		727,246	13·1	5·5		
Islington		507,752	6·6	2·8		
Lambeth		1,420,852	17·5	7·3		
Lewisham		1,820,203	37·5	15·6		
Southwark		1,510,429	20·8	8·7		
Tower Hamlets		703,432	12·1	5·0		
Wandsworth		1,831,617	29·0	12·1		
	Total	11,053,687				

		£	Gain or Loss as Rate Poundage			
			Gain		Loss	
		£	d	p	d	p
Outer London						
Paying Authorities						
Ealing		233,385			2·4	1·0
Hillingdon		44,900			0·6	0·3
Hounslow		161,371			2·4	1·0
	Total	439,656				
Receiving Authorities						
Barking		127,425	3·4	1·4		
Barnet		51,986	0·6	0·3		
Bexley		436,068	9·9	4·1		
Brent		87,779	1·1	0·5		
Bromley		603,029	9·1	3·8		
Croydon		301,505	3·7	1·5		
Enfield		195,673	2·6	1·1		
Haringey		426,161	7·8	3·3		
Harrow		301,391	6·5	2·7		
Havering		530,934	10·9	4·5		
Kingston upon Thames		15,016	0·4	0·2		
Merton		159,396	3·5	1·5		
Newham		315,971	5·4	2·3		
Redbridge		410,465	7·8	2·3		
Richmond upon Thames		142,381	3·2	1·3		
Sutton		226,144	5·8	2·4		
Waltham Forest		478,247	10·1	4·2		
	Total	4,809,511				

Net Transfer from Inner to Outer
London   4,369,915

If the scheme is to be seen as one of re-distribution rather than as an exercise in reducing rate poundages, the total re-distribution under the pre-1965 scheme may be compared with the Ilersic scheme. In 1964-5 the paying authorities contributed a net £7·8 million towards the costs of the remaining 22 metropolitan boroughs. In 1971–2 (the year when the Ilersic scheme became fully implemented), the four Inner London paying boroughs contributed more than twice this amount, of which £4·3 million was transferred to Outer London. This must be set against the transfer of responsibility for health and welfare services to the London boroughs and for education to Outer London boroughs. These are some of the major revenue expenditure generating services. Moreover, expenditure also varies considerably amongst the boroughs, according to differences in school population and numbers of aged. The Ilersic scheme attempts to compensate for these factors by giving an additional weighting to these elements in a borough's total population when assessing need. However, with the loss of the LCC's equalizing role, a greater proportion of revenue expenditure is being borne today over smaller rateable areas than prior to 1965.

The question must also be asked if such a complex scheme is really necessary for the amounts which are received by some boroughs. The Herbert Commission advocated a selective scheme which would merely reduce the range of rate poundages but would not touch upon the finances of the great majority of boroughs. This scheme was unworkable; the Ilersic scheme on the other hand attempts to re-distribute rateable value rather than reduce rate poundages.

In order that such a scheme should be seen to be equitable it must apply to all London boroughs which necessarily means a complex and time-consuming exercise for which some boroughs have little to show in return. On the other hand, those boroughs where reorganization produced some of the sharpest rate increases in 1965 benefit substantially and are enabled to have some share in the wealth of the central area. Given the existing distribution of functions and the wide disparity of resources amongst the boroughs some arrangements would have been necessary if only to assist in the financing of the education service which was proving expensive for some outer boroughs in 1965 as Table 10.5 indicates:

Table 10.5

Education – Rate Poundage before deducting general grant 1965–6

	s.	d.	p.	Change from 1963–4 d. p.
Outer London Boroughs				
Highest (Havering)	10.	6.	(52·5)	+ 20 (8·3)
Lowest (Ealing)	5.	9.	(28·7)	− 1 (0·4)
Average	7.	7.	(37·9)	
Inner London Education Authority	4.	7.	(22·9)	+ 7 (2·9)

In early 1965, some suggestions were being made that all, or some proportion of education expenditure should be met by a common rate in the pound over Greater London. This would have been unacceptable to most boroughs, but an equalization scheme by which more boroughs benefited than contributed proved acceptable. The fact that an equalization scheme had existed before 1965 in the administrative county, and that the Herbert Commission had favoured the continuance in principle was undoubtedly important in this approach being adopted.

Quite apart from the additional work which the devising of a new equalization scheme and its operation have brought to the boroughs, the importance of this discussion is that it illustrates how change in one direction can bring unintended effects in another. The reform of London Government changed the areas of local authorities and the distribution of functions between them. The maintenance of an approximation to the pre-1965 equalization scheme in Inner London had the effect of distorting the expenditure patterns of some of the Inner London boroughs. The introduction of a new scheme applicable to the whole of Greater London, although it had little effect on the finances of most boroughs, was significant for some either because they were paying out appreciably more than they were under the previous scheme (Camden, Westminster and the City of London) or because they were receiving more (eg Lewisham and Wandsworth were contributing and are now receiving authorities) or less as in Tower Hamlets where the change to the new scheme meant an increase in the rate poundage of more than 5 new pence. It is impossible to assess these effects in any precise manner since we do not know whether any individual borough would have altered its pattern of expenditure (and if so in what ways) if these changes had not been made. This is an illustration of the imponderable factors which have to be taken into account in discussing the financial consequences of the reforms.

Two further points should be mentioned in this section. The first is the transitional assistance that was given to the truncated counties, of note since it was one of the few financial questions which was discussed during the Parliamentary passage of the 1963 Act.[27] In its final form, the Act provided for any excess over a 5d rate burden imposed on the counties as a result of reorganization to be met by Greater London as a whole; in 1965-6 the excess was to be met in total, but was to diminish by one-eighth each year over eight years.[28] In fact, Hertfordshire did not qualify for assistance, the excess being less than a five penny rate; the assistance for the three other counties is set out below:

Table 10.6

	Additional rate burden	*Additional rate burden in excess of a 5d. rate*		
		Amount	*Rate in the £*	
	£,000	*£,000*	*d.*	*p.*
Essex	2,844	1,910	10	4·2
Kent	1,378	375	2	0·8
Surrey	1,621	587	3	1·2
Total Assistance paid by GLC 1965–6		2,872		

This amounted to 1s. 2d. (5·8p) rate over the whole of Greater London in 1965–6.

The second point is the arrangements made in the 1963 Act for capital expenditure by the GLC. The Council, like the LCC before 1965, was required to sponsor an annual Money Bill[29] which, once approved by the Treasury and passed by Parliament, enabled the authority to borrow capital money without seeking individual loan sanctions. The Royal Commission had favoured special facilities for the GLC and on the basis of the Ministry of Housing and Local Government's oral evidence, suggested that a similar procedure might be adopted without the complexity of securing the passage of an Act of Parliament every year. The Ministry had thus recognized that effective control of the LCC's capital expenditure rested in their annual discussions with the Council prior to seeking Treasury approval.

In the event, the Act made provision for an annual Bill to be passed by Parliament, and the form of the GLC Bill differed in few respects from that of the LCC's Bill. Following the LCC procedure three schedules were to be attached to the Bill, the first comprising the estimated capital expenditure of the Council's spending committees, the second, the estimated loans the Council would be making in the period for which it was seeking approval, and the third schedule containing an agreed figure which would permit a reserve to cover unforeseen expenditure. The advantages of this procedure from the Treasury's point of view are that the reserve can only be drawn upon with Treasury approval; the use of this latter device is thus very similar to the traditional form of loan sanction to which every other local authority is subject.

The GLC Money Bill was to include money required by the GLC for its own capital expenditure and for that of the ILEA. Thus while education in Inner London remained an ILEA responsibility it maintained the advantageous position that it had held prior to 1965 with regard to school building of being able to build once it had received approval of its plans by the DES without waiting for loan sanction.[30] Although power was also given in the 1963 Act for the GLC to lend capital sums to the boroughs, a power which the LCC had possessed but had not exercised to any extent since before the war, this has not yet been used.

The costs of reform. In this section we are mainly concerned with two questions. First, has there been increased expenditure on the provision of services which is attributable to the reforms? Secondly, what is the significance of any such increase?

In considering these questions the first aim is to identify changes in expenditure since 1965 and to compare them as far as possible with the pre-1965 situation. Next, increased expenditure which is due to factors other than the reforms (eg inflation) must be allowed for. Finally, having identified increased expenditure which is attributable to the reforms, it is necessary to judge how far such increases are justifiable.

This final stage in examining these two initial questions is the most crucial of all. One effect of reform may be to provide a stimulus for the development of particular services either through increased demand or through increased provision to meet existing demands. This will almost inevitably mean increased expenditure. Part of the answer to the question will then be to see whether services have been provided economically – whether, that is to say, the increased expenditure has been matched by increased standards of provision of services, as measured in the ways indicated earlier in the functional chapters.

There is, however, more to the question than this, and one cannot avoid value judgments about the way in which particular services should develop. Take, for example, the provision of small residential homes for elderly people; as was suggested earlier,[31] the London reforms led to or at least provided the occasion for concentrating on the provision of smaller homes to replace the large homes which some of the boroughs inherited. Such a policy is bound to be expensive and might well lead to increased expenditure over and above what could be accounted for, in quantifiable terms, in improved service. Yet the policy might be justifiable precisely because although more expensive than previous policies it was more conducive to the well-being of the elderly people who benefited from it even though such advantages could not be directly measured.

There are, in other words, two basic problems: first to identify the costs of reform, and, secondly, to try to interpret the meaning of those costs. The problems are familiar; increasing emphasis has been devoted in recent years to attempts to throw light on the relation between cost and efficiency in the provision of local government services, and on the factors which are of most importance in the provision of a high standard of service.[32] These analyses make use of developed statistical and other techniques, but it seemed neither possible nor appropriate to attempt anything as ambitious as an examination of this kind would require for the present somewhat broader inquiry. Of more immediate relevance is the study commissioned by Cheshire County Council in 1969;[33] this did examine, among other things, the increased costs of providing certain services in Greater London following the reforms and the extent to which these increased costs were not matched by a corresponding improvement in the standard of service provided. Its concern was to identify the once-for-all cost of reform. Further reference to this study is made later.[34]

The following account will largely be concerned with three aspects of the financial effects; first, total rate-borne expenditure; secondly, statistical data relating to a number of specific services; and thirdly, a broader examination of three services where reform might have been expected to have a direct effect on the cost of providing the service. The emphasis will thus be largely on questions of revenue expenditure. This is because it is here that the financial effects of the reforms are likely to be felt most immediately. Capital expenditure by local authorities is greatly affected by the general economic situation and by specific central government action which has a direct effect on local authorities through the system of loan sanction. In the period since 1965 interest rates have been at a high level, and government policy has generally operated to restrain the growth of local authority capital expenditure. Thus it is practically impossible to isolate in the capital expenditure field the effects of the London Government reforms. As was indicated earlier,[35] for example, capital expenditure by the boroughs on providing homes for the elderly since 1965 has fallen short of borough plans and a similar point is discussed later in relation to GLC responsibility for the ambulance service.[36]

There are, however, certain questions about capital expenditure which are relevant to the general discussion. There is, first, the question of the relation between capital and revenue expenditure. To the extent that authorities make revenue contributions to capital expenditure, an examination of revenue expenditure alone may not give a true picture. This will arise most obviously where the new London authorities have adopted a different policy from their predecessors. The effect of any such changes is likely to be small in the case of the London boroughs since the effect of a change of policy by one borough individually would not be of great significance in relation to total borough expenditure, and there is no evidence of any substantial change in policy by the boroughs as a whole.

The GLC is in a different position because of its very large capital programme of over £100 million annually.[37] The GLC has followed the LCC in making specific provision in the revenue estimates for a contribution to capital expenditure and in financing small capital projects from revenue. As Table 10.7 indicates, the amounts have fluctuated quite considerably over the years, but in no one year has the revenue contribution by the GLC to capital projects amounted to more than a 2d rate over the whole of Greater London.

Also relevant here is the operation since 1965 of the special arrangements for GLC capital financing through the annual Money Bill, and the effect of the 1963 Act on the GLC's capital programme. The process involved in receiving Treasury and Parliamentary approval for the Money Bill would not seem to have changed greatly since the days of the LCC. Before being presented to the GLC Finance Committee,[38] spending committee estimates are discussed with central government departments, and the degree of interest displayed by the departments is directly related to the amount of grant (if any) that is involved for any one project. The major grant receiving capital projects – highways, housing and sewerage works – are subjected to the same yardsticks or unit costs standards that are applied nationally. In other services where the GLC is meeting the majority or all of the cost themselves, it appears that more information is now being required by government departments on the nature of the individual projects for which the Council is seeking borrowing powers in the Money Bill. Not only does this reflect the increasing volume of expenditure for which Parliament is being asked to give sanction

N

Table 10.7

Revenue Contribution to Capital Expenditure by LCC and GLC

	Revenue Contribution to Capital Expenditure £		Items of Capital nature financed from revenue account £	
LCC	*Total*	*Amount allocated to Education Committee*		
1954–5	500,000		5,000	
1957–8	2,500,000		5,000	
1959–60	3,000,000	—	5,000	
1960–1	3,000,000	—	5,000	
1961–2	500,000	—	5,000	
1962–3	1,008,000	—	5,000	
1963–4	9,196,000*	4,534,500	10,000	
1964–5	2,006,000	—	10,000	
	GLC	*ILEA*	*GLC*	*ILEA*
1965–6	5,000,000	—	10,000	20,000
1966–7	1,000,000	—	10,000	20,000
1967–8	1,000,000	—	10,000	20,000
1968–9	3,000,000	—	20,000	20,000
1969–70	5,000,000*	2,500,000†	20,000	20,000
1970–1	1,000,000*	1,000,000†	20,000	20,000

* Included £5 million taken from balances.
† Estimates.

(£166·7 million in 1965 to £204·8 millions in 1970) but also the increasing interest of the Treasury in local authority expenditure as a means of national economic management. Again, the GLC is today undertaking capital projects in fields of which the Ministry of Housing has little experience as, for example, the £10 million refuse incinerator at Edmonton.

Once the Money Bill has become law, the Council can spend on its capital account up to the limits imposed by the Bill, and central government powers to reduce or cut back the Council's expenditure are more limited than for other local authorities. It is true that the LCC had in the past accepted voluntary restraint on its spending in times of financial stringency when the Government had requested a reduction in capital expenditure, and the GLC has continued this practice; but it is possible that high interest rates have had more effect in containing the growth of capital expenditure than the authoritys' willingness to comply with central government requests. Moreover, the fastest growing areas of GLC expenditure have been those which are subject to central government grants, a field in which the departments have greater influence.

Certainly, a comparison of LCC and GLC capital expenditure (Table 10.8) does not so far disclose any significant changes as a result of the reforms, except to the extent that the GLC has greatly increased both the amount and the proportion of capital expenditure on loans. The reason is simply that the major items of capital expenditure are for housing and education, and that these have remained as substantially the same responsibilities[39] under the GLC as under the LCC.[40]

In the capital expenditure field, therefore, there is little evidence of reorganization costs, that is, of additional expenditure of a temporary nature due to the reforms themselves, nor is it easy to identify any positive differences or trends of a long-term nature brought about by the reforms. The reason in both cases is essentially the same, that capital expenditure is constrained by factors beyond the control of local authorities and that consequently the effects of the reforms are submerged in the total picture.

Table 10.8
LCC and GLC Capital Account

Year	Original Estimate		Actual Net Expenditure	
	Services main-tained by the Council	*Loans to other bodies*	*Services main-tained by the Council*	*Loans to other bodies*
	£ million		£ million	
1960–61	31·76	4·85	27·42	1·69
1961–2	33·59	5·55	32·65	2·73
1962–3	37·77	3·17	37·86	3·06
1963–4	47·49	4·12	46·1	25·10
1964–5	59·26	37·25	60·27	61·37
1965–6	75·51	81·25	62·89	67·9
1966–7	77·07	51·35	65·87	47·0
1967–8	83·71	52·0	77·50	33·8
1968–9	99·34	50·0	86·39	6·7
1969–70	106·21	21·0	104·25	19·3

Source: LCC and GLC annual estimates and abstracts of accounts.

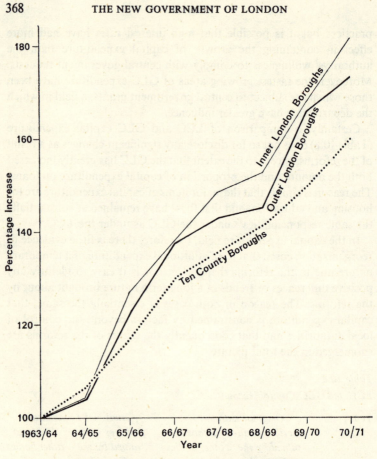

Figure 8. Rates levied in Greater London 1963–70.

In considering the effects of the reforms on the total rate-borne expenditure of the London authorities we take as our starting-point the Herbert Commission's view, already quoted, that 'our recommendations should not involve any material change in the aggregate rate expenditure in the Greater London area'.[41]

On the other hand, the Commission did suggest that whereas most boroughs in Outer London would be able to reduce their rate levies many of those in Inner London would have to increase theirs.[42] The Commission appear to have assumed that their proposals represented simply a re-apportioning of existing functions among a new set of authorities in which increased costs in some services would be offset by savings elsewhere. Such an assumption is unrealistic; furthermore, it takes no account of any additional costs which the reorganization itself might cause. It could be argued that in any event the changes made by the Government in the Commission's proposals invalidate the latter's forecasts.

Table 10.9 summarizes the rate levies for London boroughs since 1963–4, and a comparison is made with ten county boroughs outside London, not affected by reorganization, but of a similar size in terms of population to London boroughs.[43] Taking Greater London as a whole, the average increase in rates levied for 1965–6 compared with the previous year was 19 per cent whereas there was only an 8·5 per cent increase in England and Wales outside Greater London. Before attributing the difference in these increases to 'reorganization costs', the increases in the Greater London area in the preceding two transitional years should be examined.[44]

Table 10.9

Rate poundages – Averages and Index Numbers 1963–1971

	Inner London Boroughs		Outer London Boroughs		10 County Boroughs	
	Average in d. p.	*Index No.*	*Average in d. p.*	*Index No.*	*Average in d. p.*	*Index No.*
1963–4	91·4 (38·9)	100	101·68 (42·4)	100	120·3 (50·1)	100
1964–5	95·7 (39·9)	104·7	106·40 (44·3)	104·6	128 (53·3)	106·4
1965–6	116·8 (48·7)	127·8	125·62 (52·4)	123·5	141 (58·8)	117·2
1966–7	126·9 (52·9)	138·8	140·9 (58·7)	138·6	156·1 (65·0)	129·8
1967–8	136·1 (56·79)	148·9	145·40 (60·6)	143	163·2 (68·0)	135·7
1968–9	139·8 (58·3)	152·9	147·30 (61·4)	145	170 (70·8)	141·3
1969–70	153·3 (63·9)	167·7	163·30 (68·0)	160·6	180 (75·0)	149·6
1970–71	166·1 (69·2)	181·8	176·5 (73·5)	173·6	192 (80·0)	160

In the final year of the constituent authorities' existence, their rates had increased by an average of 4·5 per cent over the previous year as opposed to a national average increase in 1964–5 of above 6 per cent. In part this lower increase can be attributed to the reluctance of the old authorities to embark on new schemes, but the rate of increase of expenditure did not fall off in the same way. In order that no single authority should contribute more than its share to the new London borough, a practice was generally adopted by the old authorities in their final year of reducing their balances in their general rate fund revenue account whilst raising a small contingency reserve as a proportion of its rateable value. By such means, it was hoped that each of the constituent authorities would contribute to the same extent to the new London borough.

In some cases, the contingency fund was exhausted by the beginning of 1965–6, and the new London boroughs took over only the remainder (if any) of their transitional rate. Thus many of the new London boroughs began life with little or no working balance and above national rate levies were made in the first year to compensate for the depleted balances of the constituent authorities.

Since 1965, however, annual rate increases have exceeded those in the rest of the country, and the increase in Inner London has been higher than that of Outer London. By 1970–1 rates had increased in Inner London by 82 per cent since 1963–4 as compared with 74 per cent in Outer London and 60 per cent in the ten county boroughs outside London. Even between 1966–7 and 1970–1, whereas rates went up by 18 per cent in the county boroughs, they rose by 22 per cent in Outer London and by 31 per cent in Inner London. In other words the rate of increase is still higher in Greater London, especially Inner London although many of the initial reasons for high expenditure in London have diminished; for example, many of the appointments which were originally 'double banked' have been reduced as some chief officers have retired. Communications between a borough's departments have improved as more space has become available and reorganization completed. Management service units have been established by individual authorities as well as the London Boroughs' Management Services Unit; all have analysed the organization of departments or individual services and claimed financial savings as a result of their proposals. Moreover, in 1967 and 1968 Conservative councils were elected in the Greater London Area to 'wage war on waste' and reduce unnecessary spending by London local authorities.

In other words, even allowing, as the Herbert Commission did not, for some costs directly attributable to reorganization, it seems that several years after the reform took place the London authorities' expenditure was still increasing at a higher rate than that of authorities which had not been affected by reorganization. Two general ways in which London differs from the rest of the country may be mentioned as being relevant here. First, the proportion of total expenditure borne by rates as opposed to grants is much higher; in 1965–6 rates met 66 per cent of expenditure in Greater London against 47 per cent in the country as a whole; on the other hand changes in grant structure introduced by the Local Government Act, 1966, had the effect of shifting the burden away from rates proportionately more in London than elsewhere so that in 1967–8 the figures were: Greater London 61 per cent, England and Wales 45 per cent. Thus the relative disparity between London and elsewhere shown in Table 10.9 would have been even higher if the rate-borne proportions had not changed. Secondly, staff costs, which form the major item in local government revenue expenditure, have increased proportionately more in London than elsewhere since 1965. As the detailed note on the fire service indicates,[45] several wage agreements have resulted in an absolute increase in the London weighting allowance which has not yet been compensated for in the rate support grant formulae. Some part of the increased expenditure of Greater London compared with elsewhere shown in Table 10.9 is accounted for by this fact.

In addition, comparisons between the pre- and post-1965 situations are bound to be complicated by the fact that additional responsibilities were placed on the GLC by the 1963 Act which had not previously been discharged by local authorities. The planning responsibilities of the GLC for example, were wider in scope than those of the previous counties, and traffic management, previously the responsibility of the Ministry of Transport, was transferred to the GLC. Thus some increased cost was bound to be reflected in Table 10.9 for this reason alone; it must be said, however, that such costs are relatively small when compared with the high-spending services such as education, housing and the health and welfare services.

Table 10.10 gives an indication of the demand made by individual services upon revenue since 1965, and makes a comparison with authorities elsewhere in the country. Several features should be mentioned, although discussed in detail elsewhere.

First the remarkable increase in expenditure under the heading 'Town and Country Planning' in both Inner and Outer London boroughs. This represents expenditure incurred by borough councils and the GLC in establishing their new planning departments. However, even in 1969 this item only accounted for 2 per cent of total revenue expenditure.

Secondly, the increased expenditure by London boroughs on welfare, children's and personal health services has been above that of county boroughs. Earlier chapters have already implied that these services received a stimulus from the reorganization and that this was first apparent in the welfare services.

Thirdly, there has been an increase in the cost of refuse disposal in Outer London above the rate in Inner London and the ten county boroughs, although total expenditure per head remains highest in Inner London. This increase would seem to be a result of transferring disposal functions to the Greater London Council and the consequential charging for this service over the whole of the Greater London area.

Other significant features of the table are the higher education expenditure in Inner London and the fact that education expenditure in Outer London has been in line with that of county boroughs. The increased housing expenditure in Outer London should also be noted, although not yet at the same level as the Inner London boroughs' housing expenditure.

Finally, it should be mentioned that the Metropolitan Police precept has also increased in London at a rate above that of police expenditure elsewhere. This is despite the fact that the impact of the London Government reorganization upon the police service was negligible. It is possible that factors other than reorganization have therefore contributed to a rise in expenditure in London.

The next question to consider is in which services expenditure has increased most under the new system compared with the old. Here, the difficulties of finding adequate data become formidable. This is mainly because there is usually little or no information about the cost of county services before 1965 in areas which became part of Greater London. The main difficulty lies over the services transferred from the counties to the London boroughs. It is true that partial comparisons can be made, as for example by comparing the cost of LCC services with those of Inner London boroughs,[46] and, more approximately, by comparing Middlesex CC expenditure with that of the 9 boroughs which roughly correspond to the area of Middlesex; but no comparisons can be made for the parts of Essex,[47] Kent and Surrey which were brought within Greater London, and therefore no global figures for the whole of Greater London can be presented. The group of services for which direct comparison before and after 1965 is most easily done are those which were previously performed by the county districts and county boroughs and are now performed by the London boroughs; on the whole, however, these tend to be services where expenditure is relatively small, as Table 10.10 indicates.

Table 10.10
Average rates levied per head of population 1965–6 and 1969–70

| | 1965–6 | | | | | | 1969–70 | | | | | | % Increase | | |
| | Inner London | | Outer London | | 10 County Boroughs | | Inner London | | Outer London | | 10 County Boroughs | | Inner | Outer | 10 County Boroughs |
	s	£p	s	£p	s	£p	s	£p	s	£p	s	£p			
Education	442	(22·10)	459	(22·95)	481	(24·05)	701	(35·01)	667	(33·35)	689	(34·45)	59	45	43
Public Libraries	27	(1·35)	18	(0·90)	16	(0·80)	40	(2·00)	27	(1·35)	24	(1·20)	48	50	50
Sewerage	34	(1·70)	25	(1·25)	26	(1·30)	51	(2·55)	35	(1·75)	41	(2·05)	50	40	58
Refuse	40	(2·00)	24	(1·20)	21	(1·10)	59	(2·95)	40	(2·00)	31	(1·05)	48	67	48
Environmental Health	74	(3·70)	46	(2·30)	40	(2·00)	106	(5·30)	65	(3·05)	48	(2·40)	43	44	20
Local Health Authority	62	(3·10)	43	(2·15)	48	(2·40)	89	(4·45)	65	(3·05)	65	(3·25)	44	51	35
Children	39	(1·95)	12	(0·60)	19	(0·95)	63	(3·15)	20	(1·00)	27	(1·35)	62	67	42
Welfare	34	(1·70)	20	(1·00)	21	(1·05)	62	(3·10)	37	(1·85)	35	(1·75)	82	85	67
Housing	84	(4·20)	15	(0·75)	14	(0·70)	148	(7·40)	39	(1·95)	20	(1·00)	76	160	43
Planning	13	(0·65)	12	(0·60)	14	(0·70)	34	(1·70)	28	(1·40)	24	(1·20)	162	133	71
Highways & Public Lighting	88	(4·40)	71	(3·55)	67	(3·35)	112	(5·60)	96	(4·80)	85	(4·25)	27	35	27
Fire	28	(1·40)	17	(0·85)	19	(0·95)	38	(1·90)	23	(1·15)	25	(1·25)	36	35	32
Police	154	(7·70)	96	(4·80)	76	(3·80)	244	(12·20)	147	(7·35)	104	(5·20)	58	53	36
Other	148	(7·40)	102	(5·10)	58	(4·90)	182	(9·10)	127	(6·35)	75	(3·75)	23	25	29

Source: IMTA Return of Rates

It is thus unfortunate that many of the most important services from a financial point of view are precisely those which were transferred from the counties to the new London boroughs and for which accurate figures are unobtainable. Nevertheless, because they are so important we consider these services first.

Education. Besides being much the largest spender of all local government services, education is a special case in Greater London because of the continuity between the LCC education service and the ILEA. Thus in Inner London the reforms would not have been expected to give rise to additional expenditure. If net revenue expenditure is taken to be 100 in the base year of 1963–4 we have the following comparison:

Table 10.11

	1963–4	1966–7	1967–8	1968–9
LCC	100			
ILEA		126·3	139·3	150·1
East Ham and West Ham	100			
Newham		135·2	144·8	151·6
10 County Boroughs	100	142·2	151·6	160·4

Source: Based on IMTA statistics

Thus in Inner London as well as in Newham total expenditure has risen by rather less than in the county boroughs outside London which were not affected by reorganization. Total figures of this kind are not, however, very meaningful. On the basis of the annual statistics published by the IMTA and the SCT the following comparisons can be made:

Table 10.12

Average Net Expenditure per 1,000 Population on Education

				% increase on	
	1963–4	1966–7	1968–9	1963–4	1966–7
(1) *Total Net Expenditure*	£	£	£		
LCC	22,006				
ILEA		31,253	34,768	26	11
Middlesex	17,765				
Middlesex LB's		25,903	30,104	31	14

	1963–4	1966–7	1968–9	% increase on 1963–4	% increase on 1966–7
East Ham and West Ham	19,990				
Newham		27,116	30,827	35	14
Essex	20,887	29,689	33,679	42	13
Essex LBs		26,041	30,112	25	16
Kent	19,876	25,623	30,233	29	18
Kent LBs		26,887	31,412	35	17
Surrey	18,682	27,016	31,749	45	18
Surrey LBs		25,777	31,103	38	21
10 County Boroughs	19,537	26,220	31,686	34	21

(2) *Administration and Inspection*

	1963–4	1966–7	1968–9	% increase on 1963–4	% increase on 1966–7
LCC	876				
ILEA		1,541	1,965	76	28
Middlesex	739				
Middlesex LAs		996	1,048	35	5
East Ham & West Ham	816				
Newham		939	1,158	15	23
Essex	664	1,056	1,159	59	10
Essex LBs		962	1,140	45	19
Kent	616	877	1,057	42	21
Kent LBs		816	1,005	32	23
Surrey	498	962	1,132	93	18
Surrey LBs		1,148	1,510	130	32
10 County Boroughs	612	821	994	34	21

The table draws attention to several important facts. First, what seems like a remarkable increase in the administration and inspection costs of the ILEA compared with the LCC, although this turns out to be not so remarkable when one discovers that the ILEA introduced a change in their methods of classifying expenditure. Since 1966, however, the ILEA's administration and inspection costs have continued to increase at a rate above that of the 10 county boroughs outside London, and of most Outer London boroughs. Secondly, the increase in administrative costs in the Middlesex boroughs does not appear to have been any greater than that experienced by the 10 county boroughs, and indeed since 1966 the increase in Middlesex has been distinctly lower than other education authorities. Thirdly, there are the considerable, and in the case of Surrey, very large increases in the costs of administration and inspection in the counties which lost territory to Greater London in 1965. Surrey London boroughs also experienced high increases, a fact which would appear to be reinforced by the Local Government Operational Research Study.[48] Fourthly, there is the very small increase in administration and inspection costs in Newham,[49] the one education authority formed by amalgamation of two existing education authorities in the London reforms.

Personal health and welfare.   Some data on these services were given in an earlier chapter[50] and indicate that net expenditure per 1,000 population has risen very sharply in the welfare services when the London boroughs are compared with county boroughs elsewhere in the country; in the health services there is a somewhat different picture of relatively small increases in expenditure in the first two years of the new system, followed by much larger increases since, particularly in Inner London. The analysis presented in the chapter emphasizes the existence of very great variations between the boroughs.

Children's service.   Total net revenue expenditure can be compared in a similar way to that for education shown earlier:

Table 10.13

	1963–4	1966–7	1967–8	1968–9
LCC	100			
Inner London Boroughs		139·8	155·2	172·2
East Ham & West Ham	100			
Newham		120·0	137.7	151·6
10 County Boroughs	100	142·2	160·2	176·0

Source: Based on IMTA statistics.

Thus the increase in total expenditure in Inner London has kept pace with the increase in the 10 county boroughs, whereas in Newham the increase has been much less.

Housing.   Of the services which were amalgamated as a result of the 1963 Act, perhaps the most important is housing. Unfortunately it is also a service in which it is very difficult to make reliable comparisons of costs. In the housing chapter, expenditure on supervision and management was examined in an endeavour to see how far the reforms had directly affected administrative costs. The results[51] showed that more than half the new London authorities had increased administrative costs exceeding the national average, and on the whole these increases appeared to be greater in Outer London boroughs.

In housing, too, we encounter another difficulty which has a bearing more generally on the London situation in comparison with the rest of the country. This is the very high cost of providing housing, particularly in Inner London, resulting from high land values and building costs as well as from the high interest rates which affect all local authorities. The important point is that the relative position of London appears to be getting worse, so much so that Camden and Lambeth have recently argued that without further Government measures to help the Inner London boroughs, it will soon become very difficult for them to maintain a housing programme at all.[52] No doubt the position has been aggravated by the sustained attempt to expand the London housing programme vigorously at a time when economic circumstances have been particularly adverse.[53] We return to this point later in general discussion of the financial effects of the reforms.

Of the other services now provided by the London authorities as a result of amalgamation we deal separately with the two GLC services of the *fire brigade* and *ambulance service* and with the borough *library service*.[54] We deal here with *refuse disposal*.

Refuse disposal. The Herbert Commission had argued that the Council for Greater London should be responsible for refuse disposal, and emphasized that transport costs were the major item in the metropolitan borough's expenditure. They claimed: 'we find it impossible to believe that the elimination of the criss-crossing of refuse *en route* to the tipping grounds would not reduce the average mileage and the cost'.[55]

While the GLC established an organization to carry out this function, responsibility was delegated to the London boroughs for two years; a phased transfer of the service was undertaken, and the GLC assumed full control of refuse disposal services in April 1967. The financial year 1967–8 was, therefore, the first year in which the economies of scale forecast by the Royal Commission might become apparent. Expenditure on the service is set out below, and a comparison is made with 10 of the larger county boroughs in the rest of England and Wales.

Table 10.14

Gross[56] *Expenditure – Refuse Disposal*

	1965–6	1966–7	1967–8	1968–9	1969–70
	£m	£m	£m	£m	£m
GLC	3·5	4·4	4·9	5·5	6·7
Index	100	127	134	149	193
10 County Boroughs[57]	1·7	1·9	2·2	2·3	2·4
Index	100	112	128	130	138

Expenditure on the service has increased more rapidly in London than in the rest of the country; moreover, there has been a tendency for expenditure in London to rise more sharply in recent years. Part of the explanation for this increase in revenue expenditure lies in the comparatively high capital expenditure by the GLC on refuse disposal.

Table 10.15

Capital Expenditure – GLC 1967–71

	1967–8	1968–9	1969–70	1970–71
£ million	2·078	3·403	3·544	3·385*

* estimate.

This expenditure has mainly been on the provision of an incinerator at Edmonton, and the high costs of building this plant has been reflected in the debt charges included in the revenue account, which by 1970–1 accounted for some 16 per cent of gross revenue expenditure on the service. The plant itself will dispose of refuse from 8 London boroughs, equivalent to 14 per cent of the total refuse handled by the GLC, and it is hoped that each year some £500,000 will be recouped from the sale of electricity generated.

The costs of refuse disposal in London are therefore higher than in other parts of the country, but as the Herbert Commission pointed out, the problems in London of finding tipping space and of transport are greater than anywhere else in the country. Moreover, some movement has been made towards devising a more 'comprehensive and economical' system for Greater London[58] as the Commission advocated.

Perhaps the most conspicuous omissions from this survey of available data about local government services in London are *highways and traffic* and *planning*. These are services for which the available information does not provide an adequate basis for an effective comparison of costs.[59] This indicates once more the magnitude of the task of trying to make a service-by-service comparison of costs before and after reform, let alone to try and interpret comparative costs in terms of the effect of the reforms. All the more interesting, then, is a recent study by the Local Government Operational Research Unit which attempts to isolate the cost of the London reorganization.[60]

Local government operational research unit study. Briefly, the method adopted in this study was to examine administrative costs[61] in the geographical area of the 5 counties[62] involved in the London reorganization for a number of services and to compare them with a control group of 27 counties which were not affected by reorganization. This indicated where administrative costs in the 5 counties had risen more than in the control counties; the next stage was to assess by means of a component analysis whether any of the increased cost could be attributed to factors other than the reorganization, such as inflation or increased standards of service. The final stage, having identified reorganization costs, was to apportion these costs between the parts of the split counties (Essex, Hertfordshire, Kent, Surrey) which were included in Greater London in 1965 and those which remained outside.

In three of the four services examined (education, welfare, local health) the study found that there had been significant increases in administrative costs in the three years 1965–6 to 1967–8, particularly in welfare; in the fourth service (housing) there had either been no significant change or (as in Kent) a significant decrease in costs.[63] When allowance was made for inflation and other factors, it was again found that the significant increases in administrative costs attributable to reorganization were largely found in the welfare services; in education and local health, increases were found only in Surrey.[64] Finally, where increased costs attributable to the reorganization were found in the split counties, these fell largely on the parts transferred to Greater London in 1965 rather than on the reduced counties.[65]

In considering these findings, the object of the study has to be borne in mind. This was to see whether it was possible to estimate what the cost would be of introducing the Redcliffe-Maud proposals in Cheshire. The study was not therefore attempting to cost the

London reforms but rather to discover a means of applying relevant experience from London to the situation as it might be in Cheshire.

This is certainly not to deny the importance of the study in the context of the present analysis. On the contrary, it is the most valuable analysis which has yet been made of the costs of local government reform, and most illuminating in the light which it throws on the questions raised here. Yet in the last resort the findings of the study must be used with caution and supplemented by more subjective assessments of the situation.

There are a number of reasons for saying this. First, although it deals with some important services, the study does not claim to be at all comprehensive, either in coverage of services or in applying to the whole of Greater London. Secondly, a crucial stage in the argument is the selection of indices for the measurement of standards of service; as is well known, the selection of such criteria is both difficult and controversial. Thirdly, the apportionment of reorganization costs between the parts of counties now included in Greater London and those remaining outside has been made on a population basis, on the assumption that within a county such costs are constant when related to population; this may serve to distort the increase in costs which has taken place within Greater London since 1965.[66]

Nevertheless, there are obvious ways in which the study links up with what has been said earlier in this section. There is, for example, the increased administrative costs for education in Surrey; as was indicated earlier, the cost per 1,000 population in the 'new' Surrey is very much higher than in the 'old' Surrey of 1963; the study suggests that Surrey has indeed incurred reorganization costs, but that they have fallen mainly on the London boroughs formed out of Surrey rather than on the county council. Yet it is probable that Surrey had a particularly difficult job in reorganizing its education service after 1965 because of the way in which the London reforms cut into the pattern of divisional organization. What seems clear from all the evidence is that reorganization of education in Outer London has brought additional costs either to the boroughs or to the counties and that these costs cannot be entirely offset by measurable improvements in the standard of service.

Doubts arise over two questions; first, whether other consequences of the reforms, though not precisely measurable may also help to offset increased costs. There is, for example, the evidence of the education chapter[67] that the outer boroughs make more intensive use of specialist advisers than the previous authorities; there is thus increased cost but improved quality of service which does not appear in the indices of measurement. Secondly, there is the time factor. The study dealt only with the three years immediately following reorganization. As the London Government Act, 1963, recognized in its provisions for transitional payments to the counties,[68] there were bound to be increased transitional costs because of the reorganization, which would diminish in the course of time. In a similar way the GLC and the boroughs may have incurred additional administrative costs whose beneficial effects in the form of improvements to the quality of the services provided may only emerge over a period of time. Whether this will happen is not yet demonstrable, but it suggests that one should be cautious about the interpretation of figures of increased costs.

The basic difficulty is that the study assumes, as it has to assume, that all additional expenditure other than what can be accounted for either by inflation or by measurable increases in standards of service represents the once-for-all cost of reorganization, whereas (as was suggested on page 363 above), the true position may be much more complex.

To supplement this discussion, and before trying to draw conclusions about the whole situation, we have examined three services in more detail. They are the fire, ambulance and library services. None of these is a major service in terms of its expenditure. But they are all important 'bread-and-butter' local government services. Moreover, they are all services which might be expected to demonstrate economies of scale, two of them at GLC level and one at borough level. The object has been to consider not only what has been happening to the costs of these services, but to examine also the influences which have had a bearing on changes in expenditure. By this means it is hoped to illuminate some of the more general points which have been made earlier.

Fire service. Net revenue expenditure on the London Fire Brigade between 1965 and 1969 increased by 17 per cent compared with an increase of 28 per cent in England and Wales and 24 per cent in Birmingham over the same period. In the first of these four years the increase was higher in the GLC brigade than in the rest of the country.

Table 10.16

	1965–6	1968–9	% increase
	£m	£m	
GLC	9·6	11·2	17
England & Wales	44·6	57·0	28
Birmingham	1·0	1·3	24

At first sight this might reflect the economies which the larger authority could expect as a result of amalgamating all or part of the 9 fire services which formed part of Greater London. However, when the expenditure is examined in more detail it appears that some large increases occurred in the early part of this period, and that a number of decisions were made affecting the organization and manning of the brigade which committed the authority to operational expenditure at higher levels than the constituent brigades prior to 1965.

The practical difficulties involved in making any estimates of the costs of providing certain services in Greater London before 1965 have already been emphasized. This section will, therefore, attempt to identify the financial implications of reorganization upon the fire service and identify those areas where economies might be expected.

The GLC inherited on 1 April 1965 a varied pattern of communications, stations and equipment. A major initial problem was the organization of the new London Fire Brigade. It was decided not to operate on a divisional basis from a central control (as had been the practice in the old counties) but with a system of three commands comprising eleven divisions.

Since 1965, many of the organizational problems which dictated the original command structure have been overcome; for example, the brigade has been able to centralize its communications network. It is possible to envisage therefore the command structure being eliminated from the organization in the future. The study of the potential costs of reorganization in Cheshire as a result of the Redcliffe-Maud proposals considered the need for a new intermediate level of administration in the Merseyside metropolitan area[69] fire brigade, and decided it was unnecessary; however, they concluded 'if a command level were introduced, staffing and accommodation costs would be substantially above the estimates made in this Study'.[70] The establishment scheme initially drawn up by the GLC in 1964 was based on the assumption that there would be full manning on 85 per cent of all calls (85 per cent confidence level). This reflected the practice which the LCC had with difficulty introduced in 1961 and was a standard of attendance above the 75 per cent which the Home Office in a circular of 1958[71] indicated as acceptable. However, the LCC scheme was based on a 56-hour week, the duty rota being worked by all the constituent authorities in Greater London with the exception of Croydon County Borough which worked a 48-hour week at a 75 per cent confidence level. In order to adopt the best practice current among the constituent authorities the establishment scheme had to be calculated on the basis of an 85 per cent confidence level while taking the opportunity to introduce the basic 48-hour week (with 8 hours voluntary overtime) over Greater London. However, when the staff numbers available for the new brigade were known, it became apparent that the service could not achieve an 85 per cent confidence level, which in fact could only be achieved if recruitment rates were to increase greatly. In fact, however, shortage of staff, a national problem, has meant higher overtime payments than originally estimated.

In May 1967, following a report from the Prices and Incomes Board[72] a general increase of $7\frac{1}{2}$ per cent in firemen's basic pay was awarded with specific additions to pay in order to encourage firemen to undertake longer hours of availability. A further agreement providing for an undermanning allowance and for an addition to the London weighting, was made in November 1969, the aim being to create a smaller but more highly paid force. However, it would appear that the result has been to increase the proportion of revenue expenditure devoted to wages and salaries, since this has risen from 75 per cent to 80 per cent since 1965.

In some areas of London, fire stations were sited where they were initially intended to serve the needs of much smaller areas. The GLC took the view in 1965 that the same standard of fire cover could eventually be provided in Greater London from fewer stations if these were more rationally sited, and therefore adopted a policy of building new stations and closing others. However, capital expenditure restrictions since 1966 have curtailed the number of stations built, and since 1966 more than half of all capital expenditure has gone towards the acquisition of strategic sites when they have appeared on the market, with a view to future building. In spite of these capital expenditure restrictions, the 122 stations in use in 1965 had been reduced to 115 by 1970, entailing the opening of 8 new stations and the closure of 15. This has resulted in marginal economies in station maintenance, administrative costs, and manpower establishment. A number of stations are planned to be closed in the next five years, and since these are mainly older buildings further economies can be expected.

These, and other special factors[73] make comparison with other fire authorities of little value, but the IMTA statistics of expenditure per 1,000 population do provide an indication of the areas in which a larger authority may expect financial savings. The GLC by virtue of its high rateable values and larger budget is able to undertake all but land purchase and new buildings from its revenue account, with a consequent saving in loan charges. Many county and county borough brigades use their capital allocation for the purchase of appliances, a practice followed by some of the constituent authorities in Greater London before 1965.

The major areas in which expenditure per 1,000 population rose after 1965 in the GLC as compared with the LCC were in those items which involved expenditure as a result of reorganization, for example, uniform and communication equipment. One significant element in the IMTA statistics is, however, the larger amounts spent by the London Fire Brigade under the heading 'administration and establishment'. This not only reflects the higher administrative costs of the London Fire Service but also the high 'central administrative costs' which all GLC departments must share.

Table 10.17

Fire Service – Expenditure per 1,000 population LCC/GLC and County Boroughs

Selected headings	1963–4		1966–7		1968–9	
	LCC	CBs	GLC	CBs	GLC	CBs
	£	£	£	£	£	£
Employees						
Uniformed	749	586	762	686	824	823
Civilian	45	35	45	46	22	47
Premises	104	49	116	60	115	68
Communications	13	5	17	7	17	9
Other Equipment	30	30	37	35	29	39
Transport	25	17	32	23	55	31
Administration and Establishment	83	24	111	36	131	39
Debt Charges	21	42	41	55	51	69
Total Gross	1407	918	1430	1142	1447	1247

In evidence to the recent Departmental Committee of the Home Office on the Fire Service (Holroyd Committee)[74] the GLC claimed that responsibility for the larger area enabled them to plan fire cover more effectively without regard to administrative boundaries. However, as a result of the restrictions on capital and revenue expenditures the brigade has not yet completed these plans, and decisions on the future organization of the brigade have yet to be made, but certain economies in communications, station maintenance and staff costs are already apparent and developments already planned are likely to produce greater efficiency in the future.

Ambulance service. In the four years after 1965 gross revenue expenditure by the London Ambulance Service increased by some 26 per cent, a similar increase to that experienced by the 14 most urbanized counties in England and Wales, but slightly greater than that of the 5 largest county boroughs.

Table 10.18

	1965–6	1966–7	1967–8	1968–9
	£m	£m	£m	£m
London Ambulance Service				
gross exp.	4·2	4·6	5·2	5·3
as index no.	100	111	124	126
Expenditure (gross) (14 most urbanized counties)[75]	7·2	8·0	8·6	7·2
as index no.	100	107	120	128
Expenditure (gross) (5 largest C.B.s)[76]	1·4	1·4	1·6	1·7
as index no.	100	106	117	121

Source: Ambulance Costing Returns DHSS (annual).

The problems of reorganization and the creation of a unified service facing the ambulance service in 1965 were greater than those of the fire service. This was a result of the nature of the equipment and the variety of practices which were current amongst the 9 authorities which in whole or in part constituted the new Greater London Ambulance Service. As with the fire service, the immediate financial repercussion of reorganization was a result of standardization of rates of pay and hours of duty, for although these are negotiated and

agreed nationally, each of the constituent authorities in 1965 interpreted these to suit their own local needs. Moreover, no two authorities in 1965 adopted the same procedure relating to uniform allowances; standardization of these practices inevitably meant greater
expenditure without any corresponding benefit.

The work of the London ambulance service may be divided into
two parts. On the one hand the emergency service, dealing with up
to 1,300 calls a day, requires a fully-equipped vehicle and trained
personnel to attend any emergency in the shortest possible time. On
the other hand, the major work of the service involves some 9,000
patient journeys a day and consists of transporting out-patients to
and from hospitals or other centres. This aspect of the work is
proportionately greater for the London service compared with other
health authorities because of the presence in the capital of a large
number of specialist hospitals attended by day patients living in both
Greater London and the South East generally.

The vehicles that composed the new GLC fleet of ambulances in
1965 did not reflect this dual function of the service. Some of the
vehicles transferred to the GLC were considerably older than seven
years, the average age of replacement by the LCC; others were a
combination of stretcher and 'sitting case' vehicles which the LCC
had found to be unsuitable for the needs of the service. A programme
of replacement was therefore embarked upon in 1965, using money
from revenue account to purchase new vehicles. By June 1968, 394
vehicles had been replaced and the total composition of the fleet
changed, as is shown in the table:

Table 10.19

	Vehicle Totals for 1965–6	Vehicle Totals for 1967–8
Stretcher ambulances	570	536
Dual purpose vehicles	123	—
Sitting case vehicles	321	465
Total	1,014	1,001

Source: Ministry of Health Ambulance Costing Return.

Prior to 1965, the daily transport of out-patients for treatment
involved the movement of ambulances across health authority boundaries; this was especially so if an ambulance was required to trans-

port a patient from what is today Outer London to a Central London teaching hospital. If the carrying of these patients involved an ambulance belonging to a health authority other than that in which the patient lived, then that health authority could charge the other for the transport. Thus, before 1965 there was a considerable amount of inter-authority payments, which in 1964–5 amounted to at least £48,000 in the GLC area. This figure, however, is not the true cost of this practice since the administrative work involved in claiming these payments was absorbed in most health authorities, central administrative accounts. As a result of reorganization, payments to other authorities within the GLC area have been eliminated, and daily routes for drivers can be planned without regard to local authority boundaries, thus achieving a greater utilization of vehicle resources.

Expenditure on communications increased substantially as a result of reorganization. In 1965 the service acquired 15 control points, some of which covered substantial areas of the counties not included in Greater London. By 1969, the service was operating from 6 centres geared to the new Greater London area. Expenditure on communications has also included £275,000 from capital account designed to equip all vehicles with radio communication. This should enable the service to operate at greater levels of efficiency in the future by reducing the miles travelled by empty vehicles, besides reducing the time taken by emergency ambulances to reach the scene of an accident.

In July 1965 all health authorities were requested by the Minister of Health to revise their ten-year plans for the decade 1966–76. The plan submitted by the GLC[77] as an ambulance authority envisaged capital expenditure over the decade of £4·5 million at current prices or an average of £450,000 a year. The majority of this expenditure was to be devoted to providing new and replacing old ambulance stations to meet the changing pattern of demand. It was expected that demand for the service would show an increase on the general side as more out-patients were treated in response to the national move towards community care. The GLC believed that this could be met best by garaging more vehicles in the outer areas and reducing the numbers of those housed in the centre of London.

Capital restrictions, however, have curtailed these plans, and expenditure on the service has not approached that which was planned. In fact only one new station has been built since 1965, the major share of capital expenditure on the service being taken up by the new headquarters and control for the central area.

Table 10.20

Actual and Estimated Capital Expenditure 1965–70

	1965–6	1966–7	1967–8	1968–9	1969–70
Actual (£)	137,006	3,056	133,805	98,450	334,553
Estimated (£) (including provisional sums)	217,000	202,000	301,000	300,000	445,000

Source: GLC Annual Estimates and Accounts.

The ambulance service has, therefore, experienced particularly high initial costs as a result of reorganization and standardization. Capital expenditure restrictions have delayed the rationalization of communications and vehicles, but there are indications that the greater scale of operations in Greater London are beginning to produce certain economies in terms of utilization of staff and vehicles.

Library service. The public library service in Greater London accounted for £11 million of net revenue expenditure in 1968–9 or some 3·5 per cent of the total net revenue expenditure of London borough councils. Since 1965, expenditure on the service has risen by 28 per cent, but in England and Wales it has risen by 67 per cent in the same period.

	1965–6	1968–9	% increase
	£m	£m	
London Boroughs	8·8	11·3	28
England & Wales	25·8	43·0	67

The standard of library service for larger authorities (over 100,000 population) was laid down by the Secretary of State for Education and Science in 1965.[78] It depended on the purchase of at least 250 volumes of all kinds per 1,000 population annually.

Table 10.21 sets out the number of London boroughs which have attained this standard of book purchase since 1965. In Inner London, only two authorities failed to attain the standard in the first year, while in Outer London, 14 authorities fell short.

Table 10.21

Number of London Boroughs attaining standard of 250 volumes purchased per 1,000 population

1965–6	17
1966–7	18
1967–8	22
1968–9	22

Source: Public Library Statistics IMTA.

This reflects the development of the service prior to 1965, when metropolitan boroughs were generally regarded as providing a high quality service,[79] while in the areas of Outer London that were to be included in the London reorganization standards varied widely, and not all of the county districts were in fact library authorities. Table 10.22 sets out the second-tier authorities before 1965 which provided a library service; those that did not provide the service relied upon the county service for their area.

Table 10.22

Authority	No. of LAs included in Greater London	No. of Library Authorities	Expenditure per 1,000 population 1963–4	
			Books £	Total £
Metropolitan Boroughs & City	29	29	236	1,033
Middlesex CDs	26	14	174	818
Essex CDs	10	5	147	781
Surrey CDs	12	10	144	710
County Boroughs	3	3	120	633
Kent CDs	8	7	135	622
County Councils (average for above five counties)			143	545

Source: Public Library Statistics IMTA 1963–4.

Although library service expenditure in Greater London has risen by considerably less than in England and Wales as a whole since 1965, an analysis of Inner London expenditure since 1963 (Table 10.23) suggests that this part of Greater London has increased by rather more than the rest of the country.

Table 10.23

Net revenue expenditure as Index numbers

	Inner London*	10 County Boroughs	Newham
1963–4	100	100	100
1964–5	111	111	109
1965–6	132	124	126
1966–7	151	137	140
1967–8	192	157	152
1968–9	203	171	162

* Excluding Lambeth

The large increase in expenditure occurring in the first two years after reorganization, and the fact that this was some 14 per cent above the pattern of expenditure in 10 county boroughs outside London not affected by reorganization, and 10 per cent above Newham, where problems of reorganization were fewer, suggests that some measure of the increased costs were attributable to London reorganization. The London Treasurers in their evidence to the Royal Commission on Local Government in England suggested that this increase was due to a general development of local government services as a result of the reforms and concluded:

> 'There can be little doubt that the new authorities intend to bring standards of service in their new areas to a uniform basis and the standard sought is not unnaturally at least the best achieved by any of the previous authorities now merged in the new London borough.'[80]

Book purchase accounts for 20 per cent of the total expenditure on the library service, yet, as Table 10.21 illustrated, many of the new London boroughs made a conscious attempt to attain the recommended standard of book purchase in their first years. In 1965–6, 17 boroughs attained the recommended standard and by 1967–8, a further 5 boroughs had reached it despite large increases in the price of books. The increasing expenditure on books has been most marked in Outer London where expenditure on this item per 1,000 population increased by 63 per cent between 1963–8 as opposed to a 34 per cent increase in Inner London. But Inner London authorities were well within range of achieving the recommended standard prior to 1965, whilst none of the book funds of the old library authorities in Outer London were sufficient to attain the national standard and larger increases were therefore required.

Table 10.24

London Borough Libraries

Total Establishment

	1965	1966	1968
Professional	1,376	1,541	1,719
Other Non-Manual	2,730	2,644	2,627
Manual	763	885	793

Expenditure per 1,000 population on staff has increased by 55 per cent in Inner London since 1963 but by 66 per cent in Outer London. Although Inner London still spends considerably more per 1,000 population than Outer London boroughs, the gap has narrowed. It will be seen that in absolute terms, the number of full-time librarians has increased since 1965, which represents a movement towards the standard laid down by the DES of one full-time professional librarian per 2,500 population. Only one borough now falls short of the standard.

One feature of the staffing picture is the rise in manual staff in 1966, and the 10 per cent fall since 1966. In the first five years after 1965, the new library authorities expanded their activities and services, providing more mobile libraries and service points, as well as record lending, which might account for the increase. It is probable that the later reduction is a feature of those authorities where amalgamations have enabled certain economies in the sharing of manual and maintenance staff between two or more libraries.

The overall effect of reorganization upon the Library Service would seem to have given the new library authorities an added stimulus on top of that provided by the 1964 Public Library and Museums Act. Whilst the latter Act gave authorities a standard to which they should work the reorganization combined authorities, who in most cases tried to adopt the best book-buying and staffing policies of their constituent authorities. Thus, in the years since 1963, the overall expenditure increases have not been greatly different from the rest of the country, although the London authorities would seem to have experienced increased expenditure in the early part of this period while the reverse has been true of the rest of the country, where the effects of the 1964 Act have more recently precipitated increases in expenditure.

Conclusion. Even taking simply the statistics presented here the financial picture which they indicate is far from clear. Where comparisons are possible with the situation before 1965, there is by no means always a consistent pattern of change in the figures of expenditure on different services. Sometimes, as in the case of administrative costs in education in the Surrey boroughs, costs have risen more sharply than in corresponding areas outside London which were not affected by reorganization. But often expenditure on services has risen no more or very little more than that of authorities elsewhere, as in the case of the children's and health services; on the other hand, in the welfare services expenditure has risen on a much larger scale.

Where comparisons are made with authorities outside London in the period since the reforms, there is again no consistent pattern even in services where, superficially at least, similar problems were likely to be encountered. Expenditure on the fire service, for example, has risen less in Greater London than it has elsewhere, whereas expenditure on the ambulance service has risen at much the same rate as elsewhere. Again, the increase in expenditure on the library service has been less than half that in the country as a whole whereas the increased expenditure on welfare services in the London boroughs is almost twice that of a group of county boroughs of comparable size outside London. And overall, especially in Inner London, the rate of increase of expenditure, as measured by rate poundages, still seems to be somewhat higher than in authorities of comparable size elsewhere.

That the picture is confused is not surprising, given the great variety of services provided by local government and the great variety of factors operating within each service to influence the cost of providing them. This has been illustrated very clearly in the fire, ambulance and library services, and in some of the functional chapters.

Moreover, higher expenditure on services in London than elsewhere is only to be expected since certain costs are higher in London. It seems likely too that these costs have been increasing at a greater rate than those outside London, which would in part account for the facts set out above.

That in London certain costs are higher than in the provinces has for long been recognized with the provision in the Rate Support Grant formulae for a 5 per cent addition to the grant for which London is eligible; this is intended to cover such issues as the metropolitan weighting of wages and salaries in the London area. Yet other factors cannot be quantified with such precision. Land costs, for example, are undoubtedly higher in London than elsewhere and similarly the cost of building and construction.[81]

However, whenever London local authorities have sought additional central government relief to meet these higher costs, they have been met with a firm refusal, it being argued that the relatively higher rateable values in Greater London are sufficient to meet these higher costs. Whether this is so or not has never been accurately established, and remains a matter of dispute between London and other local authorities.

The reforms themselves in this sense provided both advantages and disadvantages; fragmentation of some services such as education, might be expected to result in increased costs which might not be compensated for by improvements in the service provided, whereas amalgamation, as in the fire service, might result in a saving of overheads with the same or even an improved standard of service.

Thus, the Treasurer of the London borough of Newham has recently stated:

'It seems to me that the review of operations necessitated by reorganization are more likely to result in improved efficiency or effectiveness than a rise in standards, but will probably result in both.'[82]

Above all, perhaps, one may note the stimulus provided by change, some of it direct, as in the need to standardize equipment and procedures, some of it indirect, providing the occasion for a review of policy. Almost all these changes would initially lead to increased expenditure on services.

But as has been emphasized throughout this discussion, it is virtually impossible to isolate only increases in expenditure which were not caused either by general factors such as inflation or by specific London factors not attributable to the reforms, such as increased building costs or increased London weighting in staff salaries; and then to discover what part of such increases were not matched over a period of time by some improvement in the services provided either directly measurable in terms of units of service or having indirect effects through, for example, changes in the relationship between the providers of the service and recipients.

The importance of the time-scale should be emphasized again. Not only may some costs be incurred immediately, whose effects can only be measured over a long period. There may be other costs which do not appear immediately but are spread over many years. This is well illustrated again from the study done for Cheshire County Council. By far the largest increase in costs attributed by the study to reorganization were the capital costs of providing new office accommodation.[83] But it is well known that capital expenditure of this nature is particularly subject to pruning and deferment according to the state of the national economy. There must be many local authorities whose plans for new town and county halls have been deferred for this reason. Certainly the new London authorities have not been able to make much progress in building new offices in their first five years. If building is spread over ten or twenty years or even longer it becomes practically impossible to attribute it specifically to a particular event such as the reform of local government structure as opposed to the many other factors which operate to influence the needs of different services over such a long period. Thus it may be potentially true that the London reforms lead to increased capital expenditure on new municipal offices but the reality may be very different.[84]

It is not however only the potential for increased costs which may be affected by other factors. Services which are now provided over larger areas than before 1965 in theory offer potential economies of scale. But, as has recently been pointed out: 'economies of scale do not emerge spontaneously as size increases; the potential must be deliberately exploited'.[85]

There is evidence from the examination of both the fire and library services that there has been a conscious effort to check the growth of expenditure and exploit potential economies, but this is not necessarily true of all such services.

In borough services there may be great disparities between the attitudes of individual boroughs to particular services which may further serve to mask the trend. This can happen not only in services where there are potentialities for economies but also in those where fragmentation might indicate a tendency towards increased expenditure. One has only to compare the very different expenditure patterns in the personal health and welfare services[86] to see this.

One should add here, too, that there are some services which must cost more to run now than they did before since they were either being performed inadequately under the old system or not being performed at all. Planning in Greater London, as has been shown,[87] now employs more and better qualified staff than was the case under the pre-1965 system, but the job of planning is rather different from what it was, so that to compare pre- and post-1965 expenditure would not be to compare like with like. Traffic management again, which employs scarce, highly-qualified staff, was not a local authority function in Greater London before 1965. Consequently, an additional cost is now being borne by the new authorities.

It will be clear, therefore, that it is impossible to put a precise figure on the cost of the London reforms. Furthermore, such a figure would be meaningless unless it was carefully qualified by reference to the particular circumstances and developments in individual functions. Thus what is needed is an examination of each service in detail relating costs to developments in that particular service and to the consequences both direct and indirect of the reforms. The studies of the fire and ambulance services offer examples of this method. The amalgamation of these services in Greater London provided complex problems which are reflected in the discussion earlier in this chapter of the reasons both for increased expenditure and for the different patterns of expenditure in the two services. In the fire service, it was shown that the maintenance of standards required relatively high expenditure in London because of the concentration of areas of high risk, but that improvement of standards which would have resulted in still higher expenditure has been held back by shortage of staff. On the other hand, changes in the organization of the service which have proved initially expensive may in time lead to more economical running which has not yet been reflected in the expenditure figures. Again, in the ambulance service initial expenditure on new vehicles and communications is likely to show effects in terms of improved service only gradually. All these developments can be attributed to the influence of the reforms.

These are not, however, the only consequences of the reforms. The London Fire Brigade and the London Ambulance Service now operate as much bigger units than they did before 1965. But in Essex, Kent and Surrey the opposite situation has arisen of these services being provided now for smaller areas than before. In Essex, the effect has been in the fire service that administrative overheads have not been reduced proportionately with the reduction in the size of the brigade. In the ambulance service, on the other hand, the area of the new Essex has proved far more suitable for running the service economically.[88] The problem then in these services is how to balance increased costs in some directions and improved services in others.

Moreover, one has to bring into the reckoning indirect consequences of the reforms, particularly on staff costs. For example, London rates of pay before 1965 applied to the Metropolitan Police District which did not include certain areas (Romford and Hornchurch) where London weighting is now payable, thus contributing to the increased cost of services there. But the effect of the setting-up of a large number of new authorities in Greater London at the same time for certain services (eg children) has also tended both to increase staff costs within Greater London and to make recruitment more difficult immediately outside Greater London,[89] because of the heightened competition for scarce staff. Indirectly, therefore, the London reforms operating first on staff costs within London have had repercussions on the ability of authorities adjoining Greater London both to recruit and retain staff, with consequential effects on their costs and efficiency.

Nevertheless, although it is not possible to draw up a neat balance-sheet of the profits and losses of the London reforms, there are some general conclusions to be drawn about the financial advantages and disadvantages of the London reforms. Perhaps the main conclusion is that the reforms have led to higher expenditure on most services. Some part of the increased expenditure can be attributed to general factors (eg inflation) and to improvements in the standard of service provided. It is doubtful whether increased expenditure not attributable to these factors is significant except in a few services, of which education is the outstanding example; the welfare services also come into this category.

Thus the Herbert Commission were wrong in predicting no increase in rate-borne expenditure. Nor did they foresee another important consequence of reform, a marked change in patterns of expenditure. This has come about in two ways; first by the need for standardization where services are now provided by amalgamated authorities (eg fire service, libraries); secondly, by differences in approach among the larger number of authorities now responsible for some services, eg the 12 welfare and children's authorities which have replaced the LCC represent 12 different ways of spending money on these services compared with one before.

Among the advantages of the reforms are, first, that the high concentration of rateable wealth in the central area is now available to a greatly increased area and population, a population which to a large measure contributes towards the creation of that wealth. However, this mainly applies to those services and functions which are the responsibility of the Greater London Council, although the equalization scheme does contribute some additional resources towards borough expenditure in Outer London. In the early years of the new London system when no equalization scheme applied to Outer London, and the GLC was in the process of adapting itself towards its new Greater London role, the new Outer London boroughs were faced with some difficulties in the financing of some of the more expensive services such as education. As the GLC's planning and development proposals become a reality, its share of expenditure on these services can be expected to increase compared with that of the boroughs, thus contributing to the more even spread of resources. Secondly, the resources of the boroughs have increased greatly in comparison with the previous situation and the London boroughs are now amongst the most wealthy authorities in the country. Only one London borough has rateable value per head below that of the average for local authorities in England and Wales as a whole.

Against these advantages, however, must be set the fact that there are still wide disparities between the resources and needs of the individual boroughs. Furthermore, large and prosperous as the new London authorities are, there is little indication that this fact has secured them a relaxation of financial control by the Central Government. The GLC, it is true, has a greater degree of freedom than any other authority in the country, but this is largely through the operation of the annual Money Act which the LCC also had. The wider sphere within which the GLC operates is counter-balanced by the fact that it has responsibility for fewer functions.

Moreover, although the London authorities have larger resources, they also operate in an area in which costs, especially staff and land costs, are considerably higher than in other parts of the country.

The analysis presented in this chapter is thus, on the whole, an encouraging one since it suggests that ratepayers of the new London authorities are getting value for their money from the London reforms. The conclusion must, however, be inevitably at this stage a tentative one. Moreover, it lends support to the view of both the Herbert Commission and the Government that financial considerations, though important, must be subordinate to more general considerations in assessing the value of reform.

Appendix

The Ilersic Scheme. In outline the scheme provides:
  (1) for each London borough to contribute at a rate of 10 new pence in the pound on
    (a) the total rateable value of its non-domestic hereditaments, plus
    (b) the rateable value of dwellings with a rateable value of over £200, minus
    (c) the value of its lower-rated dwellings (taken as below £101 in Inner London and below £57 in Outer London).
  (2) The sum contributed is then divided between Inner and Outer London in proportion to the total amount demanded from the ratepayers of Inner and Outer London in 1967–68.
  (3) The Inner and Outer London amounts are each split into
    (a) a housing pool, based on the percentage which housing costs represent of total expenditure in each area (15 per cent in Inner London and 7 per cent in Outer London); and
    (b) (the balance) a needs pool.
  (4) The housing pools are distributed between individual boroughs in the two areas by reference to the difference between their housing loan charges and Exchequer subsidy.
  (5) The needs pools are distributed by reference to their entitlements of needs element of rate support grant but with the special supplementary amounts for children under 5 and under 15 and for persons over 65 multiplied tenfold.

Following a recommendation made by a joint working party of officials of the Ministry of Housing and London Treasurers the weighting attached to the Housing factors (3(a) above) was discontinued.

11
The Greater London Council

The constitution of the Greater London Council under the provisions of the London Government Act, 1963, was in all essential respects similar to that of other local authorities. There were, it is true, some distinctive features not generally found among local authorities (eg the electoral arrangements and the provision for an annual Money Bill), but these were derived from similar arrangements which had applied to the LCC. The most distinctive feature of the constitutional provisions of the 1963 Act was the limitation of the number of GLC councillors to 100. The effect of this was that each councillor on average represented 80,000 people whereas in the old counties of London and Middlesex the figure had been about 25,000.

There were, however, obvious ways in which the GLC differed from other local authorities and from the authorities which it had replaced; it was, for example, much larger both in area and in population than any other urban authority. More especially the GLC's range of functions represented something quite new in local government since

i. it had almost no personal service functions;
ii. it combined a strategic planning function with a large-scale operational task of running services such as the fire brigade and the ambulance service.

The first of these two distinctions is particularly important in marking the GLC off from other large local authorities. All county and county borough councils have responsibilities for education, personal health, welfare, and children's services. The GLC has none of these except for the special arrangements for education in Inner London. On the other hand the GLC inherited from the LCC housing powers which county councils do not normally possess; these powers were, however, to be mainly of a temporary nature under the 1963 Act except for those relating to new and expanded towns outside London.

The GLC has constantly referred to itself in its Minutes and Reports as a 'regional authority' to emphasize its distinctiveness from other local authorities. The problem is to see how distinctive it has been in practice and whether this distinctiveness implies that it is indeed a new kind of authority and not truly a local authority at all. In particular, what have been the implications of the rather special range of functions assigned to the GLC for its internal organization and staffing? These are the questions which will be mainly considered in this chapter.

Committee and departmental structure 1965–7

The Herbert Commission did not devote much attention to the internal organization of their proposed Council for Greater London. They seem to have assumed that it would operate through the traditional local government committee and departmental system; they did, however, emphasize that it would need to have a fairly streamlined organization with only 100 councillors; and they hoped 'that it will be relieved of a great deal of the detailed routine which at present tends to clog the wheels of the county machinery'.[1]

The first question, then, is to see how far the GLC has found it possible to adapt traditional local government organization to its rather special range of responsibilities. The original departmental and committee structure, as Table 11.1 indicates, was closely modelled on that of the LCC. There were two main reasons for this: first, the great pressure of time which marked the transitional period when these matters were under discussion; secondly the prominence of LCC members and officers in the discussions as compared with those of other authorities. Nevertheless, some of the changes which were made between the LCC and the GLC were significant. At committee level, the Planning and Communications Committee was, as its name implied, intended to have responsibility for major policy questions affecting both planning and transportation.

At departmental level, there were three important developments:

i. A separate planning department was instituted whereas under the LCC planning had formed part of the Architect's responsibilities;

ii. Under the LCC the Chief Engineer's Department had been responsible for all engineering work, but under the GLC this was split between three departments, Highways and Transportation, Public Health Engineering (sewerage, refuse disposal, etc.)

Table 11.1

Comparison of the Committee and Departmental Structures of the London County Council (1964–5) and the Greater London Council (1965–6)

Committees		Departments	
LCC	GLC	LCC	GLC
Finance	*Finance	Comptroller	†Treasurer
Fire Brigade	Fire Brigade	Fire	†Fire
Parks	*Parks & Small-holdings	Parks	†Parks
Public Control	Licensing	Public Control	†Licensing
Supplies	*Supplies	Supplies	†Supplies
Health	*Ambulance	Health	†Medical Adviser
Rivers & Drainage	*Public Health Services	Chief Engineer	†Public Health Engineering
Roads	*Highways & Traffic	Chief Engineer	†Highways & Transportation
		Chief Engineer	Mechanical & Electrical Engineering
Town Planning	Planning & Communications	Architect	Planning
New & Expanding Towns	*New and expanding Towns	Architect	†Architect
Housing	*Housing	Housing Valuation	†Housing Valuation
General Purposes	*General Purposes	Clerk	†Clerk
		Legal & Parliamentary	†Legal & Parliamentary
Establishments	*Establishments		Establishments
Staff Appeals	*Staff Appeals		
Education	*ILEA	Education	†ILEA
		Schools Meals & Catering	†Catering‡
Children		Children	
Welfare		Welfare	

(Table based on committees' terms of reference and the major service responsibilities of departments.)

   * Chairman ex-LCC.
   † Chief Officer ex-LCC.
   ‡ Originally a GLC department, but later transferred to the ILEA.

and Mechanical and Electrical Engineering (a general service department responsible for the design, installation and maintenance of electrical and mechanical equipment);

iii. A separate Establishments Department was set up.

The second of these changes in GLC compared with LCC practice was due to the wider responsibilities carried by the GLC. The separate Establishments Department seems to have resulted mainly from the feeling that staffing matters were likely to be of greater prominence and difficulty in the new authority, and that the Clerk's Department would be sufficiently burdened with other duties.[2] The separate planning department, together with the higher status of the Planning and Communications Committee compared with the LCC's Town Planning Committee were a recognition of the greater importance of the GLC's planning role.

Continuity between LCC and GLC was emphasized by the fact that 11 of the 14 committees in the initial GLC structure were chaired by ex-LCC committee chairmen, and that 13 of the 17 chief officer posts were filled by ex-LCC officers. In particular, the important General Purposes and Finance Committees were chaired respectively by Mr Victor Mishcon, formerly Chairman of the LCC Fire Brigade Committee and Mr (now Sir Reginald) Goodwin, formerly Chairman of the LCC Finance Committee; and the posts of Clerk of the Council and Treasurer were held by Sir William Hart and Mr W. L. Abernethy respectively who had filled corresponding positions in the LCC. On the other hand, it is worth noting that the Planning and Communications Committee, which was to be responsible for one of the GLC's much-changed functions compared with the LCC, had as its chairman Mr Christopher Higgins who had previously been Leader of the Labour group on the Middlesex County Council; and the Director of Planning (Mr B. J. Collins) was a former Middlesex chief planning officer.[3] The other main committee concerned with the GLC's new strategic role, the Highways and Traffic Committee, was under the chairmanship of Mr Richard Edmonds, formerly chairman of the LCC Roads Committee, and the director of Highways and Transportation, Mr P. F. Stott, had also formerly been with the LCC as chief engineer.

Also important for the development of the GLC's internal structure was the size of committees. The LCC General Purposes Committee had had 22 members out of a total membership of the LCC

(Councillors plus aldermen) of 147; the GLC's General Purposes Committee had 31 members out of a total membership of 116.[4] This was, however, much the largest of the GLC's committees and bore a heavy responsibility in the early stages as the committee which had to draw up the initial plans for the Council's organization. Most of the GLC's standing committees had 12 members each, the exceptions being the Housing and Planning and Communications Committees each with 18. Corresponding committees of the LCC were generally larger, though with a greater range, from the Establishment and Roads Committees (12 members each), through the Finance, Fire Brigade and Public Control Committees (15 members) to Town Planning (21) and Housing (30). The LCC, however, had large committees for functions which were transferred to the boroughs, notably Health (40) and Welfare (30); the largest committee of all was for Education (47). Under the provisions of the London Government Act, 1963, all 40 of the GLC members of Inner London boroughs became members of the Inner London Education Authority.

It is clear then that the initial structure of the GLC was not based on any view of the need for a different kind of approach to the way in which business was organized and dealt with. The assumption was that adaptation of the familiar LCC methods of doing business would be adequate for dealing with the GLC's responsibilities.

This implication appeared to be challenged in a motion put down only three months after the GLC had taken over its responsibilities by Mr Plummer (who was later to become Leader of the Opposition and subsequently Leader of the Council) and Sir Percy Rugg (Leader of the Opposition) requesting the General Purposes Committee to 'examine what changes are needed in the relationship between officers and elected members in order to provide for more effective services'.[5] There was, however, nothing particularly new in the idea of such an examination. The LCC had several times in the course of its life carried out such investigations, the last as recently as 1962,[6] the net effect of which had been a progressive increase of delegation of responsibilities from the council to committees, and from committees to committee chairman and senior officers. This development was in line with the increasing volume of work with which local authorities had had to deal.

The Conservatives, who had long been in opposition on the LCC, took the view that none of these developments went far enough. With their emphasis on managerial efficiency they saw the LCC as still too much bound by traditional local government practice, with too much detail considered in committees, not enough delegation of responsibilities and inadequate means of co-ordinating policy. Mr Plummer's motion was in this sense not specifically directed to the new situation brought about by the substitution of the GLC for the LCC. It was rather that the new situation provided the opportunity for raising again a familiar point of view.

It was only subsequently[7] that a number of factors combined to bring about changes in the GLC's structure which considerably modified the original arrangements. There was first of all the experience gained in running the services for which the GLC was responsible and the realization that they presented certain problems which had not arisen or not so acutely in the case of the LCC. The General Purposes Committee, in its report to the Council,[8] drew attention to the somewhat rigid limits on the decisions which could be taken by committee chairmen and officers, and to the time and energy spent by both officers, and members in taking purely formal decisions; there was a need for 'more radical reform of delegation of authority' and a general examination of the Council's procedures. A Special Committee on Procedure[9] was duly set up for this purpose under the Chairmanship of Sir William (now Lord) Fiske, the then Leader of the Council, with 11 other leading members including Mr (now Sir Reginald) Goodwin, Chairman of the Finance Committee, Mr Mishcon, Chairman of the General Purposes Committee and Mr Plummer, Leader of the Opposition.

Two other factors were important in influencing the activities of the Special Committee: first, in the GLC elections of 1967, the Conservatives gained control by a large majority; secondly, the report of the Committee on the Management of Local Government appeared in May 1967. Following the 1967 election the Special Committee on Procedure was re-constituted under the chairmanship of Mr Plummer. With their more managerial approach, and as a party new to office after a long period in opposition, the Conservatives undoubtedly were prepared to go further in making changes than the cautious approach which the Labour Party had been making up to that point. Before considering the decisions which were taken, however, it is worth stressing that because of the way in which developments took place, the particular nature of the GLC and its responsibilities was only part of the motive for making changes in its internal arrangements.

This is illustrated by one of the first proposals to be made by the Special Committee on Procedure concerning the role of the Clerk of the Council. Co-ordination of a local authority's business has always been a major difficulty given the tradition of individual committees and departments responsible for particular blocks of an authority's work. The problem was a major preoccupation of the Maud Committee on Management which, among other things, suggested the need for a recognition of the Clerk explicitly as the leader of the team of chief officers. The GLC, following the recommendation of the Special Committee, agreed to re-designate the Clerk as 'Director General and Clerk to the Council; he was to be formally

> 'head of the Council's paid service, leader of the team of chief officers, and the chief executive responsible to the Council and its committees for advice on Council policy, for securing co-ordinated and integrated advice from other chief officers, and for consistency of departmental objectives and effective management'.[10]

No doubt the GLC were correct in stating that existing GLC practice already implicitly recognized the Maud view of the Clerk's powers.[11] The question was whether this was really adequate in view of the nature of the GLC's activities as compared with the LCC's. The same report which advocated formalizing the Clerk's overall responsibilities also referred to the need for forward planning of GLC resources and commitments over a period of at least five years ahead, looking at manpower as well as finance.[12] A firm of management consultants called in by the GLC to examine management information had also found a need for a chief executive with power to direct departmental chief officers. The GLC was, however, cautious in its approach and certainly was unwilling to make bold experiments in the way that some other authorities, such as Newcastle upon Tyne and Basildon were doing. Indeed but for the facts that Sir William Hart was due to retire early in 1968 and that a new Council was elected in 1967, it seems doubtful whether the formal changes in the Clerk's role would have been made.

The cautious approach was certainly evident in relation to the Maud Committee's proposals for severe reductions in the number of departments, and for the grouping of departments under senior chief officers acting as co-ordinators. The Special Committee on Procedure observed that 'as the departmental structure is still in process of evolving since the reorganization in 1965, it would be a mistake at this time to undertake any sudden or drastic changes'.

Furthermore,

'The Maud Committee's proposals for the grouping of departments under senior chief officers or co-ordinators does not really suit a regional authority such as the Council.'[13]

However, the Director General was instructed to keep departmental organization under review and some tentative moves in the direction of strengthened forward planning were made.[14]

Changes in the committee structure

Nevertheless, following a further report from the Special Committee on Procedure, the GLC in July 1968 did decide to reorganize its committee structure. The committee drew attention to the two different roles of the GLC:

'one is to provide a forum for the discussion and settlement of policy on long-term planning. The other is to carry out the extensive executive responsibilities of the Council.'[15]

They therefore proposed two types of committee, corresponding to the two roles of the GLC. Three forward planning committees were proposed:

   i *Leader's Co-ordinating Committee*: consideration and co-ordination of major policy matters;
   ii *Policy Steering Committee*: assessment of medium- and long-term objectives; broad financial planning; allocation of resources; London Transport;
   iii *Strategic Planning Committee*: Greater London Development Plan; relations with Standing Conference, Economic Planning Council and London boroughs on strategic planning.

Nine executive committees were proposed: *Arts and Recreation* (including parks); *Establishment; Finance and Supplies; General Purposes; Housing; New and Expanding Towns; Planning and Transportation; Public Services; Thamesmead*. In addition, there were to be the two special committees on *Procedure* and for *Staff Appeals* together with a *Scrutiny* committee.

The way in which these proposals were presented disguised the fact that the GLC was faced with several difficult problems in revising its committee structure. In particular, a simple division into two (or three with the special committees) types of committee did not really correspond to the realities of the situation.

First, any large and complex organization is faced with the problem of adequately co-ordinating its activities; in local government the problem is further complicated by the political nature of decision-making. The establishment of the Leader's Co-ordinating Committee and the Policy Steering Committee was the GLC's response in 1968 to this problem; and the reasons which led it to seek a solution at that particular point of time were largely the stimulus provided by the publication of the Maud Committee's report together with the new approach of the Conservatives who, in 1967, came into power for the first time.

Secondly, there is the problem of the distinctive GLC functions and how they should be handled at committee level. Here the new stage marked by the 1968 changes is the isolation of questions concerning the Greater London Development Plan for consideration by a separate committee; but apart from this, the new structure does not represent any fundamentally different way of handling the relationship between planning and other activities of the GLC, and, in particular, what was earlier referred to as the development complex of functions.[16]

Thirdly, there is the problem of committee responsibility for the more traditional local government functions – fire brigade, parks, sewerage, refuse disposal, etc. Here the 1968 changes closely followed the Maud Committee's suggestions by grouping responsibility, in this case under two committees, one for Arts and Recreation, and one for Public Services.

Fourthly, little change was made in committees responsible for general services, except to the extent that some of their important duties were transferred to the Policy Steering Committee, as happened with the General Purposes Committee and, to a lesser extent with the Establishment and Finance and Supplies Committees.

Finally, as part of the attempt to strengthen the managerial side of the Council's work there was the introduction of the Scrutiny Committee[17] charged with the duty of investigating, as and when required to do so by the Council or a committee, particular parts of the GLC's activities.

The changes in the GLC's committee structure introduced in 1968 are thus a blend of several different approaches. Much of what was done then can be paralleled from what other authorities were doing at that time. This is particularly true of the establishment of the Leader's Co-ordinating and Policy Steering Committees. The former is a committee of leading members of the majority party only; its membership of 10 consists of practically all the major committee chairmen together with the Chief Whip; its chairman is the Leader of the Council. The minutes of this committee only record the matters considered and its papers are not available to non-members. Thus the majority party caucus has become an official committee of the Council, functioning as a kind of Cabinet and having officials present at its proceedings.[18]

As with the Leader's Co-ordinating Committee, the introduction of the Policy Steering Committee marked the formal concentration in the hands of a few leading members of the majority party of responsibility for major policy decisions and the allocation of resources to the different activities of the GLC. This is a tendency which was becoming common, perhaps inevitable, among large local authorities. As a recent study has shown,[19] many county boroughs have tended in the past few years to set up policy committees, usually in combination with finance responsibilities. The GLC too, after the 1970 elections, renamed its Policy Steering Committee the Policy and Resources Committee with responsibility for the annual budget as well as for broad financial planning.[20] Such committees generally consist of leading members of the Council. The GLC's Policy Steering Committee, for example, was chaired by Mr Plummer the Leader of the Council and contained most of the committee chairmen, as well as the Leader of the Opposition. Again, this was somewhat modified in 1970 when the Policy and Resources Committee came under the chairmanship of Mr Horace Cutler, Mr Plummer's deputy; it is significant, however, that the chairmen of three of the major committees (Strategic Planning, Environmental Planning[21] and Housing) continued to be members of the new committee.

A further modification in the committee structure which was made in 1970 was to make the Thamesmead and New and Expanding Towns (renamed Town Development) Committees subordinate to the Housing Committee.

These various changes in the GLC Committee structure thus appear to be very much in line with similar developments in other large local authorities. In other words, given that from one point of view the GLC was simply a large local authority with the inevitable problems of co-ordination and forward planning which such a body gives rise to, the changes made in 1968 with the further modifications in 1970 were a reasonable attempt to meet the problems involved. But the GLC, as has been said above, was not simply a large local authority but one with an unusual range of functions. In particular, it was a strategic planning authority with responsibilities in a number of closely relevant functions (especially transportation and housing). How was this group of functions dealt with under the new system of committees and how were they related to the remaining functions of the GLC, such as running the fire brigade or refuse disposal?

The answer to the first question is that basically the 1968 changes did no more than separate off responsibility for the Greater London Development Plan under the Strategic Planning Committee without making any fundamental changes in the relationship between planning and its related functions. To that extent these changes were no more than a strengthening of what had been intended in the original committee structure when the Planning and Communications Committee had been set up with the basically similar task of preparing the development plan. The relationship between the planning group of functions and the GLC's other functions was altered by the 1968 changes but only in the general sense discussed above that priorities and programmes (allocation of resources between different GLC activities) were now to be dealt with by two new committees (Leader's Co-ordinating and Policy Steering, later Policy and Resources Committees). It is also necessary, however, to examine changes which have been made in the GLC organization at officer level.

Departmental changes

As in the case of the committee structure, the GLC was increasingly drawn towards the need for a greater degree of co-ordination of policy allied with a new approach to forward planning and the allocation of resources. Apart from obvious devices such as the use of project teams, the first tentative suggestion put forward by the Special Committee on Procedure following the re-designation of the Clerk's post was the idea of a team consisting of the Director General, treasurer and director of establishments to prepare and keep up-to-date 'a long-term programme of resources and commitments',[22] and bringing in other departmental heads as necessary.

The Chief Officer's Board, set up in 1968 as a result of this proposal, had the task not only of preparing the long-term programme but also of advising on major policy changes and on arrangements for the review of departmental performance against objectives. Significantly, it was soon enlarged to include not only the three members originally proposed but also the Architect, the Joint Directors of Planning and Transportation, the Controller of Services[23] and the Director of Housing. The Valuer and Estates Surveyor was subsequently added. A body as important as this could not be confined to the original triumvirate if it was to carry the other chief officers along with it. Effectively, therefore, the board now contains all the heads of major departments.

Further developments have taken place on the same lines, notably in the setting up of a strategic planning board, again under the chairmanship of the Director General, to consider major questions of policy in this field, and with a membership consisting of the chief officers of the departments directly affected[24] together with the executive head of the planning and transportation department.[25] Thus the chief officers' board parallels the Policy and Resources Committee and the strategic planning board the Strategic Planning Committee.

A further development in the direction of co-ordinating the work at officer level has been the appointment in February 1970 of a controller of services within the Director General's department whose function is to co-ordinate the work of the public health engineering service, fire brigade, ambulance service, parks service, licensing, the Royal Festival Hall and the scientific branch, and, also to assist the Director General in relation to matters affecting the London Transport Executive. It is not yet clear to what extent the controller will function as 'overlord' of the miscellaneous services listed nor just what co-ordination means in the context of these services. Since he is a member of the Chief Officer's Board he will presumably be in a position to state the case for these services in the general programme of the GLC, but to do so he must inevitably establish priorities within the group of services. To that extent the chief officers of the departments concerned have lost something of their independent status, although they remain fully responsible for the management of their departments.

The new Director General who succeeded Sir William Hart in

May 1968, Mr A. W. Peterson, came from outside local government, having spent his whole career in the civil service. A major aim of his was to continue further the breaking-down of the compartmentalization of departments. The development of the Chief Officers' Board and the other boards mentioned above is one indication of this approach. But more than anything, perhaps, he will be identified with the introduction of a Planning, Programming and Budgeting System (PPBS).

Much has been written about PPBS recently, and its applicability to local government has been much discussed. A number of authorities (eg Gloucestershire CC, Liverpool CB, Westminster LB) have introduced such systems but undoubtedly the GLC is one of the pioneers. Essentially the idea of PPBS is to bring together information needed by decision-makers in a more meaningful way in the hope that the quality of decision-making may be improved. As conceived by the GLC it seems to be intended to be mainly a management tool, giving precision to the definition of objectives and the measurement of performance against those objectives.

PPBS proceeds by way of identifying specific goals and objectives for an organization, evaluating alternative means of achieving those goals and the financial and manpower resources required, and monitoring progress once decisions have been made on the measures to be taken. Like all such moves towards long-term programmes, the introduction of PPBS carries considerable implications for the politicians; how far, for example, can a majority party elected for a mere three years make much impact on a programme in that time? But especially in a local government context the introduction of PPBS has profound organizational implications. The whole object of PPBS is to get away from the formulation of policies simply by reference to particular departments or committees, as has traditionally been the case in local government, and to move towards thinking in terms of the areas of policy which are important. If PPBS really can be introduced and made to work in the GLC, it is likely to have further consequences for departmental organization. At the moment, the intention is to introduce it gradually so that by 1973–4 a programme budget should replace the traditional annual financial budget.[26] So far, six programme boards have been set up, each effectively a sub-committee of the Chief Officers' Board, to prepare programmes for each of the six major areas of GLC activity which have been identified.[27] For many problems which will come up for discussion under this new system departmental boundaries will be irrelevant.

So far, however, there has only been one major, formal change in the departmental structure. This was the merging on 1 October 1969 of the Planning and Highways and Transportation Departments, together with that part of the Architect's Department which was concerned with road design and the Research and Intelligence Unit. According to Mr Desmond Plummer, the Leader of the Council, the object was

'to create an organization which is better suited to the Council's responsibilities as a strategic planning authority ... a unified approach to the problems of environmental planning, including communications, is essential'.[28]

However, 'to avoid lack of continuity' the new department of Planning and Transportation was to have two joint heads, the existing Directors of Planning and of Transportation. Nearly 18 months earlier the Special Committee on Procedure had argued that there was little scope for a major amalgamation of departments if it were accepted that 'too large a department can be beyond the grasp of any one head of department'.[29] Meanwhile, however, the situation had changed both by the appearance of a new Director General and by the introduction of changes in the committee structure.

In practical terms, the merger raised a great many problems. This was in part because of difficulties over the reorganization of the Research and Intelligence Unit;[30] in part, because of the difficulty of bringing together people with such widely varying approaches as planners, architects and traffic engineers. The oddity of a joint head for the combined department – obviously a compromise solution – has not made it easier to achieve the fundamental recasting of the department's structure called for by the new organization. This merely emphasizes that structural change in itself will not bring about the co-ordination of planning and transport policies.[31] It should also be noted that the significance of this merger lies not in the size of the combined department – the total staff is very much less than that of a number of other departments – but in the concentration in the planning and transportation activities of the GLC of many of the major policy issues which it has to face.

Other problems

Perhaps the main question raised by the various changes which have been made in the GLC's internal organization since 1965 is whether they adequately meet its specific organizational problems, which derive from two facts, those of being a large local authority and those of being an authority with an unusual set of functions to carry out. As has been indicated, the GLC has perhaps paid more attention to the first than to the second of these questions. Yet they are closely connected, and the position of the Research and Intelligence Unit which is discussed below is an illustration of the difficulties which have faced the GLC.

Two other questions which have a bearing on this discussion are the position of the Inner London Education Authority, and the extent to which service on the GLC as an elected member has proved burdensome, both financially and in terms of the time required. The ILEA occupies a curious position; it is both distinct from and yet closely linked to the GLC; formally, it is a special committee of the GLC,[32] yet the latter has no control over its membership, nor, formally, over its finances. A major problem has been that of co-ordinating the activities of the ILEA with those of the GLC as a whole. This problem is of considerable importance in view of the developments in forward planning and resource allocation described above since education is one of the biggest users of money and manpower. So far the GLC has had to rely on informal means[33] and these have worked tolerably well in practice. However, until 1970 these arrangements were greatly helped by the fact that both the GLC and ILEA were under the control of the same political party. Labour's recapture of the ILEA in the 1970 elections with the GLC remaining in Conservative hands is likely to strain these arrangements and make it difficult to achieve effective co-ordination.

The 40 Inner London members of the GLC also serve on the ILEA, thus emphasizing the problems of service on a large authority which has been set up with a basically similar constitution to that of other local authorities. Here the GLC has argued that it is a special kind of authority, but has failed to convince the Government. At a very early stage in 1965 the GLC, adopting a proposal made by the LCC, approached the Minister of Housing and Local Government and suggested that GLC members should be paid allowances on a system similar to that of the sessional allowances paid to members of the House of Lords. The GLC's argument was that 'as a regional authority' they needed to pay allowances on a more adequate basis than was possible under existing local government provisions. Although this proposal was apparently favourably received at first it subsequently met objection from the Inland Revenue and the Treasury and was finally rejected, to the indignation of the GLC, which claimed that 'many leading members spend more than 30 hours a week on the Council's work; some give more than 50 hours a week'. It was suggested that without adequate financial compensation the Council might find it difficult to find candidates of sufficient calibre.[34]

Support was given to this view by the resignation in 1969 of the Chairman of the Finance Committee, Mr Roland Freeman on the grounds that the burden on unpaid members was too great. He is reported to have said that it was 'utterly absurd that committee chairmen who give so much time remain unpaid'.[35] There was also, as The Times[36] noted, an unusually large number (26) of Conservative members elected in 1967 who did not stand again in 1970. The amount of time required for work on the GLC may have been one ot the reasons for this. Certainly, one of those members, Mr Robin Leach, in September 1969 resigned from the chairmanship of the Lee Valley Regional Authority on somewhat similar grounds to Mr Freeman.

There is little doubt that service on the GLC is very time-consuming, especially for the leading members of the Council;[37] nor that this poses in an acute form the dilemma facing other large authorities of how to attract and keep members unless some change is made in the present system of remuneration for financial losses. The changes in the committee system have done something to reduce the number of meetings which members have to attend, and this has helped leading members (chairmen, vice-chairmen, minority party leaders) through the reduction in the number of committees; on the other hand, there are still a comparatively large number of sub-committees.[38] The Government, consistent with the approach of the London Government Act, 1963, has refused to treat the GLC as a special case.

The Research and Intelligence Unit

The constraints under which the GLC has operated and the difficulties which it has experienced in establishing a satisfactory organization are nowhere more evident than in the history of the Research and Intelligence Unit. It will be recalled[39] that the Herbert Commission had attached a surprisingly large degree of importance to the establishment of a 'first class intelligence department', even going to the lengths of claiming that if local government did not itself provide a service of continuous research and information for Greater London, it would cease to exist, for the central government would then have to take over the job and 'this would be . . . the death knell of local government in Greater London'.[40]

This strongly-expressed view seems basically to derive from the Commission's insistence on the urgent need for a body to plan Greater London as a whole and for that body to be a local rather than a central government organ. If such a body was to operate effectively then clearly it must obtain and keep up-to-date relevant information about Greater London and its needs. This was precisely what was lacking in the existing situation. Hence the importance which they attached to the idea of a central Intelligence Department for the Council for Greater London was part and parcel of the case which they were making for a fundamental change in London's system of local government.[41] The Commission saw the Department, moreover, not only as a body which would collect and collate information and carry out research but also as one which would provide a service to other departments of the GLC, to the London boroughs and to the Central Government. It would also publish a regular series of statistical publications and special studies.[42]

The Government, in accepting the Commission's view of the need for some such body, seem to have gone out of their way to leave as open as possible questions about its scope and organization. Sir Keith Joseph, speaking on the Committee stage of the London Government Bill, said that it 'would not be an advisory bureau but rather an information bureau on whose information policy decisions can be based'.[43] The Bill itself did not make it obligatory on the GLC to set up such an organization, although after pressure Sir Keith agreed to amend the wording of the Clause. Under the Act, the GLC was required to:

'establish an organization for the purpose of conducting, or assisting in the conducting of, investigations into, and the collection of information relating to, any matters concerning Greater London or any part thereof and making, or assisting in the making of, arrangements whereby any such information and the results of any such investigation are made available to any authority concerned with local government in Greater London, any government department or the public'.[44]

As L. J. Sharpe has pointed out,[45] the idea of a research unit as a permanent feature of a local authority's administration was something quite new in local government. Apart from the novelty of the idea, the exact scope, location, and role of such a unit were of great significance for the GLC's administrative structure.

For these reasons alone, the Government were clearly anxious to leave it to the GLC to decide the two closely-related questions which are considered here: how far the unit should be independent of other executive departments of the GLC, and how far it should go in initiating and carrying out research. It is noteworthy and in keeping with the Herbert Commission's view of the importance of what they termed the Intelligence Department that they specifically stated their views on these questions, and that this was the only occasion in their whole report when they did make such specific recommendations on the internal organization of the new authorities. In their view:

'To this central Intelligence Department all the other departments should be able to turn for information and to it they should address requests for research and advice. Certainly other departments must in limited sectors, for example, traffic engineering and management be in a position to conduct their own inquiries and research; but we can see no hope of a coherent policy for Greater London unless this central Intelligence Department exists and functions properly'.[46]

This was a radical departure in terms of its implications for the traditional local government internal structure. Research in local government has been traditionally geared to the needs and policies of individual departments. What the Herbert Commission were suggesting was that departmental research should be restricted to more limited and technical aspects, and that the broader research which contributed directly to policy should be the responsibility of an entirely separate department.

At an early stage, however, it became clear that the GLC had no intention of going as far as this. The GLC Joint Committee set up before the new system was instituted, seems to have viewed the new organization largely as a co-ordinating body, keeping in touch with research which was going on, maintaining a library of information, and advising on the methodology of research. On the critical question of how far the unit should itself undertake research the committee obviously inclined to a restricted role with the unit undertaking only work which could not be done elsewhere. Correspondingly, the planning and other departments which were established, themselves contained branches which were concerned with research and intelligence in their own specific fields. The important work of the London Traffic Survey, for example, which had considerable implications for the production of the Greater London Development Plan, was entrusted to the Highways and Transportation Department at the end of 1964. It was also decided at that time that the proposed intelligence unit should be attached to the Clerk's department.

By these decisions the GLC clearly rejected the Herbert Commission's concept of an intelligence organization. Yet at the same time there was a great deal of doubt about what such an organization should do. So far from regarding it as 'the first requirement of all'[47] the GLC agreed with Sir William Hart that 'it would be wise for the Council to feel its way slowly towards the definition of the precise functions and scope of the intelligence unit'.[48]

In a detailed report to the General Purposes Committee Sir William made it clear that having rejected the Herbert Commission's radical proposals the GLC were in a dilemma; they had, as he put it: 'the problem of fitting such a novel unit into the traditional local government structure of separate departments'.[49]

Skilful as the report is in suggesting possible ways in which the unit might operate and develop, it is most revealing as indicating uncertainty about its role. The impression is undoubtedly left that section 71 of the London Government Act was proving an acute embarrassment to the GLC and at this stage Sir William saw the appointment of a director as the next logical step.

In the event there were no less than 315 applicants for the post designated as 'Director of Intelligence'. Dr Bernard Benjamin, a former Chief Statistician of the General Register Office and Director of Statistics at the Ministry of Health, was appointed at a salary of £6,000 p.a.[50] He was not able to take up his appointment full-time until 1 February 1966, that is, almost a year after the GLC had assumed its responsibilities and 18 months after most of the chief officers had been appointed. With virtually no staff to begin with, Dr Benjamin's first task was to survey and assess the field within which the unit (now entitled Research and Intelligence Unit) was to work.

Much of his report to the General Purposes Committee on this topic was not controversial and covered ground which had largely been agreed throughout the discussions on the scope of the unit. Providing statistical data for the London boroughs, publication of a regular series of statistics, close liaison and co-operation with the departments – these were activities which were generally agreed to be useful and appropriate to the unit. Yet soon the tensions which were latent in the unit's ambiguous position began to appear. The main question was the extent to which it should itself undertake research, and how that research should be related to policy-making. Dr Benjamin felt that there were disadvantages in the unit's subordinate position and in his own status as, in effect, a deputy chief officer.[51] But any move in the direction of strengthening the unit could only bring him into conflict with existing departmental heads who had no wish to see an independent body impinging on what they regarded as coming within their own sphere of responsibility. And clearly to the extent that the unit was more than merely a clearing-house and collecting centre for statistical information it was bound to be concerned with questions directly relevant to departmental policy. With the concentration of major research activities in the Planning and Highways and Transportation Departments it proved difficult for the unit to make a case for doing independent research although this was conceded in principle.[52]

In practice, much of the unit's research has fallen into one of two categories; there has been, first, work done for the London boroughs and, secondly, work not specifically falling within the sphere of any one existing department of the GLC. The first category has included an immense variety of tasks; examples include a survey of children with spina bifida in relation to estimates of future school populations, a survey of the professional qualifications of social workers and an examination of the system of grants to voluntary bodies. In the second category fall especially the population studies which the unit has carried out, particularly projections of population for London and special studies of the implications of population trends.

Some of the unit's research findings cast doubt on basic assumptions made by other departments. This was particularly evident in the case of population forecasts. The GLC, for example, following the South East Study, had been working on the assumption that in the period up to 1981 a planned overspill programme for approximately one million people would be needed. The Research and Intelligence Unit questioned this assumption[53] and although their arguments were eventually accepted,[54] the incident did not make for the happiest relations between the unit and the Planning Department. Yet it symbolized fundamental differences of view about the role of the unit; or, to put it more broadly a difference of approach to the question of the right way of organizing research within the GLC.[55]

The Director continued to press for a wider role for the unit. The opportunity came when, for reasons which have been described earlier, the merger of the Planning and Highways and Transportation Departments was under consideration. The question of the organization of research work in the combined department proved difficult. The solution proposed was to bring together all the research and intelligence activities of the GLC within the Research and Intelligence Unit, but to make the unit part of the new Planning and Transportation Department. Two factors were undoubtedly important in inspiring this proposal: first, the co-ordination of a research programme for the GLC to prevent overlapping and gaps had not proved easy largely because of the departmental fragmentation of research work; secondly, there was a strong argument for concentrating scarce staff resources by bringing all research work together,[56] and if this were done there was an obvious case for putting it in the new combined department which had the largest need for research of all the GLC's departments.

This was not, however, the end of the story. Unity of research had been achieved, as the Herbert Commission had advocated. But there still remained the question of the relationship between the work of the unit and policy-making. The Commission had suggested in effect a separate Intelligence Department. The GLC's 1969 solution was a unit as one branch of the department which would be mainly concerned with using the results of the unit's research. If this arrangement had lasted it would have required a high degree of co-operation between the unit and the other branches of the department, as indeed would the Herbert Commission's proposal between the intelligence department and other departments. The arrangement was, however, soon abandoned.

Shortly after the new department of Planning and Transportation was set up on 1 October 1969 Dr Benjamin resigned. Following his resignation, further organizational changes were made; research which was specifically related to the work of the department was assigned, not to the unit, but to other branches. In particular, an economic and social planning unit was set up under Dr David Eversley, the Chief Planner (Strategy) and transport research was brought under a new post of Chief Planner (Transportation). The R & I Unit became the Intelligence Unit and retained its London Borough Division and its responsibilities for advising on methodology and for co-ordinating and disseminating research and statistical information for Greater London. In June 1970 the GLC once more advertised for a Director of Intelligence 'with managerial and co-ordinating capacity in a statistical environment as well as experience of research work', but 'research' had dropped out of his title. Mr Cecil Russell, head of the Census Division at the General Register Office, was appointed to the post.

The organization of the research and intelligence function is important in a body like the GLC with a strategic planning role which must inevitably require a great deal of soundly-based data if it is to be carried out effectively. The merit of the Herbert Commission's analysis was that it fully recognized this need for effective research; at the same time the Commission may not have fully realized the implications of their precise proposals for the organization of an intelligence department. The idea of a wide-ranging and comprehensive intelligence unit which is also independent of the departments which are to use its services is very difficult to reconcile with a traditional local government structure. The tendency in local government is for departments responsible for policy to rely as much as possible on gathering their own information and research which is directly relevant to the formation of policy. As L. J. Sharpe, a strong supporter of the idea of a research and intelligence unit, pointed out in 1965, 'there is a strong case for allowing the traffic and the planning departments to have their own survey units' even though he recognized that there might in any case be pressures to restrict the scope of a research and intelligence unit.[57]

Thus the extent of and scope for research of an R & I Unit which was not located in one of the major research-using departments was bound to be more limited than the Herbert Commission had imagined. This was especially so because the unit only came into being after the GLC's initial structure of committees and departments had already been established. It would have been impossible to insert into that structure at that stage a unit with the kind of wide-ranging and independent role which the Herbert Commission had had in mind. The only possible alternative within the local government context was the one proposed in 1969, that is, a unit located in the major research-using department. The fact that this was soon abandoned is perhaps the clearest indication of the very great difficulty, in a local government context, of separating responsibility for research and responsibility for policy-making.

This discussion has not been concerned with the value of the work done by the GLC Research and Intelligence Unit which has been considerable nor does it seek to deny that the present intelligence unit has an important role to play even with its present reduced scope. It illustrates primarily how much the GLC remains in the local authority tradition; to have accepted the Herbert Commission's contention that 'the first requirement of all' was the establishment of an intelligence department would indeed have compelled the GLC to have examined much more closely when it was set up what kind of internal organization was appropriate and, in particular, what kind of relationship there should be between policy-making in planning and related functions and the organization of research. It does not, of course, follow that such an examination would have produced any radically different solution to the organizational problem, but it would have helped to clarify the whole question of how far the GLC should be regarded and treated as some new kind of authority, indeed as the regional authority which the GLC claims itself to be. The story of the R & I Unit, however, like that of the committee and departmental organization mainly illustrates how the GLC has tried to adapt traditional local government organization to its somewhat special position.

Staffing

As Table 11.2 indicates, the GLC employs a total staff of nearly 37,000, excluding the ILEA. The exclusion is significant because the education services are much the largest employers of staff of all local government services. Although the GLC employs 23 per cent of the 170,000 local government employees in Greater London (again excluding those in education) its total staff is 3,000 less than that of Birmingham, with about 40,000 staff;[58] many of the latter are employed in the social service departments which have no counter-part in the GLC.

This is one way of drawing attention to the peculiar range of the GLC's functions. As Table 11.2 indicates, relatively few departments of the GLC employ large numbers of staff and these are mostly concerned with the traditional local government functions which have been projected onto a Greater London scale, such as housing, fire brigade, medical (ie mainly the ambulance service) and parks; these are also the departments with a high proportion of manual workers among their total staff. The major departments concerned with the GLC's strategic planning functions on the whole employ comparatively few staff as do the general service departments (Director General's, Treasurer's, Establishments etc.). There has however been a tendency for these departments to increase in size somewhat more rapidly than the traditional departments, some of which have actually declined in size.

Table 11.2

(a) *Number of GLC department employees for selected dates*

(b) *Number and types of staff in GLC departments on 31 May 1970 (in post)*

(a) Department	September 1965	August 1966	In post May 1970
Valuation and Estates	1,017	1,048	1,089
Clerk/Director General	1,268	1,158	776*
Establishments	112	147	163*
Fire Brigade	5,814	5,583	5,445
Housing	10,831	10,740	9,832
Legal and Parliamentary	248	243	281
Licensing	1,662	1,835	1,582
Mechanical and Electrical Engineering	1,183	1,229	1,372
Medical	3,022	2,903	2,710†
Parks	2,712	2,554	2,836
Planning (Inc. Communications)	338	399	1,390*
Transportation (Inc. Highways)	524	653	*
Architect's	3,041	3,170	3,383*
Public Health Engineering	1,686	2,348	3,019‡
Supplies	1,565	1,812	2,094*
Treasurer's	670	664	795
Total	35,693	36,486	36,767

(b) Department	APTEC§	Other staff
Architect's	2,959	424
Director General	553	223
Establishments	161	2
Fire Brigade	301	5,144
Housing	2,055	7,777
Legal and Parliamentary	280	1
Licensing	1,547	35
Mechanical and Electrical Engineering	857	515
Medical	202	2,508
Parks	257	2,579
Planning and Transportation	1,291	99
Public Health Engineering	711	2,308
Supplies	1,013	1,081
Treasurer's	793	2
Valuation and Estates	1,064	25
Total	14,044	22,723

* Figures for May 1970 are not strictly comparable because of departmental reorganizations:

(a) Formation of Planning and Transportation Department by the amalgamation of the former Highways and Transportation Department and Planning Department, together with some staff from Architect's and Director General's Departments.

(b) Transfer of Central Reprographic Service, cleaners and messengers from Director General's to Supplies Department.

† Figures for May 1970 exclude about 200 staff employed in the School Health Service who are chargeable to the Inner London Education Authority and are no longer included under GLC staff.

‡ Figures for August 1966 and May 1970 include staff formerly employed by London Boroughs and various sewerage and drainage authorities whose work has been transferred to the GLC.

§ Administrative, Professional, Technical, Executive and Clerical.

Perhaps the most important element in the staffing situation which influenced the GLC's approach to its responsibilities was its inheritance. The GLC inherited staff from a number of different authorities, but particularly from the LCC. Thus the problem of assimilating staff which was common to all the new authorities in Greater London in 1965 was strongly influenced by the attitudes and traditions of the

LCC formed by the distinctive role which the LCC had played in the pre-1965 London government situation.

At the outset, however, the GLC significantly departed from LCC practice in setting up a separate Establishments Department;[59] in the LCC establishment matters had been dealt with by the Clerk's Department. The GLC did, however, adopt broadly the method of working of the LCC in that day-to-day management of staff remains the responsibility of heads of departments, liaison being maintained between them and the establishments department through a system of departmental establishment officers. The GLC's Director of Establishments, however, came from outside local government altogether;[60] the LCC's Senior Assistant Clerk responsible for the establishment branch was appointed Deputy Director of Establishments.[61]

The creation of a separate Establishments Department under its own chief officer, and the appointment to that post of a man with no local government background provided the opportunity for a fresh approach to personnel and staffing questions. This has been most evident in the introduction and extension of management techniques, such as work study, operational research and, on a limited scale, management by objective. To some extent these can be seen as a response to the GLC's particular situation and to the need, described earlier, to concentrate more on resource allocation, within which manpower resources are obviously of great importance; to some extent these developments follow an earlier tradition of the LCC which was a pioneer in management methods, such as O & M; to some extent they merely parallel developments taking place elsewhere in local government, particularly in the wake of the reports of the Maud and Mallaby Committees. Yet the effect has been to steer the GLC towards a greater emphasis on the more effective use of staff; and this, as has already been noted, was very much in accordance with the emphasis of the Conservative administration which took over in 1967.

Developments of this kind clearly depend on the active co-operation of the staff concerned. Here the main difficulty has been that emphasis on the more impersonal side of the GLC's activities has been criticized as leading to a more impersonal treatment of staff matters. Psychologically the root of this difficulty goes back to the GLC's inheritance from the LCC. There is no doubt that many of those who had served in the LCC regretted the change to the GLC.

They were, therefore, reluctant to accept the new emphasis which brought home most forcibly the differences between the LCC and the GLC. This is a transitional problem since in the course of time the GLC will increasingly consist of staff who have not served under the previous system. It does emphasize, however, that there has so far been a strong restraining influence on movement towards a somewhat different view of the GLC's role. On the other hand, there has already been some dilution of the GLC's inheritance of staff from the LCC and other authorities, not only through the normal wastage processes, but because of the need to recruit specialized kinds of staff, many of whom have either not previously been employed in local government or have only been employed on a much more limited scale than the GLC needs. Traffic engineers, for example, were almost non-existent and were required in comparatively large numbers by the GLC; the GLC therefore at a very early stage[62] approached a number of universities with a view to their providing places on postgraduate courses for its nominees, and these arrangements have continued since. Again, the programme office in the Director General's department, with its overall responsibility for the introduction of the PPBS, has been largely staffed with people recruited from industry and other outside sources since this is a very new field of specialization which local government is only beginning to enter.[63]

As with any organization set up largely on the basis of existing organizations, there are two main competing elements in the development of the GLC's staffing arrangements; first, the tendency towards identifying and stressing the particular characteristics of the organization and the need for staffing policies consistent with that aim; secondly, the influence of the previous organizations and the element of continuity provided by the fact that many members of the GLC staff had previously served under the LCC.

The decision of the GLC on the question of negotiating machinery for white-collar staff,[64] and subsequent developments illustrate the influence of these different tendencies. The LCC had not taken part in the national negotiating machinery for salaries and other conditions of service,[65] but had instituted its own separate negotiating machinery, notably with the LCC Staff Association. This was an arrangement which was satisfactory to the staff who thought, probably rightly, that it was to their advantage; on the whole LCC staff were better paid than their neighbours, particularly in Middlesex.[66]

This was naturally one point of difficulty in the initial staffing arrangements. But apart from this, the LCC Staff Association pressed hard for the continuation of separate negotiating arrangements for GLC staff. Other local authorities had, however, not looked with favour on the LCC's separateness; in the context of the London reforms, the new London boroughs were particularly anxious that the national negotiating machinery should apply equally to GLC and borough staffs since they feared the effect of any marked differential between GLC and borough rates of pay on their ability to attract staff in areas of competition eg for planners.

The GLC's decision was in favour of a separate GLC Whitley Council. This negotiates on behalf of the great majority of the white-collar staff, although about 500 senior officers and the chief officers conduct separate negotiations. The GLC like the LCC justified these arrangements on the grounds that its responsibilities and those of its staff differed substantially from the responsibilities found in all other local authorities. This argument was specifically rejected by the National Board for Prices and Incomes.[67] Initially, GLC senior officers had had their salaries temporarily fixed at those paid by the LCC or Middlesex County Council until a review of salary structure could take place in the light of experience of the GLC's responsibilities. An independent inquiry under Lord Heyworth recommended a range of increases in March 1966 which was accepted with two minor modifications by the Council – although officers were much more critical. Following a reference by the Minister of Housing and Local Government the NBPI supported the Council on the Heyworth Report but questioned

'whether the exclusion of the GLC from the national negotiating machinery can be defended. If negotiating machinery can embrace authorities as small as the urban district council of Llanwrtyd Wells with its population of 510, and as large as Lancashire, whose population for administrative purposes exceeds 2¼ millions, there seems no reason why it should not also include the GLC.'[68]

The NBPI recommendation was a general one in relation to GLC staff, although it had only been concerned specifically with the upper echelons of its officers. However, a review of the pay and grading structure of officers below £2,400 had been instituted in March 1965, and it was found that

'The Council's staffing needs, both in nature and quantity, differ in important respects from those of its predecessors ... (and) in the light of the results of the review, a new pay and grading structure has been devised for the Council's staff.'[69]

The new structure had the aims of simplification, more effective use of highly qualified staff, provision of an attractive career structure and the maintenance of a common pattern for administrative and professional staff.

These objectives were achieved by a continuous straight line pay structure with scale points at standard levels, by a greater use of the technical and executive classes to leave the professional and administrative classes free to concentrate on more important work, but with provision for controlled transfer between the two groups of classes; and promotion on a service-wide basis for clerical, executive and administrative classes.

This new structure is certainly substantially different from the LCC system which it replaced, in that it made provision for support staff for the administrative and professional grades by the creation of executive and technical grades as part of the overall career structure. The GLC differs from other local authorities in aiming to offer a career pattern within its service whereas the more usual career pattern in local government is for staff to move to higher positions in another authority.

Because of these differences, it is not possible to make direct comparison between GLC and national grades. The main grades and the scales payable are as follows (as at 1st July 1970):

Table 11.3

GLC		*National*	
Admin/Prof		*Admin/Prof*	
Grade A – £1167 – 1812		Trainee – £ 489 – 1515	
B – £1902 – 2865		AP1 – £1038 – 1272	
C – £3012 – 3510		AP2 – £1272 – 1515	
D – £3666 – 3936		AP3 – £1515 – 1776	
		AP4 – £1776 – 2025	
		AP5 – £2025 – 2268	
		Senior Officer – £2106 – 2751	

Table 11.3 contd.

	GLC		National	
Executive				
Grade 1	–	£ 963 – 1812	Principal	
2	–	£1902 – 2178	Officer –	£3366 – 4332
3	–	£2274 – 2466	*Technical*	
4	–	£2562 – 2757		
			Grade 1 –	£ 429 – 942
Technical			2 –	£ 942 – 1089
			3 –	£1089 – 1272
Junior	–	£ 681 – 1227	4 –	£1272 – 1515
Grade 1	–	£1284 – 1812	5 –	£1515 – 1776
2	–	£1902 – 2178	6 –	£1776 – 2025
3	–	£2274 – 2466	7 –	£2025 – 2268
4	–	£2562 – 2757		

(Addition throughout of £15 supplementary London weighting)

(In London there is also payable in addition supplementary London weighting of £90 at age 18, or over; under 18 it is £50)

The GLC has more justification for emphasizing in its pay structure its separateness from the rest of local government than the LCC had. Most of the services for which the LCC was responsible were the same as those of other county councils. The LCC differed, therefore, largely in the scale of its operations and to this extent the NBPI's strictures would have been more applicable to it than the GLC. The latter is not only very much bigger than any other local authority, but it has a very different range of responsibilities.

In its pay and staffing structure, the GLC seems to have been readier than in its committee and departmental structure to move away from the traditional local government pattern. There is no doubt that particular circumstances helped here – the readiness, even anxiety of the staff association to support such a move and the fact that the Director of Establishments was, as an outsider, not bound by local government traditions. But paradoxically it was the fact that a tradition of separateness had already been established by the LCC that gave the main initial impetus, just as it was LCC influence which largely determined the original committee and departmental structure. Given that initial impetus it was easier to move further in the direction of a staffing structure which would suit the particular circumstances of the GLC, especially with its emphasis on a unified career structure. Whether the GLC were justified in doing so is another question. Here we merely note the fact that in this one important respect alone the GLC has virtually cut itself off from the rest of local government, as though it were indeed some new kind of regional authority.

Conclusion

The implication of the arguments presented in this chapter is that the claim that the GLC is a regional authority must be rejected if that is intended to imply that the GLC is some new type of authority. Both constitutionally and in the way in which it has organized itself it is a local authority. Nevertheless, it is a local authority of a distinctive kind and this fact has been reflected in some of the difficulties which it has encountered in organization and staffing. To some extent the distinctiveness of the GLC is a reflection of its inheritance, particularly from the LCC; this is to be seen in statutory provisions, such as those for the annual Money Bill, and in part in the separate staffing and pay structure.

The GLC is not, however, simply a large-scale LCC. It has a very distinctive function to perform, that of providing the broad planning framework for the future development of an entire urban area, together with a range of other functions which are also distinctive in that they do not correspond to those of any other authority. Thus the GLC differs from a county borough in not having all local government functions to perform and also because of the much larger scale of its operations. It differs from a county mainly because it is a purely urban authority and because it lacks the social service functions. The important question which this chapter has been largely concerned with is how successful the GLC has been in devising an organization to meet its special role and functions.

The London Government Act, 1963, was drawn up on the assumption that the GLC was simply a large local authority. Government spokesmen during the passage of the Bill through Parliament emphasized this point and appeared to see no difficulty in the GLC's ability to operate through the traditional committee and departmental structure of local authorities. Only the limitation of the GLC to 100 councillors, despite its large area and population, seems to indicate a somewhat different approach, but this limitation was simply derived from the Herbert Commission's recommendations and little thought seems to have been given to its implications.

P

By and large the first GLC from 1964 to 1967, was organized and operated as though it were a larger LCC. Between 1967 and 1970 there were considerable changes in organization. On the whole, however, these changes have been more successful at dealing with the problems of co-ordination and forward planning which are common to all large authorities than with those presented by the carrying out of the GLC's particular range of functions.

This situation has arisen mainly because of political and professional pressures and growing concern generally with questions of internal management to which the Maud and Mallaby Committee reports gave impetus. In other words, the establishment of policy and co-ordinating committees and of chief officers' boards and even the introduction of PPBS are not primarily due to a recognition of the GLC's specific role so much as to the fact that it was a large authority in which this type of problem presented particular difficulties to elected members and to officers.

The particular form which these and other developments have taken in the GLC may in fact have made it harder in some ways for the GLC to carry out its functions effectively. It seems, for example, to have been a major aim of the GLC to ensure that physical planning should have close relations with the transportation group of functions. Under the original structure, this was largely to be achieved through the medium of the Planning and Communications Committee, and through ad hoc joint working arrangements between the Planning and Highways and Transportation Departments. Later changes have gone much further, particularly at officer level, with the merger of the two departments. At the same time, it seems to have been recognized that a major problem of GLC organization was to separate strategic policy issues from the operation of executive functions; so far as planning is concerned, this has been attempted, first, by means of the establishment of a Strategic Planning Committee and, secondly, by the institution of the Strategic Planning Board.

There is, however, a certain inconsistency in this approach since the combined Planning and Transportation Department is concerned both with strategic planning and with a very important group of executive functions concerned with traffic measures and highway construction and maintenance. Since this department's views must obviously be a major influence on the Strategic Planning Committee's thinking, there is a great danger that strategic policy will be unduly influenced by transportation considerations.

This issue, however, raises a question of great importance for the role of the GLC in the new system of local government. It might be argued that the 1963 Act was expressly designed to facilitate the integration of planning and transportation by the GLC and that this was in accordance with the Herbert Commission's analysis. The current organization of the GLC is, therefore, simply a recognition of the importance of this integration. The Herbert Commission did, however, also recognize the importance of other considerations in planning, especially housing. As has been indicated earlier,[70] this recognition was not carried through either into the Commission's detailed proposals or into the 1963 Act. The question is, therefore, whether the inconsistencies in the GLC's approach really go much deeper and are an indication more of failure of the statutory provisions to give the GLC adequate powers than of failure by the GLC to make effective use of the powers which it has.

It was argued in an earlier chapter[71] that the GLC could have adopted an alternative approach, particularly to the planning and related group of functions, which might have proved a more satisfactory means of carrying out its role. Nevertheless, there are considerable difficulties in the way of its achieving a satisfactory organization. These difficulties are in part due to inadequacies in the statutory definition of powers; but much more they arise from the fact that the GLC is different, a local authority with a peculiar range of duties to perform. This is why the story of the Research and Intelligence Unit is instructive, just because it focuses attention on the problems which arise within the traditional approach to organization in a new situation of this kind.

The GLC, then, is to be regarded as a body rooted in and developing out of local government as it has been traditionally understood in this county. In its approach to its organization and staffing, the GLC has fully recognized this, moving essentially towards the kind of streamlined and co-ordinated structure which most large authorities have found to be increasingly necessary; at the same time, it has not yet satisfactorily resolved the problems of its own distinctive range of functions.

12*
The London Boroughs

Introduction

The role of the London boroughs in the new system of London Government was stated by the Herbert Commission in the following terms:

'the primary unit of local government in the Greater London area should be the borough, and the borough should perform all local authority functions except those which can only be effectively performed over the wider area of Greater London or which could be better performed over that wider area.[1]

What is important in this statement is the link between the term 'primary unit' and the distribution of functions between the GLC and the boroughs. It puts the emphasis on the boroughs as the main providers of local government services in a quantitative sense. It does not follow from the Commission's view that the boroughs were to have all the most important local government functions. The services requiring to be performed by a Greater London authority might be few in number but of major importance for the functioning of local government in Greater London. Indeed, in relation to functions to be shared between the Council for Greater London and the boroughs the Commission themselves proposed that the former should have sole responsibility for making a development plan for Greater London and should have a major role in traffic and highways and in education.

The Government, in the White Paper of 1961, endorsed the Commission's view of the boroughs as the primary unit.[2] By creating larger boroughs, however, and by strengthening their powers, particularly in relation to the GLC, they put a somewhat different and narrower interpretation on the range of functions which required to be performed on a Greater London scale, and correspondingly broadened the meaning of 'primary unit'.

* The author is grateful to Mr David Southron for much detailed help in the preparation of this chapter.

The London boroughs represent a new development in English local government. They are second-tier, most-purpose authorities, differing not only from all-purpose county boroughs, but also from the county districts with their limited range of functions.[3] The London boroughs have exclusive responsibility for many important functions which are elsewhere the responsibility of counties and county boroughs, especially the personal social services and, in Outer London, education; at the same time, where they share functions with the GLC, it is not as subordinate authorities but as authorities with distinct and separate responsibilities.

Thus, two questions are important about the boroughs which were constituted under the London Government Act, 1963. First, how have they operated in practice as primary units of local government? Secondly, how they have interpreted their somewhat unusual role in relation to the GLC?

Characteristics of the boroughs

There are wide differences between the London boroughs[4] in population size, resources and social composition. Even in population size, which was the basic criterion for the amalgamation of the old authorities to form the new, the largest borough (Croydon) is well over twice the size of the smallest (Kingston upon Thames).[5]

More significant is the variety of social conditions of the London boroughs. This can be seen most obviously from census data relating to socio-economic groupings. There were, in 1966, 6 boroughs[6] where the percentage of employers, managers and those engaged in professional occupations was less than 10 per cent; on the other hand, there were 8 boroughs[7] where the proportion was over 25 per cent. The range was from 5·6 per cent in Tower Hamlets to 29·8 per cent in Kensington and Chelsea. Correspondingly, the proportion of unskilled manual workers ranged from 3·5 per cent in Barnet to 18·3 per cent in Tower Hamlets.[8] Similar distinctions can be made between those boroughs with relatively favourable housing conditions and those with relatively unfavourable conditions.[9]

There is also a close association between the social characteristics of the boroughs and their political complexion. In the 5 elections for London authorities since 1964 (3 GLC, 2 borough), 10 boroughs have always voted Conservative in GLC elections and had a Conservative majority in borough elections. Eight of these 10 boroughs have been identified above as having a high proportion of employers, managers and professional men. Only 4 boroughs have always voted Labour and all of these are in the group of boroughs with a low proportion of men in these socio-economic groupings; the other two boroughs in this group (Hackney and Islington) have voted Labour in all elections except the 1968 borough elections. Between these two groups come the marginal boroughs, in the sense that they have changed their voting allegiance at least once in the five elections.

The position may be summed up by saying that there is a sharp contrast, socially and politically, between the group of boroughs formed from the eastern and south-eastern parts of the old County of London and its fringes, and the outer ring of suburban boroughs, especially those formed from the old County of Surrey, but in a broader sense Inner London (the former County of London) has in many ways different characteristics from Outer London. The former tends to contain the boroughs with relatively unfavourable social conditions and the latter to have a higher proportion of boroughs with relatively favourable conditions.

One other point about the boroughs is important, especially in relation to the future. The Government in 1961 accepted that 200,000 population was the desirable minimum for the London boroughs.[10] The population of London is, however, declining. The effect on the size of the London boroughs can be seen from the following:

Table 12.1

Number of London boroughs in different population ranges

Population	1965	1970	1981
over 300,000	7	5	4
250,000 – 300,000	9	6	5
200,000 – 250,000	11	14	12
150,000 – 200,000	4	6	10
under 150,000	1	1	1

Sources: Registrar General's Population Estimates, 1965, 1970; GLDP Statement, s.3,T.1,p.17, *higher* 1981 forecasts.

Thus when the London reforms came into effect only 5 of the 32 boroughs fell below the minimum; five years later there were 7; and in ten years' time, if the GLC forecasts prove correct, one-third of the London boroughs will be below the population level, which was originally considered desirable.

Moreover, the decline in population is likely to be very unevenly spread. Very marked declines are projected in the GLDP for some boroughs; Hackney, Islington, Newham, Southwark and Tower Hamlets are expected to show decreases of population of 25 per cent or more between 1961 and 1981. By contrast, some of the outer boroughs are expected to show modest increases in population, such as Bromley (8 per cent), and Sutton (over 5 per cent).

If these projected changes in population size do occur, they will tend to make even more prominent the marked difference in character between boroughs in different parts of Greater London. Such differences are important for the present analysis partly because of the possible effect they may have on the performance of functions by boroughs which have widely different problems and may take different attitudes to such questions as priorities and the intensity of provision of services. But these differences are also important for their possible effect on the view which the boroughs have taken of their role in the new system of London Government. These points are considered later but before that the consequences of the London reforms will be examined in relation, first, to the way in which the boroughs have organized themselves for their tasks and, secondly, to their functioning as democratic bodies.

Internal administrative arrangements

All the boroughs initially set up committee and departmental structures of the traditional kind.[11] Nevertheless, there is some evidence that at this stage, despite wide differences between them, they consciously aimed at limiting the number of both committees and departments. The researches of the Maud Committee on Management in Local Government enable comparisons to be made of the position in 1965 (Table 12.2).[12]

Even the Outer London boroughs with responsibility for education had far fewer committees and departments than county boroughs of comparable size. For this there are two reasons: first London boroughs do not, even in Outer London, carry the full range of responsibilities of a county borough; apart from the fact that some functions (eg the fire service) are in London the responsibility of the GLC, county boroughs frequently perform services which in London

Table 12.2

Number of committees

	Average	Range
County Boroughs (200,000 – 400,000 pop.)	23	15 – 35
London Boroughs		
Inner London	13	10 – 18
Outer London	15	11 – 18

Number of Separate departments

	Total authori-ties	6 depts or less	7–9	10–12	13–15	16–18	19–21	22–24	25 & over
County Boroughs (200,000 – 400,000 pop.)	10	—	—	—	—	1	1	5	3
London Boroughs									
Inner London	12	1	3	7	1	—	—	—	—
Outer London	19	1	3	9	4	2	—	—	—

were, in 1965, subject to special arrangements (eg water supply, police, public transport). Secondly, in long-established authorities there is a tendency for committees and departments to increase unless conscious means are adopted to limit them. The London boroughs, as new authorities could at least start with the minimum number of committees which they thought desirable.

The picture then of the London boroughs in 1965 is of authorities organized on conventional lines but with fairly compact departmental and committee structures. At quite an early stage, however, some boroughs began, in advance of the publication of the Maud Committee's report, to move in the direction of modifying the traditional structure. This was particularly so in the case of arrangements for securing a greater degree of co-ordination of local authority activities, a major concern of the Maud Committee and of other authorities, such as Coventry and Newcastle, which re-examined their structure in advance of the Committee's report. Camden, for example, in 1966 set up an Advisory Committee consisting of leading members of the majority party only to co-ordinate major policy matters.[13]

The main impetus to re-examination of committee and depart-mental structure, however, in London as elsewhere was the publica-tion in 1967 of the Maud Committee's report. The London boroughs, like other local authorities, rejected the idea of a small management board consisting of members of both majority and minority parties, but many of them have accepted the need for a policy co-ordinating committee. Seventeen of the 32 London boroughs had by 1969 instituted such committees, and in all but four cases this development had followed the Maud Committee's report; 9 of these policy committees, like the GLC's Leader's Co-ordinating Committee, have majority party membership only.[14] It is interesting to compare this situation with that in the county boroughs. Of 77 CB's examined by the Institute of Local Government Studies, only 28 had by 1969 established policy committees, although a higher proportion of these pre-dated the Maud Committee's report than in the case of the London boroughs; 11 of the county boroughs had one-party policy committees.[15]

In other ways too developments in the London boroughs parallel those in the county boroughs. Many of the London boroughs have, like the county boroughs, reduced the number of their committees, but, in line with what was said above, the scope for and degree of reduction in the number of committees in county boroughs of a similar size has been greater than it has been in the London boroughs.[16] In the extent of delegation to officers, the degree to which the clerk has become responsible for co-ordinating the work of the officers, and the efforts made to reduce the number of departments the London boroughs have moved forward in line with the general trend in local government.

Many have, for example, designated the clerk as chief executive officer, often in association with a departmental reorganization, as at Sutton, and some have recruited chief officers from outside local government. Haringey, for example, which reorganized its administration in a five-man directorate under a chief executive officer, recruited two of the five (the director of public services and the director of finance) from the private sector of industry;[17] Wandsworth[18] and Southwark[19] both appointed chief executive officers with experience outside local government, in the first case a solicitor from industry and in the second a district auditor. Apart from the boroughs mentioned, a number of other boroughs have grouped departments together under directors in an effort to achieve greater co-ordination, among them Brent, Greenwich, Islington, Hammersmith and Kensington and Chelsea.

In all this, the London boroughs as a whole have shown a

readiness to adopt current ideas on the need for revision of internal management procedures, and some of them have been in the forefront of developments. The London Government reforms coincided with a period when increasing attention was being paid to questions of internal management of local authorities; they gave some stimulus to this process.

Local democracy

The Herbert Commission, it will be recalled, attached very great importance to sustaining the health of local government and they held that the existing structure which they had to examine was 'not conducive to the health of representative government'.[20] Two aspects of a healthy local government which it is important to consider here are:

i.   the quality of councillors;
ii. public participation in the democratic process.

The Herbert Commission argued strongly that there had been a decline in morale in local government in Greater London, largely because of uncertainty about the future of local government in the area. In particular, they claimed that 'the erosion of powers and the irritations arising out of the practice of delegation have seriously lowered the status of the boroughs and other county districts in the eyes of potential councillors of ability and ambition' so that there was difficulty in attracting in sufficient numbers 'the right kind of councillor'.[21] If the Commission were right in believing that the remedy lay in strengthening the boroughs' powers so that there was a worthwhile job for councillors to do, and also in removing uncertainty about the future, then the 1963 Act should have provided just the kind of opportunity which they had in mind.

To examine the effect of the reforms on the recruitment of borough councillors presents formidable difficulties. There is, first, the obvious practical point that for the first borough councils which lasted from 1964 to 1968, the problem was one of too many rather than too few candidates since the new boroughs had less councillors than the combined total of their predecessors. Even after the 1968 elections there was still a strong element of continuity of membership, as is indicated by the examples in Table 12.3. On the other hand, the fact that the 1968 election results represented a strong swing from

Labour to Conservative inevitably brought many new faces on to those councils which changed their political complexion (Hackney, for example).

Table 12.3

London Borough	May 1964		May 1968	
	(1)	(2)	(3)	(4)
	Total – Councillors & Aldermen	Council members also a mbr. of a constituent authority in 1964	No. from column (1) still members of council	No. from column (2) still members of council
Enfield	70	53	30	23
Haringey	70	54	20	19
Kensington & Chelsea	70	53	35	28
Tower Hamlets	70	63	51	45
Westminster	70	64	36	34

Attracting the right kind of councillor has to be seen first, therefore, in the context of the continuity between old and new authorities. but even allowing for this it is extremely difficult to make any worthwhile judgment about the quality of councillors before and after the London reforms, largely because of the absence of any agreed criteria which can be applied. The Herbert Commission, it is true, indicated what kind of a person they would expect a good councillor to be; he must 'know his people' and not simply those who had voted for him, must use professional advice but not be a slave to it, have close relations with the officials but not interfere in the detailed executive work, and act as an intelligent link between his party and the council. Although they did not expect all councillors to fulfil all these requirements, 'a high standard of intelligence, experience, personality and character should be aimed at and can be legitimately hoped for'.[22]

This may represent an admirable declaration of faith on the part of the Herbert Commission based on the numerous discussions which they had had with elected members as well as on the more formal oral evidence which they had heard. But their report was written at a time when there was little information available even about the characteristics of councillors in Greater London or elsewhere, let alone how effective they were at carrying out their tasks. Since then individual studies have been made,[23] and a great deal of information relating to councillors generally has become available from the researches done for the Maud Committee on Management in Local Government. Even these, however, throw very little light on the question of what makes a good councillor and under what conditions he is likely to be attracted to local government. More recent research has shown that there are no easy answers to either question.[24] But even more it has shown that a great deal more work needs to be done before one can make any meaningful assessment of the extent to which the London boroughs have attracted councillors of high calibre.

There is, meanwhile, some information which bears on this question. The Maud Committee's researches showed that there was almost universal difficulty in recruiting candidates to stand at local government elections.[25] The problem has become easier in Greater London. The reason may be that as the Herbert Commission believed,[26] the London boroughs with their greatly strengthened powers compared with the previous authorities, especially in the old LCC area, are more likely to attract candidates since they provide scope for the exercise of abilities and the satisfaction of ambitions. But, as a recent study has suggested,[27] there does not seem to be any strong connection between a local authority's powers and its ability to attract potential councillors. The improved position is more likely to have arisen partly from the fact that fewer councillors are needed under the present system, and partly from the increased involvement of the political parties in London borough elections. As the Maud Committee researchers noted: 'competition for seats is strongest in urban areas and appears to bear a relationship to the presence of the party organizations'.[28] To the extent that, as noted in an earlier chapter, reform of London Government compelled the parties to devise new organizational machinery, it can clearly be shown that the reforms did contribute to a sharpening of competition for seats on the new London borough councils. The almost complete elimination of uncontested elections from London borough elections compared with those for the previous authorities is a measure of the change.

The Herbert Commission also attached importance to authorities not being too large, since, if they were, either councillors would represent so many people that they would find it difficult to keep in touch with people's needs, or the council would need to be so big that the transaction of business would tend to fall more and more into the hands of the party caucuses.[29] They included a table in their report showing for each authority in the Review Area the population per councillor. In view of the importance which they attached to this question it is surprising that they made no suggestions about the size of the council of their proposed Greater London boroughs, although they did discuss this question in relation to the proposed Council for Greater London. None of the previous boroughs and urban districts had more than 60 councillors, and if we assume that the Commission had in mind that this was about the maximum size for a manageable council,[30] then their ideal was probably that each councillor should represent a population of about 3,000, given that most of their proposed boroughs fell in the 150,000 – 200,000 population range.

Under the London Government Act, 1963, the London boroughs were restricted to a maximum of 60 councillors each despite the opposition of some of the Government's own supporters.[31] The effect has been to reduce the disparities between different authorities in the number of people per councillor but also to increase the number under the present as compared with the previous system. Metropolitan boroughs for example varied, from Holborn with a population of 509 per councillor to Wandsworth with 5,647, and Middlesex boroughs from Brentford and Chiswick with 2,110 to Hendon with 5,611 whereas the London boroughs vary only from Kingston upon Thames with 2,408 to Lambeth with 5,488. The great majority of the previous authorities fell within the range 1,500 to 3,000, whereas the London boroughs fall mostly within the range 3,500–4,500.

In the light of later research,[32] however, it seems doubtful whether such variations in the number of people per councillor are of as much significance as the Herbert Commission supposed. Clearly there are significant differences between a GLC councillor with 70,000 or 80,000 people in his constituency[33] and a London borough councillor with 4,000, but whether a councillor has 3,000 or 5,000 people in his area is unlikely to affect to any great extent his effectiveness as a democratic representative.[34]

Public participation in the democratic process, like the quality of councillors, proves to be a somewhat elusive subject on close examination. It is now generally accepted that the one measurable index which is readily available, the percentage of people voting in local elections, does not by itself tell us very much about the extent to which people feel interest and involvement in local government. The survey of electors carried out by the Government Social Survey for the Maud Committee on Management in Local Government did indeed reveal that 89 per cent of those interviewed thought it important to vote, although significantly a majority (56 per cent) agreed with the view that 'The people you vote for say they'll do things for you, but once they're in they forget what they've said'.[35] Apathy towards and ignorance of local government in general are the theme of a number of recent surveys;[36] a majority (58 per cent) of councillors interviewed by the Social Survey for the Maud Committee also thought that the public were not interested in local government and the proportion was as high as 72 per cent among metropolitan borough councillors.[37]

In the light of this situation, interest has tended to concentrate more on other means of participation. It has been argued recently, for example, that the reform of local government by creating local authorities 'truly responsible for their own decisions' should encourage the growth of local interest groups with a real chance of influencing the decisions of local authorities.[38] Little evidence has been revealed by the researches undertaken for the current study that this has happened in the London boroughs,[39] or indeed that the reforms have so far led to any significant changes in public participation in the democratic process. The subject is one of such general interest and importance, however, that the Greater London Group hope to include a study of it in a further research project on democracy and politics in Greater London.

How far, then, has the creation of the London boroughs justified the Herbert Commission's views on the health of local government? Apart from the somewhat negative evidence of the figures of turnout in borough elections,[40] there is little evidence either in support of or opposed to those views. That the London boroughs have a worthwhile job to perform need not be doubted; that this job is more important than the job which the previous authorities had to perform is also evident, particularly in the case of the metropolitan boroughs. The London boroughs, with the partial exception of Croydon and Newham, are now responsible for a greater range of services and they command greater resources than the previous authorities. What is more doubtful is whether this situation is healthier in the sense that it has had a beneficial effect either on the calibre of councillors or on the participation of the public. The Commission offered no real evidence on these questions, relying basically on their faith in local government, derived as much from their own informal discussions with councillors as from any formal evidence submitted to them. They may well have been right to take this viewpoint, and to believe that the London boroughs would offer a better chance than their predecessors of reviving a healthy local democracy. Unfortunately, in the present state of our knowledge about what constitutes 'a viable system of local democracy' there is little positive that can be said about the London boroughs in this context.

Performance of the boroughs

The evidence of the functional chapters raises two points about the performance of the boroughs in those functions which the 1963 Act assigned exclusively to them, that is, the health, welfare and children's services, and in Outer London, education, together with the library service to which reference is made in the finance chapter.[41] The first point of note is that, generally speaking, the boroughs have proved successful as providers of these services; the second is that there are marked differences in the performance of different boroughs and these differences correspond roughly to the broad distinctions which have been mentioned earlier in this chapter and, more particularly, to a division between Inner and Outer London.

The transfer of county services to the boroughs inevitably produced certain disadvantages. The boroughs have not, for example, been able to provide specialization whether of men or of equipment and buildings on the same scale as the counties formerly did, and in certain fields, notably further education, the smallness of the new authorities has been an evident disadvantage. On the other hand, the fact that the boroughs are smaller has led, as in the case of education, to specialist advisers being more widely available than they were before; it has also made senior staff more accessible than before.

These consequences, like the greater variation in standards of provision of services, were to be expected from the creation of 33 authorities in place of 9. What was less to be expected, however, was the considerable effort made by the boroughs to develop and improve the services which they inherited. This applies not only to the raising of standards in services like libraries, which were inherited from the previous county districts; there have also been marked improvements in what were previously county services, eg in the provision of meals for old people and of accommodation for children.

The success has not, however, been uniform. The information presented in the functional chapters is not sufficiently detailed to enable comparisons of performance of individual boroughs to be made, but it does indicate that there is a good deal of difference in the way in which boroughs have approached their new responsibilities. In particular, there seems to be an important dividing-line between the Inner and the Outer London boroughs. This is most noticeable in the children's service, where the inner boroughs are better staffed and have generally developed better provision than the outer boroughs; but there are other examples too of services which have been developed more fully by the inner than the outer boroughs (eg the meals service for the elderly).

As was indicated earlier, the distinction between Inner and Outer London does not exactly correspond to the most marked contrasts in social and political conditions between boroughs; nevertheless, there are certain broad differences between Inner and Outer London which suggest that social and political factors play an important part in their respective performances. Inner London contains a larger proportion of boroughs with unfavourable social conditions and of boroughs which tend to have Labour majorities; Outer London has a greater preponderance of boroughs with relatively favourable social conditions and mainly Conservative majorities. There are some notable exceptions – Westminster in Inner London, for example, and Newham in Outer London; there are also boroughs in both Inner and Outer London which are intermediate in social conditions and marginal in political terms, such as Lewisham and Waltham Forest. But these variations do not detract from the broad pattern of explanation.

There are clearly other factors at work, too. There is, for example, the importance of the inheritance of the boroughs, both in men and institutions, to which Bleddyn Davies has drawn attention.[42] There is, in the case of the children's service in particular, the fact that the Outer London boroughs have much smaller numbers of children in care which is clearly a factor in the difficulties which these boroughs have had in providing a good standard of service. There is also a direct link with the specific provisions of the London Government

Act, 1963, in the fact that education, with its large financial demands, is in competition with other borough services for available funds in the outer boroughs but not in inner London. Though not directly measurable, this factor is likely to have an influence on the relative development of services such as health, welfare and children, in particular tending to limit their development in the outer London boroughs.

No mention has yet been made of the services shared between the GLC and the boroughs. There is again a good deal of evidence of the capacity of the boroughs to deal adequately with the provision of these services within their areas. The boroughs employ a high proportion of the qualified planners within Greater London, for example, and many of them have established large and effective planning departments; in their first five years the boroughs under-took large programmes of slum-clearance and house-building. Again, wide variations in the performance of different boroughs have been a feature of the situation, most notably in the programme of house-building.

Yet these functions which are shared between the GLC and the boroughs for that very reason raise broader issues. It is not simply a question of whether the boroughs are able to operate satisfactorily within their own areas, but what their role is in these functions within a Greater London context, and in particular, what their relationship is to the GLC and to each other.

Co-operation and co-ordination: the London Boroughs' Association

The relationship between the GLC and the boroughs in a number of important shared functions is, in theory at least, not that of a master and subordinate but one of partners dealing with different aspects of the function. This is something new in English local government; it implies not only a particular relationship between each borough and the GLC but also some means whereby the boroughs as a whole may put forward their views. At the same time, the boroughs as 'primary units' tend to see themselves as largely independent units of government.

It is therefore relevant to the working of the new system to ex-amine how far the boroughs have seen a need to look at problems jointly, the means they have adopted for doing so and what has been achieved in practice. It is particularly important here to see on what issues there has been general agreement among the boroughs. There have been some issues where the boroughs individually have taken different attitudes to proposals made by the GLC. The proposed transfer of GLC housing to the boroughs was perhaps the most notable example,[43] but a very similar situation arose over the transfer of parks. Under the London Government Act, 1963, the parks and open spaces belonging to the London and Middlesex county councils were transferred to the GLC but provision was made for the GLC to consult the boroughs and submit proposals to the Minister by 1 April 1970 for transferring some or all of these to the boroughs. When the GLC duly drew up proposals, they were welcomed by the majority of the boroughs. Tower Hamlets and Southwark, however, were opposed to them, largely on financial grounds.[44]

In considering the extent to which the boroughs have made common cause, the activities of the London Boroughs' Association merit first consideration. The LBA was formed in June 1964, as the London Boroughs' Committee. The substitution of 'Association' for 'Committee' in the title was made in 1966 and is of some significance when viewed in relation to the objects of the Association as set out in its constitution:

(1) 'To protect and advance the powers, interests, rights and privileges of the constituent councils and to watch over those powers, interests, rights and privileges as they may be affected by legislation, or proposed legislation or otherwise;

(2) To discuss questions of London Government and to advise and assist the constituent councils in the administration of their powers and duties;

(3) To express the views of the Association and to consult with appropriate bodies or persons whenever deemed advisable. Provided that the Association shall not have power to bind or commit a constituent council.'

The first of the objects of the Association is closely parallel to that of the major local authority associations, such as the AMC and the CCA, all of which exist 'to protect the interests, rights and privileges of their members, but not specifically, their powers; the addition of the latter word in the LBA's constitution is derived directly from the constitution of the LBA's immediate predecessor, the Metropolitan Boroughs' Standing Joint Committee. The latter body had a long history, having been formed in 1912, and it is quite clear that the

London Boroughs' Committee was set up very largely on the initiative of prominent members of the MBSJC, to carry on its work in the larger context of Greater London Government. This was very much in accordance with the wishes of the Herbert Commission who had warmly praised the work of the MBSJC.[45]

There are, however, notable differences between the spheres of responsibility of the metropolitan boroughs and those of the London boroughs. The London Boroughs' Association reflects this difference to some extent in its title. The MBSJC was set up as a joint committee under the Local Government Act, 1948, and provided a useful forum for discussion of common problems by the metropolitan boroughs particularly in relation to the LCC. The LBA felt itself to be much more of a local authority association on the lines of the AMC or the CCA but representing the specific interests of the London boroughs. The question is how this role has operated in practice.

It is, first, interesting to note that a formal agreement was made at an early date between the AMC and the LBA under which the latter were to speak for the London boroughs on all matters concerning London Government or particularly affecting London, while the AMC agreed to make arrangements for the views of the London boroughs to be made known in other matters, such as wage and salary negotiations. The London boroughs thus have this dual and somewhat ambiguous role; on the one hand they are boroughs, members of the AMC[46] and represented by the AMC on matters which affect boroughs generally; on the other hand, because of their distinctive status and powers they stand apart from the AMC in their own separate organization whenever specifically London questions arise.

There is no parallel to this situation elsewhere in the country. There are, it is true, local associations of authorities which provide a forum for discussion of common interests but these have never aspired to a representative role to the same degree as the LBA has done. On one level, therefore, the LBA has functioned very much as the AMC or any other association of local authorities particularly as a channel of communication between the boroughs and central government departments – and this function forms a major part of its work.

Closely allied to this role as a channel of communication with the Central Government is the role of the LBA as representing the views of the boroughs to the GLC and of dealing with GLC requests or proposals affecting the boroughs as a whole. This role is very similar to that of the Metropolitan Boroughs' Standing Joint Committee, a principal motive for whose creation was the desire to put a metropolitan borough point of view in negotiations with the LCC. The activities of the LBA in this direction were particularly prominent in the early years of the new system when there was a certain fluidity in the definition of powers between the GLC and the boroughs (eg in the negotiations over the development control regulations).

First and foremost, therefore, the LBA has been, as its constitution implies, a defence association, acting to protect the powers and interests of the London boroughs primarily in relation to the Central Government and the GLC. To see what it has done beyond this it is necessary to consider the different kinds of issue which have arisen affecting London as a whole. There were, first, a number of purely transitional problems such as the need to make arrangements for inter-authority accommodation in the children's welfare and mental health services or to continue the LCC practice of arranging an annual holiday for mentally subnormal children attending London Training Centres. These were problems on which decisions had to be taken, and which sprang directly from the transfer of county services to the boroughs under the London Government Act, 1963. For the most part these questions were settled relatively easily. Some, however, proved much more intractable; a particular example was the issue of grants to voluntary bodies and especially those operating in the social services. Because it illustrates not only the constraints within which the LBA has operated but also some of the unintended consequences of the reforms this issue is discussed more fully in the following paragraphs.

Before 1 April 1965, the LCC and other counties and county boroughs made grants under the National Health Service Act, 1946, and the National Assistance Act, 1948, to various voluntary organizations providing health and welfare services within their areas. Under the London Government Act, 1963, these powers were given concurrently to the GLC and the London boroughs. One problem was that many of these bodies which operated on a national or regional scale now had to seek help from a much larger number of authorities, each of which naturally tended to consider such requests from the point of view of its own needs. There was considerable uncertainty and financial difficulty as a result for some of these voluntary bodies, such as the National Council for the Unmarried Mother and her Child.

The LBA, aware of these problems, commissioned the GLC Research and Intelligence Unit to undertake a review of the whole question in 1968. Dr Benjamin, the then director of the unit, pointed out in his report that apart from the effects on the voluntary organizations themselves, the existing arrangements were administratively cumbersome; he, therefore, made a number of suggestions for simplifying the system, eg for increasing the number of organizations to whom grants would be paid by the GLC to a total amount specified by the LBA and recoverable by precept on all London boroughs. Following this report, the LBA in June 1969 proposed a clause for inclusion in the GLC's annual General Powers Bill to provide that if at least three-quarters of the London boroughs desired it the GLC could make contributions to voluntary organizations 'specified by resolution of the LBA' up to a rate of one-tenth of one penny in any one year and recover it by precept on all the boroughs.[47]

There are several points of interest about this development. There is, first, the statutory recognition of the LBA as the spokesman for the London boroughs. Furthermore, in spite of the proviso in the Association's constitution that it cannot 'bind or commit' constituent councils, here is an instance where an Association resolution perhaps supported by only 24 of the 32 boroughs could bind all London borough councils to expenditure, although admittedly on a very limited scale. The main significance of the LBA's role in grants to voluntary organizations lies in this latter development which goes beyond the normal function of a local authority association. The break-up of county services and their transfer to the London boroughs posed this dilemma. Either the boroughs would have had to admit that the question of grants to voluntary organizations was a Greater London question and therefore appropriate for the GLC as the Greater London authority or they had to show that they could themselves co-operate to find an answer. The first alternative was ruled out (even had the GLC been willing to do it) precisely because it would have called in question the boroughs' ability to provide services which had been assigned to them; the boroughs therefore had to find an answer and the LBA was the obvious, indeed the only, means by which this could be done.

Nevertheless, the long negotiations and discussions[48] illustrate the difficulties facing a body such as the LBA when questions arise which impinge upon the individual boroughs' policy-making powers. There is a conflict between an individual borough's desire to decide what voluntary organizations it should support and how much it should contribute, and the collective needs of these organizations operating on a regional or national basis. Even more obvious is the conflict which may arise in more politically sensitive areas of policy.

Housing is probably the best example here. Sir Keith Joseph, it will be recalled,[49] had put considerable emphasis on the need for the boroughs to agree on a London housing programme. This has not happened, mainly because there are fundamental differences of interest and hence of policies between the London boroughs. The LBA, because it is essentially a defence organization, is not in a position to resolve such fundamental differences.

Perhaps this fact emerges even more clearly from two specific housing issues which have come before the LBA: the standardization of qualifications for admission to housing waiting lists; and a common rent policy for Greater London. On the first, progress was made, although not without considerable difficulty; the second was rejected by the LBA. It is not hard to see why. Rent policy is one of the most sensitive of all local political issues. Moreover, it is even more sensitive in the London context because it involves the issue of the relationship between the boroughs and the GLC who in fact first put forward the idea of a common rent policy. For the LBA, discussion of a common rent policy would have resulted either in a recommendation which many boroughs would have rejected, thus defeating the whole object of the proposal, or in some kind of compromise recommendation which again would have had to be so flexible as to be useless.

The LBA then, is not and cannot be within the present structure of London Government, the body to resolve major policy issues which touch on the capacity of the boroughs to perform adequately the functions assigned to them. This is not to dismiss the work of the LBA as negligible. On the contrary, although it is by no means 'the key articulating element in the structure' claimed by Mr Derek Senior,[50] it performs a useful, even an essential role in the new system in a way which has no real parallel with apparently similar bodies found under the previous system. It has been highly successful as the champion of the 'powers, interests, rights and privileges' of the London boroughs in negotiating with the GLC, the central departments and other bodies; it has, for example, represented them in the Rate

Support Grant negotiations and the discussions on the revision of capital expenditure and loan sanction procedure; and it is invited by the Central Government to make recommendations for appointment to national bodies such as the Central Health Services Council.

Again, a glance at any agenda of an LBA meeting will indicate the very wide range of matters raised and discussed, from legislation of particular interest to the London boroughs to negotiations with street traders over increased charges or the difficulties over coke supplies to education establishments. As a forum for discussion of common problems which do not carry profound policy implications the LBA has thus performed a useful and valuable role. It would indeed be true to say that in this kind of way the London boroughs, acting through the LBA have a much closer relationship with each other than would normally be found in a group of contiguous authorities, and particularly, among authorities as large and powerful as the London boroughs are. This reflects the unique status of the boroughs and their special relationship within London Government which has produced a conscious effort on the part of the LBA to evolve a distinctive organization for this purpose.

Co-operation and co-ordination: the London Boroughs' Training Committee (Social Services) and the London Boroughs' Management Services Committee. Apart from the LBA, two other organizations exist with the specific purpose of furthering inter-borough co-operation; they are the London Boroughs' Training Committee (Social Services) and the London Boroughs' Management Services Unit (LBMSU) which operates under the aegis of the London Boroughs' Management Services Committee.

As the names of these organizations imply, they both have specific and limited tasks to perform, and they both exist because it was felt that more could be achieved by pooling resources than by each borough acting alone. But their different origins, functions and degree of success illustrate the extent to which inter-borough co-operation is possible.

The London Boroughs' Training Committee[51] was set up as a direct consequence of the London reforms. The county councils, and in particular the LCC, had had training schemes for officers in health, welfare and children's departments; the LCC suggested during the transitional period that means should be found for continuing its existing training arrangements under the new system. The London

Boroughs' Committee (as it was then) encouraged the idea, and proposed that it should be extended to all London boroughs who were invited to become members of a newly-constituted London Boroughs' Training Committee. By the middle of 1965 23 boroughs had joined and by March 1970 only Haringey[52] and Harrow were not members. The Committee is financed by equal contributions from member boroughs.

The Committee has proved a successful organization in its three main spheres of activity: co-ordinating arrangements on behalf of the London boroughs for professional training by outside bodies; providing special courses for staff of member boroughs;[53] and advising and assisting boroughs in their own training programmes. But two questions, each pulling in a different direction, may be raised about the organization: first, if it is a good idea to have co-operative arrangements for the staff of social service departments, why not also have similar arrangements for other staff, eg planning? Secondly, if the real reason for the initial success of the Training Committee was the boroughs' lack of experience in the social service field, may we not see the boroughs increasingly inclined to build up their own training departments as their experience grows and less inclined to take part in joint enterprises?

The need to examine afresh the position of the London Boroughs' Training Committee (Social Services) has arisen because of developments since the London Government Act, 1963, and, in particular, the establishment of the Local Government Training Board. In 1969–70 the Board accepted contributions to the LB Training Committee as admissible for grants, but has obviously found some difficulty over the London arrangements generally. Not only are there other bodies with training responsibilities (eg the Greater London Whitley Council), but there are over 50 training officers in post in the London boroughs. Indeed the latter development has been stimulated by the LGTB's general policy of encouraging the employment within authorities of qualified training officers.

These developments are likely to bring out sharply the characteristic dilemma of the London boroughs: on the one hand there are obvious advantages in pooling resources, on the other hand there is the natural tendency for all local authorities to emphasize their capacity to deal exclusively with the services assigned to them. So far, in the social service training field, the advantages of co-operative arrangements have been accepted by the great majority of boroughs, but the Training Committee may not survive in its present form, although it seems likely that some body with general training responsibilities may well continue to be needed by the London boroughs.

The LBMSU has had a very different history. It derived directly from the Metropolitan Boroughs (O & M) Committee set up in March 1951 as a joint committee to enable the metropolitan boroughs to take advantage of the then relatively new technique of Organization & Methods. By 1965 the Committee had in membership 23 of the 28 metropolitan boroughs and had branched out into the fields of work study and computers besides retaining its O & M activities. In 1965 the Committee was re-named the London Boroughs' Management Services Committee, and became responsible for the work of the LBMSU under its director.

By 1970 the Committee had 18 members; whereas all but three of the twelve Inner London boroughs joined, only 9 of the 20 Outer London boroughs did so.[54] Each borough contributed £5,000 a year to the work of the unit besides paying for individual projects or services. The work and the staff of the unit increased greatly from 1965 onwards,[55] but this was not entirely accounted for by the growth of work for the London boroughs since the unit also undertook work for authorities outside London.[56] Much the biggest part of the LBMSU's work for the London boroughs was a quinquennial review of each borough, including an assessment of the manpower requirements of every service and examination of the committee structure and inter-departmental relations.

Thus the LBMSU, in spite of being a large organization, has never managed to attract much more than half the London boroughs, and even among its existing 18 members some are only lukewarm in their support. There are a number of reasons for this situation, but undoubtedly the basic one, and the one which most concerns us here, is the fact that some boroughs were dissatisfied with the work done and believed that they could do the job better themselves. Greenwich, for example, which in November 1969 announced its intention of withdrawing from membership of the LBMSU in 1972, decided to enlarge its own O & M team.

This tendency stems directly from the London reforms which created larger and more powerful boroughs, better able to provide their own management services. The tendency is most obvious in the installation of computers; many boroughs have their own computer installations and the LBMSU's role is largely confined to operating two joint computer schemes.[57] Nevertheless, there are arguments, as in the case of training, for pooling scarce and specialized resources for the benefit of all London authorities. In the case of management services they have not proved strong enough in the eyes of many boroughs to outweigh the advantages of acting independently.

The LBA, the London Boroughs' Training Committee (Social Services) and the LBMSU all contain elements which specifically reflect the London situation. But their primary interest here has been in demonstrating the limits which exist to co-operation among the London boroughs. It is true that there are closer relationships between the London boroughs than are generally to be found among powerful neighbouring local authorities. But the existence of those relationships does not mean that the 32 London boroughs have evolved or could evolve a common policy on major issues. The differences between the boroughs are more profound than their similarities.

Conclusion

The London boroughs have proved generally successful in performing the functions which were assigned exclusively to them under the London Government Act, 1963. Two qualifications to that success have been noted. There is first, the fact that the success has been uneven; some boroughs have developed services with far more vigour than others, and the differences between boroughs in this respect correspond broadly to the differences between them in social and political circumstances. Secondly, there are limitations on the ability of the boroughs to provide the more specialized services, particularly where these depend on larger catchment areas than single boroughs can provide, as in further education. Similar problems arise over the provision of welfare services, for example, for those with relatively rare disabilities.

Possible ways in which these disadvantages might be overcome or diminished are the creation of larger boroughs, the transfer of some services to the Greater London Council or joint arrangements between boroughs. There are obvious attractions in the first of these possibilities, but it is by no means clear that the advantages of greater opportunities for specialization would on balance outweigh the disadvantages which might result from the creation of larger boroughs, eg in terms of accessibility of senior officers and the intensity of use of specialist provision. The fact, however, that the population of Greater London, and hence of many of the boroughs, is declining does suggest that the whole question of the size of the boroughs will need to be reconsidered sooner or later.

The second possibility, that of transferring some services to the GLC has been discussed in an earlier chapter[58] in relation to further education. There are few other borough functions where a transfer of this kind is feasible. On the other hand, no case has been made out for giving the GLC responsibility for services such as the personal social services, which would be one way of overcoming the disadvantages of the present system. It would, however, only achieve this at the cost of creating even greater disadvantages for the running of these services. It seems clear, therefore, that with the possible exception of further education, there is little to be gained from strengthening the GLC at the expense of the boroughs so far as the running of the present borough services is concerned.

The third possibility, that of increased co-operation between the boroughs, is also of limited application. As has been shown in this chapter, the boroughs have in various ways set up machinery for joint action; in addition, there are examples of ad hoc arrangements, usually between neighbouring boroughs, for specific purposes such as the sharing of educational advisers.[59] But it is not to be expected that this kind of co-operation will go more than part of the way towards solving the difficulties.

The boroughs, then, have the ability to perform their functions satisfactorily. The limitations on that ability have to be seen in perspective. No local government structure is likely to produce authorities which are ideal for performing all the functions assigned to them. A balance has therefore to be struck between the various advantages and disadvantages. The London boroughs on the whole represent a reasonable balance.

This very success does, however, raise other questions about the role of the boroughs in the new system. Fundamental to the working of the new system is the idea of a partnership between the GLC and the boroughs. Each would have its own sphere of responsibility and neither would be superior to the other. To the extent that functions are distinct and separate and can be assigned either to the GLC or the boroughs, there are clear advantages in such an arrangement. The strength and success of the boroughs, however, in performing functions assigned exclusively to them have in practice raised doubts whether this is the most effective way of handling functions which have both a Greater London and a more local concern. This is particularly so in the case of planning and housing. In both these functions the boroughs have the resources and ability to deal with their local aspects but the question is whether too much emphasis was placed on these aspects in the 1963 Act. This question has wider implications than the subject of this chapter and will, therefore, be discussed more fully in the following, concluding chapter of this volume. It is, nevertheless, the most important question raised by consideration of the role and functioning of the London boroughs in the new system.

13

The New Government of London: An Appraisal

In *The Government of London: the Struggle for Reform* and in the preceding chapters of the present volume the origins of the London Government reforms have been discussed and their effects analysed The present chapter evaluates the reforms as a whole and, in particular examines:

i. the extent to which the objectives of reform have been achieved in practice;
ii. the satisfactoriness of the new system of local government in Greater London.

The first analysis is concerned with the purpose of the reforms, as set out most specifically in the Herbert Commission's report but modified by the provisions of the London Government Act, 1963, and with the effects and consequences of the reforms; the second looks more broadly at the question of whether the new system is adequate for the tasks which now face local government and at the problems of organizing local government in a metropolis.

Enough has been said in earlier chapters about the difficulties of an appraisal based on only five years' operation of the new system. The present study makes no pretence of aiming at a broad historical judgment, which could only be made after a much longer period. Nevertheless, from the material presented in this volume the main trends are already clear. In this chapter the ways in which the new system is developing, its effectiveness and the contrasts with the previous system will be drawn upon to assess the nature of the reforms and their justification.

The objectives of the London reforms

The Herbert Commission proposed two criteria or tests for judging a local government system – administrative efficiency and the health of representative government; these twin criteria provided them with justification for the view that a new structure of local government was needed in Greater London.[1] The Government's endorsement of the Commission's analysis[2] also provided the justification for the rather different system which was introduced as a result of the London Government Act, 1963. The objective of the London reforms was, therefore, to create a better system of local government – better because it would be administratively more efficient and because it would improve the health of representative local government. However, a great deal turns on how these two concepts are used in practice to produce a particular structure of local authorities with a specific allocation of functions.

In the Commission's analysis their two criteria were closely linked to produce their proposed structure with two complementary parts – a Greater London authority and 52 Greater London boroughs. The division of local government functions between the two depended on the Commission's twin tests but with somewhat different emphasis in each case.

The principle on which functions were assigned to the boroughs was explicitly that of reviving the boroughs and contributing to the health of local government. They were to be primary units of local government and they were to perform all functions except those which could only or could be better performed by a Greater London authority; in this way they might be expected to attract able councillors and officers since there would be a worthwhile job for them to do. At the same time two considerations helped to determine the size and number of the boroughs. There was first the argument from administrative efficiency, that areas of 100,000 population should be adequate to perform effectively the functions assigned to the boroughs. Secondly, there was the view that boroughs should not be too large, deriving specifically from arguments about the health of local government.

The main arguments for the establishment of a Greater London Council were, however, derived from the Commission's view of administrative efficiency; there were certain tasks which were either not being done at all for Greater London or were being done inadequately; it was, therefore necessary, to have a Greater London authority to carry out these tasks. Their general faith in a healthy local government made them argue that this authority should be a local authority rather than an ad hoc body or an organ of the Central Government.

Thus the structure of local government proposed by the Commission was indeed consistent with their twin objectives of administrative efficiency and a healthy local government, but only because those objectives were very widely drawn. Furthermore, their application in practice may be summed up by saying that administrative efficiency was the justification for the Greater London Council in the Commission's plan, and the health of local government the justification for the Greater London boroughs. This corresponds closely to the Commission's analysis of local government functions; those which in their view were most defective under the existing system of local government were precisely the ones in which there was need for a Greater London authority, particularly planning and traffic powers. In considering other functions they examined a rather different question, namely, what would be the best way of organizing these functions in a reformed system; in other words, the efficiency of functions such as the children's services or the personal health services were not the reason for reforming local government. Rather, their organization as borough functions was determined largely on the second of the Commission's two criteria. Only education, as so often did not quite fit this analysis; of all the Commission's proposals those for education seemed least plausible in reconciling administrative efficiency with the health of local government.

The importance of the connection between broad objectives and specific proposals becomes more obvious when the system established by the 1963 Act is compared with the Herbert Commission's proposals. That system had the same objectives as the proposals of the Herbert Commission, and in broad outline the structure was similar to that of the Commission; yet because both the number of boroughs and the division of powers between them and the GLC differed from that proposed by the Commission, the new system raises questions both about the detailed aims which it was intended to achieve and about the means for doing so.

It is not enough simply to state that the objectives of the new system were to create a better system of local government, nor even to state that this was to be achieved through the establishment of a Greater London Council and a number of London boroughs. What is also important is that this structure was intended to have quite specific effects in terms of individual local government functions.

Under the Herbert Commission's proposals, the Council for Greater London was to have exclusive responsibility for a number of services such as the fire brigade, but a number of other important services were to be shared with the boroughs. Among these shared services, the Council was to have a dominant role in planning, transportation and education, and a more limited role in housing. At the same time powers were to be shared with the boroughs in such a way that the latter were not to be in any sense agents of or subordinate to the Council for Greater London. Under the 1963 Act, the GLC was given a more restricted role than the Commission proposed in planning and no responsibilities in education apart from the special arrangements for Inner London; correspondingly, the boroughs were given a strengthened role in these services; but again the relationship between the GLC and the boroughs was to be one in which each had a separate group of powers and responsibilities.

The difference between the two systems, that proposed by the Herbert Commission and that introduced by the 1963 Act, might be seen simply as a difference of view about the right organization of certain important functions, since in many respects the two systems were alike. Yet this disguises the fundamental point of difference, which was really about the importance of and emphasis to be given to different aims. For although the Herbert Commission's proposals on the whole, with the major exception of education, do hang together and follow from their principal objectives, yet it is none the less true that they had to reconcile what at first might seem to be incompatible aims. These were the need for a strong Council for Greater London and the need for strong boroughs, each of these aims following, as has been shown, from different principles. In the Commission's scheme the first aim was to be achieved by giving the GLC adequate powers for the Greater London tasks which they had identified; the second aim was more complex because it had to balance the necessity for boroughs of adequate size and resources for their tasks against the necessity to make them accessible and responsive to local needs.

The new system introduced by the London Government Act, 1963, implicitly put much less emphasis on the need for relatively small boroughs and therefore on one strong element in the Commission's argument for a healthy local government. It placed much more emphasis on the need for strong boroughs largely on grounds of administrative efficiency, and correspondingly limited the role of the Greater London Council, particularly by giving education to the boroughs in Outer London and by giving boroughs the power to produce their own development plans. Thus although the broad objectives of the London reforms were the same as those put forward by the Herbert Commission, their detailed aims were nevertheless somewhat different. Basically, the aim was to create a system of local government in which strong boroughs would be the main providers of services, and this aim was given precedence over the creation of a strong GLC; in particular, the planning function was conceived as one in which the GLC and the boroughs were given complementary powers rather than the GLC being given over-riding powers, except in so far as the latter were required to follow the guidelines laid down by the strategic GLDP.

To sum up then: basically, the reforms were designed to replace the multiplicity of authorities of various classes within Greater London by a single system specifically designed for that area; the justification of this change was that it would bring benefits and improvements in local government in the area of Greater London; and the essential part of the new structure was the creation of two levels of local government in which each level would have separate and defined tasks. Before answering the question whether the new system has achieved the objectives of reform, it is necessary to examine what the effects of the reforms have been, and then to see how far these match up to the declared aims.

The new system and the old

(a) *Performance of functions.* The starting-point of much of the analysis, and particularly of the functional analysis, in this volume has been a comparison of performance under the new system compared with the old. In practice, it has been impossible to limit the analysis only to such a narrow front, mainly because different functions do not lend themselves equally well to this treatment. Some functions were transferred from smaller to larger authorities (eg fire brigade, libraries), some from larger to smaller (eg welfare services, education in Outer London). Although the problems involved in such transfers were often complex the function itself was not changed in character: a borough's responsibility to provide welfare services in its area was not essentially different from the previous county's responsibility to provide those services in its area.

Q

In principle, therefore, it is possible to compare directly the performance of these transferred functions by the old and the new authorities.
Some functions are, however, different in character under the new
system. Planning obviously falls into this category as also does much
of the transportation function of the GLC; the reforms in these cases
did not simply shift a pre-existing function from one set of authorities
to another; they also brought together a different combination of
functions for the new authorities to perform. The task of planning
and the task of transportation were under the new system largely
different jobs from those carried out by the previous local authorities;
the GLC was not simply to do on a bigger scale what the LCC or
Middlesex had done before. Direct comparison of performance
under the old and new system is, therefore, less easy and of less
relevance in the case of these functions.

Enough has been said earlier to indicate the difficulties of making
comparisons of performance in this way. It is not possible to draw up
a comprehensive functional balance-sheet of comparisons between
the old and the new systems covering the whole range of functions.
It is possible, however, to assess the way in which the new system
differs from the old and this is an important part of the evaluation.

Of the functions which were the subject of a simple transfer from
one set of authorities to another, personal health and welfare, the
children's services and education[3] have been examined here in detail.
They share one common characteristic; their administration under
the old system was not found by the Herbert Commission to be
inefficient nor were there any grave defects in the performance of the
counties and county boroughs which required them to suggest
changes in the administration of these functions. Even in education
where the Commission were extremely critical of the operation of
delegation arrangements, especially in Middlesex, they did not suggest that education was not well or at least adequately administered.
It has, therefore, been an important part of the analysis of these
functions to see whether they are performed now at least as well as
they were performed before. Much of the opposition to the transfer
of these functions to the boroughs sprang from the argument that
it was wrong to break up well-established and efficient services.
Although both the Herbert Commission and the Government
naturally stressed the advantages of the new proposals, the truth was
that examination of these functions alone would have provided
little justification for a large-scale reform of London Government.

In the light of this we can now consider the findings of the three chapters dealing with these functions. In the personal health and welfare services the main advantages of the new system compared with the old have been the greater opportunities for rational planning and integration of these services; the main disadvantage has been the lack of means for dealing with the more specialized aspects of the services especially in catering for numerically small groups such as the young physically handicapped. On the whole, these services have been progressing at a faster rate than they have elsewhere in the country, but in detail there are marked variations, such as the relatively slow growth of expenditure on health services in boroughs which were formerly in Middlesex compared with a fast rate of growth of expenditure on welfare services in these areas. In the process the former Middlesex mental health services have become 'a casualty of the reforms'. Again, some problems which it was hoped might be more easily tackled as a result of the reforms have proved intractable: the problem of homelessness is a notable example. On the other hand, there has been a marked expansion of community welfare services and considerable progress in such fields as the registration of the handicapped.

The chapters on the personal health and welfare services and on the children's services both draw attention to the success of the London Boroughs' Training Committee (Social Services) in contrast to the fears which were expressed at the time of the reforms that the boroughs would not be able to sustain an adequate programme for training professional staff. The chapter on the children's services does, however, point out that one effect of the reforms has been to create in Outer London many authorities with small staffs for child care work and this is a serious drawback to the efficient running of services. Provision of residential accommodation for children has, however, increased more rapidly than would have been the case if there had been no reforms; but within the total provision the London boroughs have not been so successful in providing the more specialized and experimental types of accommodation. The generally improved communications between borough departments which have been beneficial to the development of the social services have been somewhat offset in Inner London by the fact that education is not there a borough function.

In education, the Outer London boroughs have generally succeeded in providing a good standard of service. A drawback of the new system is the inability of the boroughs to provide a full range of specialist advisers and other specialist staff such as educational psychologists. At the same time, there is a more intensive use of the specialist advisers who are available. Another drawback to the new system is the lack of co-ordination of special education over the whole area of Greater London. The most important disadvantage of the new system is, however, the difficulty of providing adequate further education in Outer London and especially through institutions offering higher level courses. In contrast to their success in running primary and secondary education, the boroughs are generally too small for further education purposes largely because catchment areas for further education institutions go well beyond borough boundaries and because each borough has too few institutions to be able to plan coherently.

The findings of the three chapters may be summed up by saying that on balance there is little to choose between the two systems in the performance of these functions. Some things are done better under the new system, some were done better under the old, but neither is markedly superior or inferior. At the same time, two important general points arise from these chapters. First, there are difficulties of administration under the new system for parts of services which need or require a larger organization than a single borough, such as further education or the provision of some specialist skills in the health and welfare services; secondly, there is the greater diversity of provision of services under the new system compared with the old and some evidence that this diversity is related to certain broad divisions between the boroughs in social and political characteristics, and especially to a division between Inner London (the old LCC area) and Outer London (the remainder of Greater London).

These consequences might simply be seen as springing directly from the fragmentation of provision for these services under the new system compared with the old. Because under the old system services were run by large and, in the case of the London and Middlesex county councils, very large authorities, these authorities had the resources and the necessary catchment areas to provide for a high degree of specialization. At the same time the standard of provision was determined for county areas, so that although standards might differ (eg between London and Middlesex or between Croydon and Surrey) there was less scope for variation over Greater London as a whole than under the new system where 33 authorities (21 in the case of education) have responsibility for services previously run by 9; more specifically, 13 authorities (including the City of London) have replaced one in the County of London and 9 have replaced one in the County of Middlesex.

The consequences of fragmentation are by no means all disadvantageous. Certainly the question of specialization is a serious one and merits further consideration; it is bound up with the question of whether in the long run some reduction in the number of boroughs may be desirable, a question which is discussed later.[4] In the case of the problem of further education there are arguments for making this a responsibility of the Greater London Council for the whole of Greater London.[5] Further education is, however, a special case and it is not suggested that other parts of the services examined in these chapters could with any marked advantage be transferred from the boroughs to the GLC.

Variations in the standard of service provided is another matter. It has to be considered in relation to other consequences of the reforms. In particular the reforms provided the boroughs with a unique opportunity to develop the services which they inherited. One reason for the marked differences in standards of provision which have arisen is that some boroughs have taken more advantage than others of the opportunity; some have forged ahead whereas others have fallen behind compared with what might have been expected to happen to these services if the previous authorities had continued to be responsible for them.

It has not been possible to explore in detail in the functional chapters the reasons for this situation. Nevertheless they do reveal the strong influence of professional and political pressures. New chief officers had in many cases come from counties where they had been in subordinate positions and were anxious to prove their ability to run and develop the services for which they now had responsibility. They frequently found a sympathetic and ready response from the new councils which also saw the reforms as an opportunity to run services geared more closely to local needs. To the extent then that the result was that variations in standards between boroughs reflected differences in local needs, a response by elected authorities to the local situation, this was an advantage of the new system in permitting a greater flexibility in the development of services.

There is doubt, however, whether the variations which have appeared in practice can be entirely accounted for by variations in needs. And to the extent that they are not, is there any other alternative means of organization which might prove satisfactory for this purpose? Obviously a reduction in the number of boroughs as was referred to earlier, would reduce the scope for variations; if this were taken further it could lead to the view logically that these services should be made the responsibility of the GLC. But either of these possibilities raises questions which go wider than these particular services. They are, therefore, considered further later in this chapter.

Another possibility is that the Central Government should impose not simply minimum standards of provision of these services but some higher standard which is judged to be desirable. To do so would in some degree conflict with one of the basic arguments for local government reform, that is, that the value of local self-government lies precisely in its capacity to settle issues locally with the minimum of intervention by the Central Government.

In effect, this discussion illustrates what is meant by saying that on balance the new system is not markedly superior or inferior to the old in the performance of the social services and education functions. The opportunities arising from transfer of services from the counties to the London Boroughs have to be weighed against the disadvantages where marked variations develop in the standards of provision.

What of the other functions examined here? There are peculiar difficulties in assessing the effect of the reforms on the performance of the housing function, largely because the period since 1965 has been one in which a transitional system has operated with the GLC retaining powers inherited from the LCC. Viewed from the point of view of housing activity – the number of slums cleared, the number of new houses built, etc. – the years 1965–70 were at least as successful as the years before reform. But, as was pointed out in the chapter on housing, this activity is not easily separable and identified as being due to the reforms. Moreover, the problem is more to see how the new system is likely to operate once the transitional period is ended. To make the boroughs the primary housing authorities within Greater London raises two questions, first, whether this is likely to provide adequately for London's housing needs, and secondly, whether there is a role for the GLC within Greater London and, if so, what it should be. Neither of these questions had been answered by the end of 1970, but a number of reports published about that time[6] raised these questions in an acute form. Indeed, they could be seen as part of a wider debate about what the housing function should be in large conurbations or metropolitan areas, since it has links both with social service functions and with planning. The success of the house-building programme since the reforms has to be weighed against the failure to bring housing adequately into the Greater London planning process. Moreover, as the chapter on housing has indicated, it is difficult to see the system established by the 1963 Act working effectively unless the GLC retains permanently the powers which it inherited temporarily from the LCC.

Of the functions which have been examined here and in which the GLC has a major role, planning provides the greatest difficulty of all in making comparisons. There is, it is true, one obvious sense in which the new system differs from the old by providing for the making of a Greater London plan. The problem is: what kind of plan? The chapter on planning argued that the way in which the statutory provisions of the 1963 Act and Regulations were drawn up made it difficult, if not impossible, for the GLC to make an effective plan for Greater London and that this difficulty has been increased by the tentative way in which the GLC has interpreted its opportunities. Moreover, it was further argued that it was in any case doubtful whether the kind of structure plan suggested in the Act and Regulations was appropriate for Greater London and that such a plan could more appropriately be devised for the larger metropolitan region. Furthermore, the GLC had not co-ordinated planning housing and transportation effectively in the preparation of the Greater London Development Plan, thus failing to provide what the Herbert Commission had regarded as a major advantage of instituting a Greater London authority.

Planning, then, on this analysis has not yet in practice fulfilled the hopes of those who looked to the creation of the GLC to provide a major advantage of the new system compared with the old. This contrasts with the other major group of functions which the Herbert Commission saw as requiring a Greater London authority, those concerned with transportation. The 1963 Act brought together under the GLC powers relating to highway planning, traffic measures and main roads. Indeed, the legislation has since gone further by bringing under the GLC a large part of London's public transport. Thus, there has been a major change and a concentration of powers within local government, and this has proved effective and successful in the sense that the GLC has gone ahead vigorously with traffic measures and a road programme in a way which was not possible under the previous system. Indeed, the criticisms which have been made of the GLC are that it may have gone ahead too vigorously and that it has become identified with one set of solutions to London's problems to the exclusion of a balanced view of its responsibilities. This is not, however, to deny that, in comparison with the previous system the new system has enabled transportation questions to be dealt with more effectively.

The functions which have been examined in detail in these six chapters by no means exhaust the list. A great many other services were affected by the reforms and for the most part the new system has proved superior to the old for the performance of these unspectacular but vital functions. Brief reference was made in the finance chapters to the fire, ambulance, refuse disposal and library services where, generally speaking, the effect of the reforms has been to give an impetus to improving standards of service, and it is likely that detailed study of these and other services such as refuse collection, sewerage and sewage disposal would indicate the advantages deriving from the new scale of organization of these services.

Thus to sum up the effects of the reforms on the performance of functions:

i. so far as services which are the sole responsibility of the boroughs are concerned, there have been improvements, particularly in services which were previously performed by the metropolitan boroughs and county districts; in services which were transferred from the counties improvements have to be set against certain disadvantages of the new system;

ii. so far as services which are the sole responsibility of the GLC are concerned, there is again evidence of improved performance under the new system;

iii. so far as the major services which are shared between the GLC and the boroughs are concerned, only the group of transportation functions seems to show any marked advantages; over both housing and planning there hang certain question marks.

These conclusions, however, and the evidence of the functional chapters suggest further reflections on the working of the new system. The most important is that whereas the boroughs have proved more successful than might have been thought, the GLC has been less so. The evidence on the boroughs, it is true, is limited largely to the health, welfare, children's and housing services, and, for the Outer

London boroughs, education. In planning and transportation less has been said about the boroughs because of the need to examine more fully the GLC's role in these services.

Nevertheless the functions examined constitute a most important part of the task which has been assigned to the London boroughs, and the evidence is that they have handled this task successfully. Even in education, where much current thinking suggests that larger authorities are desirable than the 200,000–250,000 population level of most Outer London boroughs, these boroughs have succeeded in providing a good service, though with certain limitations which have already been noted. Inevitably, there are qualifications to be made; it has been suggested, for example, that many of the Outer London boroughs have too few children in care to be able to provide an adequately-staffed children's service. But against such disadvantages, there are positive advantages in comparison with the previous system in the ability to forge closer links between borough departments than was the case in the previous counties, or in the greater accessibility of senior officers.

The position of the GLC is more complex. It had a much more difficult task than the boroughs, largely because it was given a new and untried role. It was the essence of this role that the GLC should combine broad and important planning powers with a responsibility for a group of executive functions, some of which, notably transportation, were directly relevant to the strategic planning function whereas others were functions which were thought to benefit from organization on a large scale (eg refuse disposal, main drainage).

The question facing the boroughs when they were set up was whether they had the resources and the will to develop the services for which they had the responsibility. The GLC had in addition to consider the nature of the new kind of planning job which it had been given, and its relationship with its other functions. Thus questions of powers and internal organization have figured much more prominently in the GLC's thinking simply because the GLC had these novel problems which the boroughs did not have to face.

The GLC has the capacity for and has successfully carried out the executive functions for which it has a clear responsibility. Uncertainty about its role and about its powers have combined to make it less successful in its planning function and in relating planning to its other activities. These are serious criticisms of the new system but they must be set against the fact that the GLC was expected to operate as a new and different kind of local authority with little in the way of precedent to guide it.

(b) *The health of local government.* Difficult as it is to assess the effect which the reforms have had on the performance of functions, it is far harder to say what the consequences have been for the functioning of representative local democracy. This is largely because the criteria for judging the health of local government have not yet been firmly established. This can be illustrated from two aspects which have received considerable attention.

There is first the question of attracting councillors of high calibre to serve on local authorities. How does the new system in London compare with the old in this respect? The calibre of councillors is notoriously hard to define in any but subjective terms. Even allowing for this, there is the problem of how far councillors and potential councillors are attracted to local government by the particular powers and status of local authorities. Certainly, the boroughs have greater powers and status under the new system than the metropolitan and non-county boroughs under the old, and in general there has been no lack of candidates for the boroughs. But the connection between these two facts and the further question of the respective abilities of councillors under the two systems would require a much more detailed study of individual boroughs to determine.

There is a different problem with the GLC. The question is not simply about the calibre of councillor but about the type of councillor needed for this new kind of authority. Some GLC councillors, particularly those who had served on the LCC, regretted the loss of the personal social services which figure prominently among the major interests of councillors in recent surveys. But should not the GLC be attracting councillors whose interest is not in the personal social services, or at any rate who can adapt to the circumstances of the GLC without regret? It is not an easy question to answer. Clearly, many new men and women came into the GLC in 1967 because of the big political swing in the elections in that year. But it was still the case that a large proportion of GLC councillors after the 1970 election had first come into local government before 1965. This was particularly true of key figures such as Mr Desmond Plummer, the Leader of the Council, and Mr Horace Cutler, the chairman of the Policy and Resources Committee; the former had served on the LCC, the latter on Middlesex County Council. Again on the opposition side Sir Reginald Goodwin, the Leader of the Opposition, and Mr Ashley Bramall, the Leader of the ILEA, were both former LCC councillors.

The GLC has therefore been largely run by men and women whose experience was gained under the previous system of local government. It may be that this, in part, accounts for some of the hesitation over the organization and policies of the GLC. There is, in particular, a problem here for the political parties since all GLC elections are contested by the main parties. It is for them to consider what kinds of candidates they should be seeking for the GLC; perhaps in time there will be developments in this direction, although there is little sign of it at present.

All this suggests that there is room for much more research into the London situation.[7] At this stage it seems clear that the Herbert Commission took much too simple a view of a very complex matter in their discussion of the problem.

The second aspect of the health of local government which has received attention recently is the extent to which there is greater awareness of and participation in local government by the ordinary elector. Here again, little that has emerged from the present study enables this question to be answered with any certainty. The voting figures for the GLC elections of 1964 and 1967 did, it is true, show a marked increase in turn-out in Outer London areas, but the 1970 figures suggest that this may prove to be a temporary phenomenon; in any case, voting figures for borough elections do not reveal any comparable changes associated with the reforms. Leaving aside the ambiguities of the voting figures, there is no doubt that as a GLC survey indicated in 1967, there exists considerable ignorance of and apathy towards the new system of local government in Greater London. This is a common phenomenon and we have no means of knowing whether it is more or less widespread than it was under the old system or than it is elsewhere in the country. On the other hand, certain aspects of what local authorities have been doing in Greater London have aroused considerable interest and concern. This is most evident in connection with the GLC's motorway proposals which led, among other things, to a campaign in the 1970 GLC elections on the slogan 'Homes Before Roads'. The GLC too, in connection with its GLDP proposals, held a series of public meetings in different parts of London which aroused a certain amount of public interest.

(c) *Political consequences*. The political consequences of the reforms are of considerable importance but with one exception they were scarcely discussed in the pre-reform period. The exception was the allegation that the reforms were deliberately engineered by the Conservatives in order to break Labour's long hold on County Hall.[8] This is, however, only one part of what is meant by the political consequences of the reforms which may be divided into two parts:

i.   the consequences for the political parties;
ii.   broader political effects.

Under i. would be included the changes in the political parties' organizations which, as described earlier in this volume, were at least hastened even if they were not a direct consequence of the reforms. Such changes were only one part of the adjustment which the parties had to make in order to come to terms with the new situation. In this sense the political consequences include changes in the way elections were conducted and the issues which were raised in the election campaigns. They also include any effects attributable to the reforms on voting turn-out.

From this point of view the most significant political consequence of the reforms is that political contests are now to be found in practically all the elections for Greater London authorities. This is most obviously true of the GLC where in the three elections which have so far taken place not only has every seat been contested but every seat has been won by a candidate standing either as a Conservative or Labour Party supporter. All LCC and practically all Middlesex CC elections were contested under the old system. But some of the other county areas which were incorporated in Greater London frequently had uncontested elections. In the Surrey CC elections of 1958 and 1961, for example, there were uncontested elections in all the divisions in these areas except Kingston; and in the borough of Sutton and Cheam 3 of the 4 county seats were uncontested in both elections.

This change may be regarded as the inevitable and indeed the logical response of the political parties to the new system, just as it was practically inevitable that the introduction of the new system would provide the occasion for both the Conservative and the Labour Parties to abandon the rather special and distinctive organizations which, for historical reasons, they had developed in the County of London, in favour of more orthodox Greater London organizations. For the political parties the GLC has become very much a major prize to be contested with the greatest vigour in much the same way as the LCC had been under the pre-1965 system.

There is some evidence of similar but less marked tendencies in London borough elections. Uncontested elections have not been eliminated but they have been much reduced in number; smaller parties and independents are represented on London borough councils, but they are far fewer than under the previous system. To some extent these changes may be accounted for by the fact that there are far fewer borough councillors now than there were, so that it is easier for the parties to contest the elections. It may be, too, that the increased resources and scope of the London boroughs makes them more attractive to potential councillors. Perhaps the main reason, however, is that the boroughs represent greater political prizes than their predecessors.

The various political consequences, though of considerable importance in themselves, form only part of the wider question of the effect which the reform had on the political situation. Here several major issues arise, and particularly the effect on policies and the development of services. The LCC had for 30 years before the reforms been under the control of the same party, Labour, whereas the other counties in Greater London over the same period had been predominantly Conservative although with some alternation with Labour control in the case of Essex and Middlesex. The GLC in its first five years[9] had two years of Labour rule and three of Conservative. The fact of change of political control of the GLC compared especially with the long period of one-party control of the LCC may be healthier in democratic terms but may at the same time have contributed to the difficulties of defining the GLC's role in London Government.

Certainly one effect of political change in the GLC has been to put a rather different emphasis on its role in addition to bringing changes in policy. The first GLC tended to operate as though it were the LCC on a larger scale, although perhaps with increasing recognition that changes were needed. Under the Conservatives there has been greater emphasis on a managerial approach to the GLC's tasks. At the same time some notable changes of policy have occurred particularly in housing where the Conservative GLC proposed to sell council houses on a large scale and to increase greatly the scale of support to housing associations. In the event the first of these proposals brought the GLC into conflict with the Labour Government, but by the early 1970s £25m was being allocated annually for loans by the GLC to housing associations.

One particular consequence of the reforms should be noted here, the possibility of political conflict between the GLC and the ILEA. Until 1970 the same party controlled both the GLC and the ILEA; the 1970 elections gave the Conservatives the GLC but Labour the ILEA. The policies followed by the ILEA and its finances are not subject to the GLC's control. The prospect of a considerable increase in the rate precept for education in Inner London in 1971–2 led to demands for a review of the situation. Conservative boroughs like Westminster wanted education to become a borough service as in Outer London; although there is little prospect of this happening in view of the ILEA's history,[10] there may well be strong pressure to give the GLC some formal control over the ILEA.

In the boroughs the fact that the majority, especially in Inner London, were Labour-controlled in the period 1964–8 contributed to the initial impetus to develop social services and also the rapid expansion of the house-building programme. The years 1968–71 was by contrast a period in which Labour held only 4 boroughs in the whole of Greater London. The effect of Conservative predominance in the boroughs was evident in many ways; boroughs which changed from Labour to Conservative control in 1968 tended to be less eager to push ahead with the reorganization of secondary education on comprehensive lines, and less anxious to sustain a large house-building programme.

Nor was it only a simple question of party political control. The division of powers within Greater London between the GLC and the London boroughs was itself a matter of profound political importance not simply in terms of the party political battle but in terms of how the new system would work in practice. The interests of the boroughs differ in important ways from the interests of the GLC and, among the boroughs there are again differences in particular between inner and outer boroughs, especially over housing. These differences may be accentuated by but do not exactly correspond to party political differences.

This political situation cannot be directly compared with the pre-1965 situation, simply because, as so often in this analysis, the establishment of a new system of local government for Greater London created an entirely new set of relationships between the GLC and the London boroughs and among the London boroughs themselves.

Nothing like it existed under the old system whether one considers the relationship of the metropolitan boroughs to the LCC or of the various other borough and district authorities to their county councils. Here, if anywhere, the characteristically different nature of the new system is to be found.

It might be thought, therefore, that it would have been important in framing the reform proposals to have paid regard to the political consequences, and to have considered whether the political circumstances were likely to exist to make the division of powers between the GLC and the London boroughs workable. The division of powers, however, seems to have been arrived at without much consideration of these questions. As has been suggested, this resulted in too little attention being paid to the need for stronger powers for the GLC in planning and housing.

(d) *Financial consequences.* Equally closely bound up with the reforms were the financial effects. Unlike the political consequences, however, discussion of the reforms did include their financial consequences, but in a curiously ambiguous and unsatisfactory way. The Herbert Commission argued that their proposals would lead to more efficient and more economic administration. This argument depended largely on the general proposition that the best guarantee of efficiency and economy will be found in a healthy, active and lively local authority which attracts good councillors and officers.[11] At the same time they argued specifically that their proposals would be economical in the sense that they should not involve much change in the total rate-borne expenditure of London authorities.

The Commission's approach made it easier for the Government to avoid committing themselves on whether reform would increase the cost of providing local government services. By claiming that their aim was to introduce a system of local government which would enable services to be provided more efficiently, they were able to shelter behind the Commission's equation of efficiency with economy without too much difficulty. But what of the other sense of economy? The Commission's argument that their proposals should not involve 'any material change' in the total of rate-borne expenditure was extremely implausible, as has been argued earlier.[12] It would be astonishing if a reorganization of local government on this scale had not increased expenditure.

In fact, the reforms introduced by the London Government Act,

1963, have certainly increased the cost of providing local government services in Greater London. The important question, however, is how much of that increased expenditure is directly attributable to the reforms and how much is due to an improved standard of service. This question, as the chapter on finance has indicated, is exceedingly difficult to answer. It is difficult enough to answer in measurable terms; it is even more difficult to answer in terms of the less tangible gains and losses involved. Nevertheless, certain tentative conclusions do emerge from the chapter; first, that a large part of the increased cost of providing services in Greater London since 1965 has gone into providing a higher standard of service; secondly, that there has nevertheless been a distinct increase in the cost of certain services which has not been matched by corresponding benefits. Certainly, the reforms have not fulfilled the expectations of those critics who foresaw large increases in costs with little in the way of benefits. Indeed, it is perhaps more surprising in some ways that the reforms were not more costly in this sense. This may be largely because some costs, and particularly capital expenditure costs, are spread over a relatively long period of time. On the other hand, some of the benefits of the reforms may also take time to emerge.

Another consequence of the reforms has been that as a result of the introduction of a new equalization scheme the wealth of the central areas of Greater London is now spread over the whole area. A perhaps surprising negative financial consequence of the reforms is how little evidence they provided either of the economies to be obtained from large organization or the diseconomies of small organizations.

How do the effects and consequences of the London reforms in practice relate to the objectives of the reforms which were discussed earlier? The first and foremost conclusion is also paradoxical. A major objective of the reforms and indeed one objective which made it essential to have reform, was the creation of a Greater London authority which would be able to produce a comprehensive plan for Greater London, taking into account, in particular, housing and transportation needs; yet the most conspicuous failure of the new system on the evidence presented in this volume has been the failure of the planning process so far to produce a satisfactory plan of this kind. But apart from this certain other conclusions must be brought into consideration.

First, to the extent that administrative efficiency provided one of the principal aims of reform, the evidence is that most local government functions are either better performed under the new system or at least as well performed as under the old system. There are, however, serious doubts whether the organization of housing under the 1963 Act in the long term (as opposed to the transitional arrangements which have operated so far) will be effective in dealing with London's housing problems, for reasons which are discussed below.

Secondly, so far as the aim of improving the health of local government depended on revising and strengthening the second-tier authorities, the reforms have been successful. Powerful boroughs have been created and have on the whole proved effective in carrying out the wide range of functions assigned to them. However, it is not clear whether this strengthening of the boroughs has had the effect of attracting elected members of high calibre.

Thirdly, to the extent that the reforms were designed to strengthen local government by increasing the powers of local authorities they have succeeded; the GLC and the boroughs between them have responsibility not only for those functions which the previous authorities possessed but also for others which were the responsibility of other bodies, notably in dealing with traffic and transport.

The situation may be summed up by saying that the reforms represent an improvement on the previous system but with severe limitations in regard to housing and planning.

In housing too great reliance was placed on the boroughs by both the Herbert Commission and the Government in the 1963 Act. However well the boroughs as a whole are able to deal with the housing situation in their own areas the particular circumstances of Greater London with severe housing problems largely confined to relatively small areas require the GLC to have stronger powers to devise and carry out a metropolitan housing policy.[13] In planning the situation is more complex: again, considerable powers were given to the boroughs and there is no doubt that they have the ability and resources to plan for their own areas. Planning for Greater London, however, raises three issues; i. whether Greater London is the right area; ii. whether the GLC has adequate powers; iii. whether the GLC has used the powers which it does possess effectively. Question i. was not considered at the time of the 1963 Act and is discussed further below; to questions ii. and iii. the arguments presented here give the answer 'no'. The GLC was severely constrained by the fact that plan-making powers were, under the 1963 Act, to be shared between the GLC and the boroughs. At the same time the Act failed to provide adequate machinery for dealing with some important planning issues; in particular it divided responsibility for planning central London among a number of boroughs instead of concentrating it at GLC level.[14]

The provisional conclusion, therefore, from the review of the evidence of earlier chapters is that amendment of the 1963 Act by strengthening the GLC's housing and planning[15] powers would go a long way towards eliminating weaknesses in the new system. Before concluding that this is so, however, it is necessary to examine further some of the problems and difficulties of a two-level system of government of the Greater London kind, which is without precedent in this country. Is it really feasible to divide local government powers so that there are distinct powers at each level and so that one level of authority does not become subordinate to the other? This is really the crucial question raised by the reforms as they have operated so far; both the Herbert Commission and the Government claimed that it was feasible, but, as has been shown neither took into account the political realities of their proposed systems. It is time to look at these and other questions which bear on the capaciy of the reforms to meet the needs of local government in London.

Local Government in London

(a) *Greater London as an area of local government.* In considering the satisfactoriness of the new system of London Government for the task of performing local government services, two major questions which have been touched on in the preceding analysis must be examined further:

i.   is Greater London the right or the most appropriate area for the organization of local government functions?
ii.   within a reformed London system how should local government be organized?

The first question is concerned with the whole *raison d'être* of the reforms. The unity of Greater London was stressed by the Herbert Commission as the major reason for the need to organize services on a Greater London basis. They accepted that there were important questions about the relationship of Greater London to the remainder of the South East region, but apart from the fact that these wider questions were outside their terms of reference, they did not see this fact as in any way destroying or weakening their arguments for the treatment of Greater London as a unit. What is at issue here is whether they were right to take this view or whether the London of the 1963 Act was already out-dated, representing a concept of the city which may have been appropriate in 1939 but was no longer so for the latter part of the twentieth century.

One preliminary point must be disposed of first. The Greater London of the 1963 Act is a good deal smaller than the Greater London proposed by the Herbert Commission. There can be little doubt that on the criteria which the Commission used – extent of the built-up area, proportion of urban land-use, density of population, etc. – the wider area ought to have been the one used in the legislation.[16] It is in this sense that Greater London is used here to denote broadly the continuous urban area within the Green Belt, and not the area defined in the 1963 Act after the Herbert Commission's proposals had been whittled away as a result of political pressures.

Essentially what the Commission did was to take the traditional idea of the city as a continuous built-up area and apply it to the London of the late 1950's; hence the Green Belt formed a convenient point at which to draw their boundary. However, as the chapter on planning in the present volume has indicated there has been increasing emphasis in recent years on the need to take account of social and economic changes which have extended the city's influence far beyond such traditional boundaries. Thus the Redcliffe-Maud Commission gave as one of the basic reasons for a change in the boundaries of local government areas that they should conform more closely to modern patterns of living in which there was an interdependence between town and country.[17] If applied to London, such a concept might imply giving to London boundaries corresponding to those of the London metropolitan region, an area of nearly 4,500 square miles containing 13 million people.[18]

Two distinct questions arise here. The first is whether there are valid arguments for extending the boundaries of London in this way; the second is whether such arguments weaken or invalidate the arguments for organizing local government on a Greater London basis. As to the first question, a strong case has been made out in this volume[19] for treating the London metropolitan region as a unit for the purpose of producing a strategic or structural plan which would also include strategic aspects of the housing and transportation functions. These arguments must clearly serve to weaken the case put forward by the Herbert Commission for confining such broad planning to the built-up area of Greater London. They do not, however, destroy the argument for a Greater London basis of organization for local government.

The appropriateness of Greater London as a local government unit does not entirely depend on these wider planning considerations. It is certainly true that if the London region were to be the area of a new planning authority, difficult questions would arise of whether there was a need also for a Greater London authority for planning purposes and, if so, what function it should have. Even apart from this, however, there are other services for which the area of Greater London is an appropriate basis of organization. In transportation and housing, for example, although there are strong links with planning at the strategic level, there are also important tasks which are appropriately linked to a more limited urban area; the co-ordination and implementation of traffic measures, for example, and certain aspects of housing policy would need to be considered in a Greater London context. So too the organization of the fire service or of main drainage or refuse disposal seem appropriate for an area delimited by the extent of urban growth.

These questions are, of course, highly debatable. On the face of it they raise similar questions but on a larger scale to those which the Herbert Commission considered. To answer them satisfactorily would require another inquiry into London Government with the prospect of another major upheaval affecting not only Greater London but the surrounding areas too. It might indeed require an examination into the problems of local government in the whole of the South East region. Among other questions which such an inquiry would have to examine would be whether an authority should be set up for at least the London metropolitan region, and, if so, what kind of a body it should be; what other authorities should also be set up in the area; and what should be the division of functions between a regional authority and these other authorities.

It is certainly not the object of the present study to embark on such an inquiry. These questions are raised to draw attention to the practical problems which would arise in trying to move away from a Greater London basis of local government organization. These problems would have to be faced if it could be shown that Greater

London was entirely inappropriate as an area for the organization of local government. But this is not so.

Certainly a superficial parallel can be drawn between the situation which faced the Herbert Commission and the situation now. Then, as now, it could be argued that there were problems which the existing basis of local government organization made it difficult or impossible to resolve. The difference is that whereas for the Herbert Commission a solution to these problems required a complete re-casting of the local government structure, it has been argued here that much could be done to reduce the main defects in the system set up by the 1963 Act by relatively modest amendments to the Act.

The difference here represents very much the difference between the ideal and the practical, between a theoretically best solution and a practically workable one. With the benefit of hindsight we can see the possibility of a solution to the problems of London's administration on a much broader scale than the Herbert Commission's examination. But the Commission, even if their terms of reference had allowed them, could not with any hope of success have put forward proposals for reorganizing local government on a larger scale than Greater London. Even today it is by no means certain that such a proposal would be politically acceptable. As it was, to have recognized that the boundaries of London in 1960 needed re-drawing and to have carried that argument through was in itself a major achievement. That these boundaries may not wholly meet the needs of London's development in the latter part of the twentieth century may be true. But they are more satisfactory than what went before; and there is still a sense in which Greater London is a unit and an appropriate area for the organization of local government services.

(b) *Division of powers in a two-level system.* The second question raised at the beginning of this section, the proper basis of organization of local government within Greater London, raises issues of some complexity. To begin with, it seems that there are two extreme theoretical possibilities – to have a single authority responsible for all local government services in Greater London, or to have a series of independent county boroughs, each responsible for all local government services within its own relatively small area of Greater London.

The first possibility was not seriously discussed at the time of the London Government reforms. For the Herbert Commission it would have conflicted so profoundly with their views of what constituted a healthy local government that they could not seriously entertain it. Yet since the Commission did their work the Redcliffe-Maud Commission have extolled the virtues of a unitary system of local government in which all services are concentrated in one authority responsible, in some cases, for quite large areas of government with populations of up to one million or a little more. Certainly there are advantages in such a concentration not least in the fact that no doubt arises as to where the responsibility lies for taking decisions on the development and operation of particular services or the relative priorities for developing different services. The main disadvantages, however, leaving aside the health of local government argument, are that in very large areas where complex planning and development questions arise, there would be serious managerial problems for a single authority; at the same time certain local government functions might suffer from being organized on a scale much larger than was ideal or necessary for the purpose. These disadvantages were recognized by the Redcliffe-Maud Commission in their proposals for a two-level system of local government in conurbation areas.[20] The disadvantages would be even more marked in the very much larger area of Greater London. Quite apart from these considerations, a proposal in 1960 to create a single Greater London authority would have received little, if any, support. It would certainly have been far less acceptable than the Herbert Commission's proposals. As it was, the latter were criticized on the grounds that a Council for Greater London would not be a *local* authority at all.

It is easier to see why a proposal for a contiguous series of county boroughs responsible for local government services in Greater London was rejected, although it was proposed to the Herbert Commission by the Association of Municipal Corporations.[21] Such a proposal makes no effective provision for determining questions which affect Greater London as a whole. It would have been no more likely, therefore, than 'the Surrey Plan'[22] to have dealt satisfactorily with the problems which troubled the Herbert Commission.

But if both these extremes are to be rejected, then inevitably a system of local government must be devised in which there are two levels of government, a Greater London level represented in the present system by the Greater London Council and a more local level represented by the 32 London boroughs and the City Corporation. It is at this point that much the most difficult decisions have to be made about which functions should be assigned to the two levels and the relationships between them. These decisions centre on questions about co-operation and conflict within the system which arise precisely because two levels of government imply two sets of interests. To examine how satisfactory the new system has proved in this respect it is necessary to consider first how the present division of powers was arrived at.

The Herbert Commission based their proposals for division of powers between the Council for Greater London and the Greater London boroughs on two principles; first, that the boroughs should have all powers except those which could only or could be better performed by a Greater London authority. Secondly, that as far as possible the Council and the boroughs should each have 'separate and distinct' functions and powers as opposed to the more usual situation to be found in other great cities with two-tier systems, such as Tokyo, Warsaw and Moscow, where the lower tier is subordinate to the upper tier. These principles, as applied in the London Government Act, 1963, have created certain difficulties, which raise the question whether it is possible to apply the principles in practice and achieve the effective working of the whole system.

Within a single-tier system of local government, as in the present county boroughs, conflicts of interest and views on how particular services should be run and what priorities should be given to different services have to be resolved within the authority itself. Where a two-level system is in operation the answers to these questions will depend on the nature of the system and the way in which powers are distributed.

If the system is such that all or most major powers are allocated to one level or if powers are shared in such a way that one level is subordinate to the other, then, generally speaking, the answer will be that these things are decided by the first or upper tier authority. This was the case in the County of London before 1965 where the LCC had exclusive responsibility for all major functions except housing and even in housing it played the major role. It was less clearly the case in the counties, where the system of delegation and the greater range of powers of the county districts compared with the metropolitan boroughs, meant that county councils, although they were clearly the major authorities, were dependent to a greater extent than was the LCC on the co-operation of often large and powerful boroughs.

The peculiarity of the new system of local government in Greater London from 1965 is that it is a two-level system of government which combines the principle of no subordination of one level to the other with the sharing of certain functions, including some very important ones, between the two levels.

This may be satisfactory enough for certain functions where quite clear divisions can be made between wider and narrower tasks. Refuse disposal as distinct from refuse collection is the most obvious example in the London context. Here the nature of the task to be performed can be regarded as being separate and distinct and it is perfectly feasible to do as the Herbert Commission proposed. Even in these functions, however, it is noteworthy that successful performance depends on a considerable degree of co-operation between authorities. For the GLC to carry out its refuse disposal duties properly, it needs to know of changes and developments in the boroughs' handling of refuse collection; equally the boroughs need to know of the GLC's plans in developing new methods and techniques of refuse disposal in case these have repercussions on their own operations. In a largely technical function such as refuse collection and disposal co-operation works reasonably well, and the same is true of sewerage and sewage disposal and the provision of parks.

The remaining functions which are shared between the GLC and the boroughs, however, do not lend themselves to such neat divisions. It is true that roads can be divided, much like parks, into major and minor and responsiblity for construction, maintenance, etc., divided correspondingly between the GLC and the boroughs. However, responsibility for roads is not an isolated function but needs to be linked with other traffic and transport measures as part of the whole transportation function; in this wider context, therefore, there is not as clear cut a division as in the case of the more technical functions such as refuse collection and disposal. With the remaining divided functions, planning and housing, it is even more doubtful whether a clear division of functions is possible without in some sense subordinating the boroughs to the GLC.

In all three of these functions there is a basic conflict of interests between the GLC and individual boroughs, as well as, in some cases, between the GLC and the boroughs as a whole. The boroughs who, under the 1963 Act, were given extensive powers in making plans and in development control, have constantly argued for restricting the role of the GLC in planning; the GLC's attempts to get the Outer London boroughs to accept responsibility for meeting Inner London's housing needs run counter to the boroughs' view of their housing responsibilities as being primarily to serve the needs of their own area; GLC traffic schemes or road proposals may conflict with individual boroughs' views of what is right for their residents, especially on environmental grounds; GLC projections of borough populations in 1981 in the GLDP were queried in many cases because they did not accord with the boroughs' own views of their future development.

Conflicts of this nature would exist whatever system of government was in operation; the important question is how they are to be resolved. All of them involve a conflict between a Greater London view and a more local view; one or the other must prevail or there must be a compromise, but the way in which decisions are arrived at will affect the effectiveness and character of the system. These are among the most difficult questions which have to be settled by those who are framing a new system. The Herbert Commission, by insisting on the need to restrict the powers of a Greater London authority, seemed to be tilting the balance in favour of the local view. In practice, their proposals for planning, traffic management and highways and education would have concentrated these powers in the GLC; the Greater London Council would, by and large, have had the effective power of deciding major policy. There were exceptions even so; the right of the boroughs, for example, to determine whether a planning application involved a departure from the GLC's development plan, could conceivably have put considerable power into the boroughs' hands to modify the plan's provisions.

Only in housing were the Commission's proposals doubtful; here they were in an acute dilemma. They held strongly to the view that housing was essentially a local function linked with the personal social services; at the same time they recognized that within Greater London there were great variations in the need and availability of land for housing in different parts of the metropolis. They thought, therefore, that the GLC should have power to build houses in one borough on behalf of others who could not solve their housing problems in their own areas. But they immediately qualified this, first, by the hope that the boroughs would be able to agree among themselves without the need for building by the GLC and, secondly, by requiring the Council for Greater London to get the approval of the Minister of Housing and Local Government if it did wish to undertake such arrangements. Above all the Commission emphasized the need for consultation and co-operation between the Council and the boroughs.[23]

Perhaps one of the clearest lessons to be drawn from the relationship between the GLC and the boroughs under the 1963 Act is that inherent conflicts of interest cannot be resolved by consultation and co-operation in the absence of a clear definition of where responsibility lies for the ultimate decision. The housing provisions of the 1963 Act, which closely followed the Herbert Commission's arguments, provide the clearest example. If, as the Commission and most subsequent studies have argued,[24] the acute problems of London's inner boroughs require some building of houses in the outer boroughs to meet their needs, how is this to be achieved? The 1963 Act is ambiguous in the extreme; on the one hand the boroughs were made the housing authorities in Greater London, but on the other hand the GLC was required to assess the housing needs of Greater London; the GLC was permitted to build within Greater London but only[25] with the consent of the borough concerned or, failing that, the Minister, but at the same time for a transitional period the GLC was to retain the LCC's housing powers.

From the Minister at the time of the Act (Sir Keith Joseph) onwards the emphasis has been on co-operation in solving London's housing problems; the note has become more urgent as time has emphasized the growing disparity between Inner and Outer London; 'a new initiative' in co-operation was called for by the Standing Working Party on London Housing in 1970 and endorsed by Professor Cullingworth.[26] But in the period 1965–70 there was as much co-operation between the London authorities as could be expected in a situation of such diversity and conflict of interest. Is it likely that, within the existing provisions of the 1963 Act, any greater degree of co-operation will be forthcoming, however much exhortation the authorities receive? And in the absence of greater co-operation there must surely be a prima facie case for amending the 1963 Act by strengthening the powers of the GLC.

Planning, too, under the 1963 Act, illustrates the difficulties of effective action where powers do not indicate how conflicts of interest are to be resolved. The Act made both the GLC and the boroughs' local planning authorities but did not make it clear what the relationship between them should be. Regulations made under the Act specified what the different plans should contain, and although the borough plans were to be produced in conformity with the Greater London Development Plan, the fact that the boroughs were plan-making authorities gave them every incentive to try to get the GLDP made as general and unspecific as possible. Conversely, the more specific the GLC made the GLDP proposals the more likely they were to encounter objections from individual boroughs. Once again, therefore, it was not co-operation between the GLC and the boroughs which was really the question at issue but whether a Greater London view of planning could emerge from the system established by the 1963 Act rather than 32 (33 with the City of London) separate views and plans.

And once again, the question is raised whether amendment of the Act is required to make clear the decisive role of the GLC in formulating a plan for Greater London. This indeed seems to be the implication of the Minister's announcement in November 1970 that the boroughs would in future no longer be required to produce structure plans, even though the reason given for the change was the administrative complexity and time-consuming nature of the system set out in the 1963 Act. Details of the change, which will require legislation, have not been made public at the time of writing but they may go far to remove the major defect in the planning provisions of the 1963 Act. It is unfortunate that the change was made too late to affect the form of the first GLDP which was already the subject of a public inquiry at the time of the Ministerial announcement in 1970.

Only with the transportation function under the 1963 Act as subsequently amended – especially by the Transport (London) Act, 1969 – has this question been resolved. There is no doubt both under the legislation and as the system has worked in practice that the GLC can decide major road proposals and traffic schemes which it considers necessary in the interests of Greater London as a whole. Some of these proposals are highly controversial and have been the subject of much dispute, not only with the boroughs but with many outside bodies. Within the system, however, the decision rests with the GLC, even though clearly they must not only consult but pay some regard to the views of the boroughs.

The argument may therefore be summed up as follows: a system of local government for Greater London based on two levels of government cannot operate effectively unless one of the two principles – no subordination and sharing of functions – is abandoned for the major, controversial functions. In regard to planning and transportation one possibility would be for the GLC to have all these powers; in housing the GLC and the boroughs could have concurrent powers, to build within Greater London; but however functions are split, the GLC would be in a position of being able to decide the major questions of policy.

The reason for the present situation is essentially that too much importance was attached to the need to strengthen the boroughs and make them more effective units of local government than under the previous system. Considered solely from the point of view of those functions which were assigned exclusively to the boroughs it was reasonable to strengthen the boroughs. The 32 London boroughs have, by and large, shown their ability to perform these functions successfully. But it does not follow from this that they can or should be given strong powers in relation to those functions which have implications for Greater London as a whole. Given that, as argued earlier, a realistic system of London Government has to be a compromise between giving all power to the GLC and giving it all to the boroughs, the need for shared functions and the precise division of functions where powers are shared become not matters of detail but critical points in the whole system. This is not only because of the practical difficulties of devising a workable system of shared powers, but because the functions where sharing is necessary are among the most important of all local government functions.

As has been indicated earlier[27] similar questions have been raised in connection with more recent proposals for the reorganization of local government on a two-tier basis in the conurbations in the rest of the country. It was pointed out that a major difference between the three sets of proposals[28] concerned the nature and powers of the first tier (metropolitan area or county) authorities. At one extreme the Labour Government's proposals would not only have made the authority a strong planning and transportation authority, but would have given it important personal functions in education and housing. At the other extreme, the Conservative Government's plan will put the metropolitan counties in an almost identical position to the GLC as indeed is its intention. On the other hand, all three were alike in proposing second-tier (district) authorities which were much more variable in size and on the whole larger than the London boroughs.

These differences between the various proposals put forward in

1969–71 and the system established in Greater London in 1965 are important in focusing attention on two questions which are relevant to the future of the Greater London system and to the reform of local government in the provincial conurbations, namely:

i.   what should be the size of the second-tier authorities?
ii. what is the nature of the first-level authority?

The first question arises mainly from changing views on the right size of authority to perform certain functions. There has undoubtedly been a tendency among expert and professional opinion in recent years to argue the need for larger local authorities to carry out effectively services such as education and the personal social services. The Redcliffe-Maud Commission suggested that authorities of at least 250,000 population were needed to perform these functions with a desirable maximum population of around one million.[29] This view was subsequently endorsed by both the Labour and Conservative Governments, and the latter argued that only in special circumstances should these services be provided by authorities with populations below 250,000.[30] In fact 11 of the 34 proposed metropolitan districts fall below this minimum, and these are the only authorities providing these services which do fall below the minimum.

Nevertheless, in spite of the implication that metropolitan areas may provide exceptional circumstances, the very high proportion of London boroughs (21 out of 32) which have less than 250,000 population suggests that, if current thinking is correct, many of the London boroughs are too small. On the other hand the evidence of the functional chapters in this volume suggests considerable scepticism about whether larger London boroughs would necessarily be more efficient. Certainly larger boroughs would have advantages, for example, in dealing with the more specialized aspects of education and the social service functions; but these advantages would have to be weighed against the evidence that, at least in the London context, boroughs of 200,000–250,000 population have proved generally successful in running these services.

This does not, however, entirely dispose of the matter. One difference between the London situation and that of the provincial conurbations is that the former has been reformed and the latter still await reform. It may well be that if one were looking at an unreformed London today one would conclude, as the Redcliffe-Maud Commission concluded, that it was desirable to have boroughs of

250,000 population or more. But given the present post-reform situation in London, the further upheaval involved in creating larger boroughs would have to be justified on much stronger grounds such as, that the London boroughs were proving inadequate to their tasks, and that services were positively suffering.

This is not the present situation in London, but it does draw attention to an important general point, namely, that reform should be given a reasonable chance to prove itself but should not be regarded as fixed and sacrosanct for all time. In the case of the London boroughs there is a specific argument from the fact that the population of Greater London is declining and that, by 1981, a considerable number of boroughs are likely to be below the minimum population level of 200,000 which was accepted as desirable in 1961. This, coupled with the development of ideas on the appropriate size of authority for the performance of certain functions, suggests that the whole question of the number and size of the boroughs should be reviewed after a reasonable interval, perhaps 10 or 15 years after the reforms.

The point has general implications in so far as it is no longer possible to assume that a system of local government will continue to meet changing needs over a period as long as the 75 years which elapsed between the setting up of the county system and the operation of the London Government Act, 1963. At the same time, too frequent changes or possibilities of change are likely to be unsettling and bad for the morale of local government. What is needed is a greater degree of flexibility so that if, for example, rapid changes in population make some boroughs unable to handle their responsibilities as well as they should, amalgamations or boundary adjustments can be carried out relatively easily.

The second question raised above, on the nature of the first-level authority, is also of fundamental importance. Too little attention was paid in the 1963 Act to the consequences of setting up the GLC as a very large authority but with a restricted and peculiar range of functions. How could and how should such an authority operate is a question which is relevant not only to the GLC but also to the proposed metropolitan county authorities in the provincial conurbations.

The GLC has had considerable difficulty in establishing its role in the new London system for three main reasons. There is first the important question of whether it had adequate powers to carry out the tasks assigned to it; this question has already been discussed at length. Secondly, there is the question of whether and how the GLC can attract the right kind of councillors; again, this has been discussed earlier. Thirdly, there is the question of ensuring that the GLC provides itself with the right kind of professionally-skilled officers to be able to function effectively; it was suggested earlier, for example,[31] that one reason for the GLC's difficulties over planning was that it had too many planners with traditional skills and not enough with the broader skills in environmental planning which the new tasks of the GLC required.

Thus, the satisfactoriness of the new system of local government in Greater London turns on two related questions, the nature of the GLC and its relationship to the boroughs in certain important functions. The analysis presented here has shown that generally speaking the new structure of local government has proved satisfactory for the performance of local government functions. Where it has not, and this applies in particular to the housing and planning functions, the position of the GLC is crucial.

The reasons for this situation are to be found basically in the political process by which the reform proposals of the Herbert Commission were translated into the statutory provisions of the London Government Act, 1963. The desire to strengthen the boroughs on perfectly justifiable grounds to enable them to perform certain functions more effectively, the desire to attract support for the reform proposals from the boroughs to counter the strong hostility of the counties and other opposition, the fear that the GLC might come to dominate the new system rather as the LCC had come to dominate local government in the County of London – all these were pressures inducing the Conservative Government in the early 1960s to pay more attention to the need for strong boroughs than to the desirability of making the GLC effective. In doing so, they were forced to make the unrealistic supposition, as Ministerial pronouncements during the course of the London Government Bill asserted, that the boroughs would be perfectly capable of working in harmony with the GLC to solve London's problems. Perhaps the truth is that within all the conflicting pressures neither the politicians in the Government nor the civil servants in the departments were entirely clear how the GLC should or would operate in the new system nor how the planning and housing functions should be organized and this is reflected in the lack of clarity of the statutory provisions.

The success and limitations of reform

In this final section various themes which have been explored earlier are brought together and the reforms set in their wider context. The two most important characteristics of the new system of London Government which have earlier been discussed at length are:

  i. that it is an urban region rather than a metropolitan area system – this follows from the fact that the limits of the new system are the boundaries of Greater London defined more or less in terms of the built-up area within the Green Belt;

 ii. that within the system the balance of power between the GLC and the boroughs is at certain key points ambiguous and unclear; in particular the relatively strong powers of the GLC in transportation are balanced by its relatively weaker powers in planning and housing.

It follows from the latter characteristic that within the new system, there is no institutional means for dealing adequately with Greater London issues in planning and housing. Furthermore, there is the secondary consequence or characteristic that the GLC's powers are unbalanced and this is one factor in its failure to give proper weight in the planning process to housing and other considerations as against transportation considerations.

These seem to be the salient characteristics of the new system and they stem most of all from the contrast with the previous system. In that system there was no institutional means of dealing effectively with Greater London questions at all and this fact alone makes the most obvious point of contrast between the two systems. But the way in which the new system was actually set up, and in particular the division of powers between the GLC and the boroughs is the second major influence on the characteristics of the new system.

Drawing attention to these questions is not to deny other important consequences and characteristics of the reforms which have been discussed earlier. But in structural terms, the change from a fragmented system of local government within Greater London to one in which Greater London is treated as an entity for local government purposes provides the best and clearest measure of the degree of change and of its significance. It is for this reason that so much attention has been given to the fact that the reforms did not give the GLC adequate powers. This fact was indeed evident when the London Government Bill was published and formed the subject of representations by the Greater London Group to the Minister of Housing and Local Government at that time.

Structural reform is not, however, the only factor which needs to be considered; in any case there is a need to keep in mind the limitations in practical, political terms of what can be achieved by way of reform or indeed of what can be formulated as reform proposals. Above all, one should appreciate, as the Herbert Commission did, that reform of structure is concerned primarily with the removal of barriers, with providing opportunities for solutions rather than the solutions themselves.

These observations may be illustrated from the stages through which the London reforms passed. Even at the first stage of formulating reform proposals there are, as has been seen, great difficulties in reconciling different and to some extent conflicting objectives.

At the second stage, that of the formulation by the Central Government of proposals which are politically acceptable and translatable into statutory terms, the conflicting demands – of those who resist all but the minimum of change, of those who want only certain specific changes, and of those who want to go the whole way – frequently lead to further compromises, further blurring of the aims of the reform programme. This too was very evident in the London reforms introduced by the 1963 Act, in the arrangements for education, in the strengthening of the boroughs and in the 'escape' of 9 areas from Greater London, to take only the most obvious examples.

The third stage is the practical working of the new system in which political pressures and the way in which the new authorities interpret their role serve to give the new system its practical as opposed to its theoretical characteristics. In Greater London these influences have so far served on the whole to reinforce rather than counteract the characteristics to which reference was made above. At this stage too it is important to remember the national context within which the London reforms operated in their first five years; as the sketch in the introductory chapter to this volume indicated, the political and economic climate of the years 1965–70 had an important influence on the way in which local government operated in those years. But these various influences were not inevitable and may not permanently characterize the London system.

What we are here concerned with is essentially the appraisal of London Government, with the quality and effectiveness of the policies pursued by the London authorities. Will they, can they, solve

R

London's traffic problems, have they the means or the capacity to alleviate the distress caused by bad housing or homelessness, can they plan effectively and yet with humanity for tomorrow's London? These and similar questions are the urgent issues which cause concern; the answers to them will demonstrate the success or failure of London Government. Success will depend not only on the structure of local government and on the powers which authorities possess, but also on their resources and on the character and calibre of their personnel, whether elected members or officials. Thus the structure of London Government is only one factor although an important one in the complex of factors which will determine the outcome.

The reform of institutions is always, to some degree, a leap in the dark. In pursuing limited and often conflicting objectives, however well-defined, we often ignore or discount other consequential effects of our actions. It is wise to do so if we are not to be frightened into inactivity. Later, looking back from a lofty historical perspective, we can see how certain things followed from certain others to make an intelligible chain of events. This does not absolve us from trying to see as well as we can what the likely consequences of our actions will be, but it does or should serve to make us cautious in our claim for what we are trying to do.

This is why this analysis has laid so much emphasis on the Herbert Commission's view of what they were trying to do, on their view of their task as essentially the removal of obstacles. That is why, too, the importance of the creation of a Greater London authority stands out as the essential mark of the new system. The obstacle above all others which the London reforms identified and sought to remove was the lack of any municipal organ capable of recognizing and dealing with Greater London problems.

Admittedly, the London reforms had other consequences, as has been shown; and for this reason alone the criterion of judgment of the new system cannot simply be whether it has satisfactorily resolved the question of the powers and responsibilities which a Greater London authority should have. Nevertheless, it is a serious criticism of the new system that in creating a Greater London authority it failed to endow it with effective powers. But it is important to remember that failure to carry through a major objective must be set not only against the ideal reform but against the alternative of no reform at all. So far as planning is concerned there is no reason to suppose that under the old system any effective new plan for Greater London (as opposed to carrying on a modified Abercrombie) would have been possible any more than that London's transportation problems could or would have been resolved. Failure here is a relative matter, just as it is in relation to the question of whether the area of Greater London is meaningful for at any rate the major functions assigned to the GLC.

In all these questions we face the same basic dilemma. For advocates of change, institutional reform never goes far enough; for those who work in the institutions subjected to reform, the changes often go too far. Governments, even those elected as crusading radical reformers, must usually steer a middle course in practice. There was no prima facie reason to suppose that the Conservative Government elected in 1959 would be the first to tackle the reform of local government structure for seventy years. By 1960 the structure of local government seemed to have hardened into a pattern first devised in Victorian times. Reform might be talked of by professors in their ivory towers but was simply not to be thought of as practical politics. Even when the Herbert Commission reported in 1960 there was little sign that the winds of change were anything more than the gentlest of breezes so far as local government reform was concerned.

That the Government at that time should have undertaken and carried through the reform of London Government is therefore remarkable. That the shape which the reform took should fall short not only of what the Herbert Commission proposed but what many reformers advocated is far from surprising. The wonder is that it happened at all and that as much of the Herbert Commission's plan survived intact. It is only right and proper that an academic body like the Greater London Group which had so much influence on the original reform proposals should examine critically the system introduced in 1965 and try to assess how it has worked in practice. But it would be quite wrong for us to give the impression that we do not acknowledge that the reforms, despite some shortcomings, marked an important advance in the development of London's Government.

Perhaps, ironically, the London reforms in the end came too early. This first major change in the local government system established in the late nineteenth century itself contributed to a re-examination of the structure of local government in the rest of England and in Scotland, from which a new approach to the organization of local government in the large conurbations seems likely to emerge. Yet for London it would be unthinkable to have another major upheaval so soon after the last one. London perhaps must pay the price of being the pioneer. It is also true that London *is* different, is on such a vastly bigger scale that its governmental no less than its other problems create unique difficulties. So we return to the question which ended the historical survey of how the London reforms came about – did the Conservative Government do the right thing in 1963?[32] They were certainly right to undertake the reform of London Government; they were right too in the broad lines of their approach to Greater London. Yet equally with the benefit of hindsight we can see the qualifications which must be made, about the relationship of Greater London to the South East region, above all about the nature of the GLC and the powers which it should have. These are important and difficult questions which must be resolved if the new system is to realize its potentialities. That they should be raised at all is a measure of the advance which the reforms represented. London Government must continue to evolve. The reforms of 1965 have provided the occasion and the opportunity for it to do so.

Notes

1 Published in 1970 by Weidenfeld and Nicolson.

2 *Local Government Reorganization: The First Years of Camden* (Camden L. B. 1972).

3 Below p. 127

4 See my 'Heart of Greater London', Greater London Group paper Number 9 published by LSE.

5 See W. A. Robson, *Local Government in Crisis* (second revised edition, 1968) for a brief account of the situation.

6 Cmnd. 4040, paras. 324–58.

7 Largely because the Redcliffe-Maud districts were based on existing large urban authorities such as Liverpool or Wolverhampton.

8 Cmnd. 4276, para. 22.

9 Ibid., para. 24.

CHAPTER 1

1 Greenwich returned 2 Conservatives and 1 Labour: Labour challenged the validity of the election of one of the Conservatives who thereupon resigned; in the subsequent by-election Labour won the seat, thus increasing their councillor representation to 19.

2 See chapter 3, p. 57–65.

3 Cmnd. 2764, HMSO 1965.

4 HC Deb., Vol. 732, cols. 627–66.

5 Public Expenditure in 1968–69 and 1969–70 (Cmnd. 3515).

6 Cmnd. 3888, HMSO 1969.

7 This August increase was not well received by NALGO members who instructed their negotiators to re-open negotiations; this led to an interim increase in February 1970 before the substantive increase of June 1970 (see *Local Government Chronicle*, 7 February 1970, p. 276; 6 June 1970, p. 1114).

8 Proposals for reorganization of local government in Wales were put forward by the Government separately.

9 *Report of the Royal Commission on Local Government in England, 1966–1969* (Cmnd. 4040, HMSO 1969) and *Report of the Royal Commission on Local Government in Scotland, 1966–1969* (Cmnd. 4150, HMSO 1969).

10 See preface.

11 *Staffing of Local Government* and *Management of Local Government* (5 vols), HMSO 1967.

12 See below, p. 000. The appointment of Mr Anthony Crosland in October 1969 as Secretary of State for Local Government and Regional Planning was an indication of the increasing interest in the links between these activities; under the Conservative Government elected in June 1970 a Department of the Environment was set up incorporating the Ministry of Housing and Local Government, the Ministry of Transport and the Ministry of Public Building and Works.

13 Cmnd. 4584, paras. 35–7.

14 *Local Government Finance in England and Wales* (Cmnd. 2923, HMSO 1966).

15 Under the Local Government Act, 1966, many specific grants were abolished and a rate support grant introduced to be calculated on the basis of needs and resources.

16 Now published as *The Future Shape of Local Government Finance*', Cmnd 4741, July 1971.

17 Cmnd. 4276, para. 63.

18 *The Future Structure of the National Health Service* (HMSO 1970).

19 Statutorily enforced in the case of children's services by the Children Act, 1948.

20 *Report of the Committee on Local Authority and Allied Personal Social Services* (Cmnd. 3703, HMSO 1968, esp. paras. 612, 617).

21 Following a reference by the Minister in 1970 of the whole question of the organization of water supply to the Central Advisory Water Committee, the Government announced at the end of 1971 that from 1974 it would be organized on a regional basis. A single region will cover most of the South East and the MWB will be absorbed into this new regional authority.

22 *Report of the Departmental Committee on the Fire Service* (Cmnd. 4371, HMSO 1970).

23 *Report of the Ministry of Housing and Local Government, 1965 and 1966* (Cmnd. 3282, HMSO 1967), p. 27.

24 *The Future of Development Plans* (HMSO 1965).

25 *Report of the Ministry of Housing and Local Government, 1967 and 1968* (Cmnd. 4009, HMSO 1969), p. 27.

26 For example, Mr P. F. Stott, the GLC's Director of Highways and Transportation, was a member of the Planning Advisory Group.

27 The Government's views were conveyed to local authorities in Circular No. 10 of 1965 of the Department of Education and Science; one of the first acts of the new Conservative Government elected in June 1970 was to withdraw this circular.

28 White Paper, *South East England* (Cmnd. 2308, HMSO March 1964) para. 2.

29 *The South East Study, 1961–1981* (HMSO 1964).

30 Cmnd. 2308, para. 10.

31 *Strategy for the South East* (HMSO 1967): the area covered was much smaller than that of the South East Study, East Anglia having been split off under a separate Economic Planning Council.

32 *Strategic Plan for the South East* (HMSO July 1970).

33 Below p. 316.

34 *Offices: A Statement by Her Majesty's Government, 4th November, 1964.*

35 *The Housing Programme 1965 to 1970* (Cmnd. 2838, HMSO November 1965).

36 HC Deb., Vol. 769, 22 July 1968: 16 of the 32 London boroughs were among the 34 authorities in the whole country which were invited to apply for new grants.

37 *The Government of London* (Weidenfeld & Nicolson, 1970).

CHAPTER 2

1 For discussion of the Regional and Area organizations of both parties see R. T. McKenzie, *British Political Parties: the Distribution of Power within the Conservative and Labour Parties*, 2nd edition, London, 1964.

2 *Interim Report of Committee of Inquiry into Party Organization*, Labour Party, 1967, para. 68.

3 Report in *The Times*, 7 July, 1894.

4 Quoted in *The London Municipal Society 1894–1954*, London, 1954, p. 10. Whilst the full story has yet to be told, it should be noted that this claim has not gone uncontested. As the *Municipal Journal* remarked, 'Mr Balfour's Bill is not on the line of the scheme which the London Municipal Society proposed . . .' *Municipal Journal*, 7 July 1899.

5 Op. cit., p. 17.

6 Op. cit., p. 12.

7 *Report of the Royal Commission on Local Government in Greater London*, Cmnd. 1164, 1960, paras. 211–12.

8 F. Smallwood, *Greater London: the Politics of Metropolitan Reform*, New York, 1965, p. 82.

9 The present writer is preparing a full-length account of the history of the London Municipal Society.

10 The extent of the Society's publishing efforts were quite remarkable. The report of the Society's Literature Committee for 11 February 1909 reveals that over 58,000 pamphlets and leaflets had been distributed during the previous quarter. Four new pamphlets and leaflets had been published during that period, and progress was reported on the preparation of a further eight. Nearly 2,000 copies of the Society's book *The Case against Socialism* had been sold in the four years since it first appeared and 1,114 copies of the companion *Socialism in Local Government* had been sold in two months. (Report in Minutes of Executive Committee of the Society, 15 February, 1909.)

11 F. Smallwood, *Greater London: the Politics of Metropolitan Reform*, New York, 1965, p. 83.

12 L. J. Sharpe, *A Metropolis Votes* (Greater London Papers LSE), p. 3. The 'heyday' of the Society could be said to have lasted from 1903 to 1948 at the longest. This was the period of the active political life of Captain (later Lord) Herbert Jessel, a brilliant political organizer who led the LMS first as Chairman, then as President.

13 Among the indications of greater local activity are the change of title of the Society's newspaper from *The Ratepayer* to *The Londoner*, to be published on a monthly basis rather than the former quarterly.

14 Mr Hare became Chairman in 1948, whilst still an LCC Alderman. He resigned in 1952 to take up a Vice-Chairmanship of the Party Organization at Central Office. Whilst in the Commons he rose to Cabinet rank with the post of Minister of Labour in Mr Macmillan's government. As Lord Blakenham he became Chancellor of the Duchy of Lancaster and Deputy Leader of the House of Lords. He was appointed Chairman of the Party Organization in October 1963.

15 The Society, op. cit., p. 30.

16 A claim made in *The London Municipal Society 1894–1954*, p. 23.

17 Ibid, p. 27.

18 *The Government of London*, p. 47.

19 Miss Gelli, the Society's last secretary, served for over half a century with the Society, becoming Secretary on the same day that Mr Macleod was appointed as the Society's first Director. The final issue of *The Londoner* claimed that the tone of the Society had become 'stamped with her personality'.

20 Mr Freeman succeeded Mr John Hay MP, who resigned from the Directorship on being appointed Parliamentary Secretary at the Ministry of Transport and Civil Aviation on 16 January 1959. Mr Freeman was appointed on 12 May 1959. He was at that time Conservative Leader on the Wandsworth Metropolitan Borough Council, and subsequently became chairman of the Finance Committee of the Greater London Council.

21 See for instance *Report of the Royal Commission on Local Government in Greater London*, Cmnd. 1164, 1960, para. 212.

22 Mr Norton was known in London as the former Agent for the Battersea South constituency.

23 Sir Eric Edwards, Chairman of the National Union 1956, President 1967, Chairman National Executive Committee 1957–65. Joint Treasurer of Conservative Party 1965 onward.

24 Sir William Urton. Conservative Party Agent since 1930. Appointed General Director of Party 1957. Retired 1966.

25 *The Londoner*, October 1963.

26 The Society was considerably richer than its Labour counterpart, having an income something like three to four times that of the LLP.

27 Director of the London Municipal Society 1950–2. From October 1961 to October 1963, Mr Macleod was chairman of the Party Organization. Appeals were made to Mr Macleod to use his influence on the Society's behalf during 1962, but he was unmoved.

28 Paul Thompson: *Socialists, Liberals and Labour: the Struggle for London 1885–1914*, London, 1967, p. 265.

29 Ibid., p. 266.

30 London Trades Council, *Minutes*, 12 March 1914, quoted Thompson, p. 283. Due to the war, the election did not take place until 1919.

31 For an account of the early days and the rise of the London Labour Party to its position of dominance, see Herbert Morrison, *An Autobiography*, London, 1960, pp. 56–68, 72–89.

32 'Factual statement' by Peter Robshaw, Secretary of the London Labour Party, presented for consideration by the LLP special committee on London Party structure, 24 November 1964.

33 McKenzie, op. cit., p. 532.

34 R. L. Leonard, 'Morrison's political bequest', *New Society*, 4 January 1968, pp. 5–6.

35 The Labour Party increased their majority on the LCC from 33 to 83. This success brought on to the Council a large number of younger and relatively inexperienced members.

36 The *Guardian*, 2 February 1959.

37 Reported in the *Tribune*, 23 January 1959, p. 4.

38 See especially 'Tin pot totalitarians', 23 January 1959, and 'I will not tolerate this offence to liberty', 30 January 1959, and subsequent articles on 20 February 1959, 29 May 1959, 27 November 1959, 18 December 1959, 1 January 1960, 22 January 1960, 12 February 1960.

39 *Spectator*, 5 December 1958.

40 Reported by Leonard, op. cit., p. 6.

41 For which see Lord Windlesham, *Communication and Political Power*, London, 1966.

42 Reportedly after a meeting with Alderman Watton of Birmingham.

43 *Report of 58th Annual Conference of the Labour Party, Blackpool*, pp. 100, 113.

44 This fundamental identity of aim was constantly stressed in the *Tribune* articles contributed by dissident members of the LCC.

45 In an interview with the author. This sobriquet was also used to denote the liberal group by the *Spectator*'s Westminster correspondent.

46 Quoted and discussed in *The Londoner*, June 1961. See also F. Smallwood, *Greater London: the Politics of Metropolitan Reform*, New York 1965, p. 305.

47 Others, however, have privately described Sir Isaac's control over decisions as notably less strict than that of his predecessor, Charles Latham.

48 The London Labour Party had in 1961 an affiliated membership of 423,000 and an annual income of £15,000. *London News* had its heyday in the period between the publication of the Government's White Paper and the first elections to the GLC, during which it ran front page articles by such major figures as Lord Morrison, Michael Stewart, Freda Corbet, Marjorie Mackintosh and Lord Longford. It suffered a decline thereafter and ceased publication in 1964.

49 See the discussion of the opposition to the Government plan in *The Government of London*, pp. 127–9.

50 Quoted in *The Government of London*, p. 223.

51 See Mr Stewart's speech, quoted in *The Government of London*, p. 169.

52 A last-minute attempt to reserve certain functions in the child care service to the GLC was made by the incoming Government, but found little support within the child care profession, or among the authorities concerned, and in the event, no changes were made. See *The Government of London*, p. 224.

53 Thus LCC affairs were the concern of the LLP executive as a whole, whilst Middlesex local government affairs were discussed in a subcommittee of the executive.

54 For examples of the manifestations of this attitude, see *The Government of London*, pp. 166–7.

55 It should also be noted that Councillor Illtyd Harrington, a noted figure on the liberal wing of the party, became unofficial assistant to the Leader of the Council.

56 I am indebted to Mr Chris Game of the University of Essex, for supplying some of the information used in the following section.

57 London Labour Party, *Annual Report 1964.*

58 *Regional Structure in Greater London,* London Labour Party, March 1965.

59 The LLP secretariat served only the LCC Labour Group, that of the GLC being served by the Secretary of the Co-ordinating Committee, Mr Len Sims. After 1 April 1965, therefore, Mr Robshaw and the LLP had not even a service role to perform.

60 Correspondence with Mr Len Williams, General Secretary of the Labour Party, June 1966.

61 London Labour Party memorandum to National Executive Committee of Labour Party, June 1966.

62 General Secretary, Foundry Workers' Section, AEF.

63 Minister of State, Home Office.

64 Lord President of the Council and Leader of the House of Commons.

65 Area Secretary, National Union of Mineworkers.

66 Acting General Secretary, Transport and General Workers Union.

67 London Labour Party, *Minutes* of Executive Committee, November, 1964.

68 The LLP's claim was for a Greater London Labour *Party*, with affiliation to the national party and representation at its Annual Conference; a General Secretary appointed by the London Party itself; a substantial affiliation fee; and the right to initiate changes in its own constitution.

69 During the late 1950s, the LLP and the national party were somewhat out of step on the question of municipalization of privately-owned rented property. The outcome of this, however, was that the LLP with their great experience of municipal housing, were able to make their view prevail against the suggested single vesting-date proposed by Party Conference.

70 *Labour Party: Interim Report of Committee of Inquiry into Party Organization,* September 1967, para. 76.

71 Report in *Minutes* of Executive of London Labour Party, February 1968.

72 *Minutes,* February 1968.

73 The *Tribune*, 22 March 1968: 'London Labour fights Transport House grip'.

74 The *Tribune*, 5 April 1968: 'London Labour fight ends in a draw'.

75 Reported in the *Tribune*, 20 September 1968.

76 When appointing Mr Dick Delafield to succeed Mr Len Sims in 1970, the NEC simply endorsed the choice made by a four-member selection committee consisting of two members of the NEC, and two of the GLRC.

77 London Labour Party, *Minutes* of Executive, 26 September 1968.

78 M. Ostrogorski, *Democracy and the Organisation of Political Parties*, London 1902, vol. I, p. 526.

79 McKenzie, op. cit., pp. 231–41, 293.

80 Labour Party *Interim Report of Committee of Inquiry into Party Organisation*, 1967, para. 70.

81 Joseph Chamberlain was perhaps the best known of the founder members of the Society and played some part in its affairs during the 1890s. The *Literary Review* (27 October 1896) attributed to him the inspiration of the LMS programme.

CHAPTER 3

1 *Report of the Royal Commission on Local Government in Greater London*, Cmnd. 1164, 1960, para. 851.

2 Ibid., para. 858.

3 Ibid., para. 856.

4 Within the County of London the boundaries of the Parliamentary and county divisions frequently cut across the boundaries of metropolitan boroughs. In other counties electoral divisions consisted either of one or more county districts, or were parts of a county district. See Local Government Act 1933, S.11(6).

5 See pp. 57–65.

6 See London Government Act 1963, Schedule 2. In fact elections were held on a borough basis for the first three elections of 1964, 1967 and 1970.

7 Letter to Home Secretary from Eric Lubbock MP, 6 December 1968; see also *HC Deb.*, 1967–8, Vol. 770, 14 October 1968, col. 36.

8 For discussions of the merits and demerits of the 'simple majority' system of elections, see E. Lakeman, *How Democracies Vote*, London 1970; F. A. Hermens *Democracy or Anarchy*, Notre Dame, Ind. 1942; F. A. Hermens, *The Representative Republic*, Notre Dame, Ind. 1957; P. Pulzer, *Political Representation and Elections in Britain*, London 1967.

9 The disparities between votes cast and seats won are far more striking in these examples than has been the case in the single-member Parliamentary elections. In the 1959 General Election, the Labour Party won 43·8 per cent of the votes, and 41·0 per cent of the seats; the Conservatives respectively 49·4 per cent and 58·0 per cent; the Liberals 5·9 per cent and 1·0 per cent. The 1970 GLC election showed a lesser exaggeration of the winning party's gains than on the previous two occasions, the distortions being clearly minimized when the parties receive roughly similar proportions of the votes cast. It may also be the case that the pattern of grouping of constituencies into London boroughs slightly favours the Labour Party.

10 Boundary Commission for England, Second Periodical Report, Cmnd. 4084, 1969, pp. 9–27. These observations relate to the Greater London Boundary established by the 1963 Act. The Greater London, Kent and Surrey Order 1968 redefined parts of this boundary and the changes reduced the theoretical entitlements of the London boroughs of Bromley and Croydon. This did not lead, however, to a departure from the number of seats provisionally recommended.

11 Ibid., p. 10–11.

12 *House of Commons (Redistribution of Seats) Act 1949*, s.2(5).

13 HC Deb., 1968–9, Vol. 785, 19 June 1969, col. 729.

14 The relative publication dates of the Boundary Commission report and that of the Redcliffe-Maud Commission are interesting, for the former was signed in April, the latter only at the end of May. The Home Secretary played down the implications of unwarranted delay in the publication of the Boundary Commission report, but given the Government's reasons for rejecting it, the timing may well have been significant. The Wheatley report on Local Government in Scotland was not available at that time, but the Government similarly postponed action upon the report of the Scottish Boundary Commission.

15 Ibid., col. 746.

16 *Times*, 28 June 1969.

17 The House of Commons (Re-distribution of Seats) (No. 2) Bill.

18 *Times*, 21 June 1969.

19 Speech to Conservative Party Conference, reported in the *Guardian*, 11 October 1969.

20 Reported in the *Guardian*, 4 October 1969.

21 HC Deb. 1968–9, Vol. 788, 14 October 1969, cols. 234–356.

22 HC Deb. 1968–9, Vol. 788, 14 October 1969, cols. 249–50.

23 Under the *London Government Act 1963, s.* 86. For the detailed discussion of these see *The Government of London*, pp. 198–200 and K. G. Young, 'Local Authority Amalgamations: the role of voluntary joint action', *Justice of the Peace and Local Government Review*, 21 March, 28 March 1970.

24 Croydon, Ealing, Enfield, Hounslow, Lambeth, Waltham Forest and Wandsworth.

25 Eg Barking with 11 four-member wards and 1 five-member.

26 Under the Local Government Act 1933, boroughs outside London are required to have either three councillors per ward or a number which is a multiple of 3 (*s.*25(5)).

27 London Government Act 1963, Schedule I, part III, para. 7.

28 Under the 1933 Act, only a resolution passed by a majority of the borough council to present a petition to the Sovereign can initiate action on such matters as re-warding or changing the number of councillors; there is then a statutory obligation on the Home Secretary to appoint a commissioner to prepare a scheme after holding local inquiries as necessary (s. 25).

29 In Bradford, for example, until proposals were made in 1967 for re-warding, there had been no changes for 30 years despite the considerable movements of population within the city during this period. As a result of this 1 three-member ward had less than 5,000 electors whereas another had over 16,000. (See report of inquiry, *The Times*, 11 April 1967.)

30 London Government Act, 1963, Schedule 3, paras. 15, 17.

31 In the period 1952–62. From 1919–49 the elections were held in the same year but separated by six or seven months. Between 1900 and 1919 the borough elections were held in the year preceding the LCC elections.

32 Standing Committee F, 21 March 1963, cols. 980–6.

33 HC Deb. 1966–7, Vol. 727, 28 April 1966, col. *926*.

34 HC Deb. 1966–7, Vol. 732, 18 July 1966, col. *35*.

35 HC Deb. 1966–7, Vol. 733, 4 August 1966, cols. 666–7.

36 HC Deb. 1966–7, Vol. 736, 15 November 1966, col. 232.

37 Ibid., cols. 232–366.

38 HL Deb. 1966–7, Vol. 279, 19 January 1967, cols. 228–32.

39 HC Deb. 1966–7, Vol. 741, 13 February 1967, cols. 235–94.

40 See above, p. 58.

41 Minutes of the Greater London Co-ordinating Committee of the Labour Party, 20 September 1965.

42 Ibid., 8 November 1965.

43 Ibid., 31 January 1966.

44 Unpublished report written by Hon. Secretary, dated 21 November 1966.

45 This unanimity was later disputed, but all the available evidence indicates that the Special sub-committee were in unanimous agreement.

46 Ibid., para. 4.

47 Ibid., para. 5.

48 Greater London Council Minutes, 8 February 1966.

49 During the Second Reading of the Bill, HC Deb., Vol. 736, 15 November 1966, col. 361.

50 See above, p. 51.

51 London Boroughs Committee. *Simultaneous elections in Greater London: Note by Working Party of Advisory Body of Town Clerks.* February 1966. The identity of the authors of reports of this nature is not customarily revealed, as the officers sit in a professional, rather than a representational capacity.

52 Ibid., para. 2.

53 Ibid., para. 8.

54 It is interesting to note in this context that the AMC in 1970 could

see no special difficulties following from holding elections on the same day. Indeed, they supported this suggestion of the Redcliffe-Maud Commission for metropolitan areas, subject only to the reservation that 'some' electors would suffer confusion, and that 'careful thought' would need to be given to changes in the administrative arrangements for the conduct of elections, to minimize the risk of error. *Observations of the Association of Municipal Corporations on White Paper*, Cmnd. 4276, 1970. Published in the *Municipal Review Supplement*, September 1970.

55 Minutes of the Greater London Co-ordinating Committee of the Labour Party, July 1966.

56 And was specifically cited as such in the Co-ordinating Committee minutes of 20 September 1965.

57 Representations were also made to the London Labour Party, for instance by Shoreditch and Finsbury CLP. The Executive of the London Labour Party planned to have the question of an election postponement discussed at the Annual Conference of the London Labour Party in September 1966, but in the event the Government had already announced their intention to legislate by that date. The Government spokesmen were quite correct in saying that no official representations were made by the LLP as such.

58 The 1971 London borough elections took place too late for analysis here.

59 The Greater London Group also hopes to undertake a further research project into political decision-making in a number of major issues which have arisen in Greater London.

60 See p. 35.

61 Maurice Green, *Daily Telegraph*, 3 March 1964.

62 *Observer*, 5 April 1964.

63 Above, p. 37–49.

64 Doubtless an echo of Labour's 1964 slogan 'Homes before Offices'.

65 Frank Roberts, *The Times*, 6 April 1970.

66 That is, excluding GLC and electorate of uncontested county divisions.

67 For example, in 1967 Lambeth had nearly 219,000 electors, more than the total electorate of Oxfordshire and of the electorate in contested divisions in Berkshire.

68 See also Figure 2(a), p. 76.

69 Hammersmith had the distinction of increasing its turn-out in 1970, having the highest figure of any borough in that year; there were, however, special circumstances in that because of the death of a candidate the election was postponed from 9 April to 23 April.

70 See also Figure 2(b), p. 77.

71 Except for Penge UD.

72 The exception in Inner London was Camden; the exceptions in Outer London were Barking and Brent.

73 Peter Fletcher in *Voting in Cities* (ed. L. J. Sharpe), Macmillan, 1967, p. 299. In a further study of 2,526 ward elections Mr Fletcher has elaborated his views and concludes that variations in turn-out 'result mainly from the stimulus provided by party political activity' (*Political Studies*, XVII, 4 December 1969, p. 502).

74 See *GLC Minutes*, 13 December 1966, 719–20.

75 For example, the survey carried out for the Royal Commission on Local Government in England (*Research Study 9*, 'Community Attitudes Survey', HMSO 1969, Section C).

76 This would be in keeping with Fletcher's general view that it is the degree of political party activity which is the main reason for increased turn-out (see footnote 73 above).

CHAPTER 4

1 Cmnd. 1164, paras. 523, 524.

2 Cmnd. 3703.

3 Written evidence from local authorities, miscellaneous Bodies and Private Individuals, Vol. V, p. 494.

4 Cmnd. 1164, para. 629.

5 Though he points out this was not true of Middlesex.

6 *Health, Welfare and Democracy in Greater London*, D. V. Donnison, Greater London Paper No. 5. (London School of Economics and Political Science, 1962).

7 This follows the practice elsewhere in the country; although in principle such posts may go to any person who is suitably qualified and experienced.

8 S. K. Ruck, *London Government and the Welfare Services*, Routledge & Kegan Paul, 1963.

9 See Townsend and Wedderburn, *The Aged in the Welfare State*, Occasional Paper in Social Administration, 1965 and Amelia Harris, *Social Welfare of the Elderly*, Government Social Survey, HMSO S.S. 366, 1968.

10 Op. cit.

11 Ibid., p. 113.

12 Cmnd. 1164, para. 619.

13 Ruck, Table *VIII*, p. 112.

14 Local Health and Welfare Authority Services in London, Joint Working Group.

15 Annual Report of DHSS for 1968 (Cmnd. 4100).

16 Quarterly return to DHSS.

17 IMTA Welfare Service Statistics.

18 Cmnd. 1164, para. 616.

19 Ruck, p. 23.

20 Though it should be noted that the comparison with the former LCC and MCC is incomplete to the extent that expenditure by the former metropolitan boroughs and county districts is excluded.

21 See Table 4.

22 Amelia I. Harris, Government Social Survey SS 366: *Social Welfare for the Elderly*, HMSO 1968.

23 *The Home Help Service in England and Wales*, Government Social Survey, published HMSO 1970.

24 Social Welfare for the Elderly, SS 366, Vol. 1,

25 Cmnd. 1973, p. 18.

26 Survey of services for the elderly provided by voluntary organizations, National Old People's Welfare Council, 1969.

27 Amelia Harris, SS 366, Introduction, p. iv.

28   K. M. Slack, *Old People and London Government*, Occasional Papers on Social Administration No. 36 (1970).

29   D. V. Donnison, *Health, Welfare and Democracy in Greater London*, Greater London Paper No. 5, LSE 1962, p. 18.

30   Penelope Hall, *Social Services of England and Wales*, ed. Forder (Routledge & Kegan Paul) 1969, p. 275.

31   Awarded by the Council for Training in Social Work set up by Act of Parliament in 1962 following the recommendation of the Working Party on Social Workers in Local Authority Health & Welfare Services (Younghusband) 1959.

32   Of these 30 had been granted the declaration of recognition of experience.

33   Numbers are approximate because a number of part-time staff are employed, especially among the most highly qualified.

34   Report of MOH, *The Health of Middlesex*, 1964.

35   Gertrude Williams, *Caring for People*, published for National Institute for Social Work Training by George Allen & Unwin, 1967.

36   J. C. Fletcher, *Mental Health Hostels*, Buckinghamshire County Council, 1970.

37   *Child Welfare Centres*, HMSO, 1967, p. 35.

38   Some boroughs, however, are greatly handicapped by their long distance away from the mental hospital in whose catchment area they fall.

39   B. Davies, *Social Needs and Resources in Local Services*, Michael Joseph, 1968.

CHAPTER 5

1   *Report of the Committee on Local Authority and Allied Personal Social Services*, 1968, Cmnd. 3703.

2   For a fuller discussion of this see the Home Office White Paper, *Children in Trouble*, 1968, Cmnd. 3601.

3   See for instance *Workloads in Children's Departments*, Home Office Research Unit, report No. 1. 1969: 'As many as a quarter of full-time officers in post in March 1966 joined the service for the first time within the preceding twelve months, and half of these new entrants had no relevant training. The high proportion of recent

recruits is due to rapid expansion in the number of field officers (which increased by a third between 1964–66) combined with high wastage rates. For example, during 1965–66 15% of the officers in post in March 1965 were, for a variety of reasons, lost to the service,' p. 5.

4 See for instance the general discussion in *Caring for People* (Williams Committee Report), NCSS, 1967.

5 Nationally about 50,000 children are received into care each year and 50,000 discharged.

6 *Workloads in Children's Departments*, op. cit., p. 31.

7 *Royal Commission on Local Government in Greater London*, Cmnd. 1164 (HMSO 1960), para. 570.

8 See for example, R. A. Parker, *Decision in Child Care*, Allen & Unwin, 1966, and J. Wakeford, 'Fostering – a Sociological Perspective', *British Journal of Sociology*, Vol. XIV, No. 4, December, 1963.

9 London Borough of Lambeth, *Third Annual Report of the Children's Officer*, for the year ending 31 March 1968, pp. 20–1.

10 Royal Borough of Kensington and Chelsea, 'A report on the first three years' work of the Children's Committee, 1969, p. 5. In the first three years the borough in fact doubled the number of its homes and increased places by 92%.

11 M. G. Powell and A. R. Hammond, 'Boarding out: A Statistical Survey', in the *Quarterly Bulletin of the Research and Intelligence Unit of the GLC*, December 1968, No. 5, p. 28.

12 Indeed one of the difficulties of small children's authorities is in actually identifying 'real' trends in their work when numbers are small. This has implications for planning.

13 Borough of Lewisham, Children's Department, *Report of the Children's Officer*, 1 April 1966 to 31 March 1967, p. 35.

14 Source *IMTA statistics* as for table 5.4.

15 All the inner boroughs place 20 per cent or more of their children in voluntary homes (1969) whereas in only 6 outer boroughs is this proportion reached – *IMTA* as above.

16 See discussion by D. V. Donnison in *Health, Welfare and Democracy in Greater London*, Greater London papers, No. 5 (LSE 1962), pp. 18–19.

17   See the Health and Welfare chapter for a more detailed discussion of the work of the Training Committee. The conclusions drawn there hold for the children's services as well.

18   Research studies No. 2. *Royal Commission on Local Government in England* (HMSO 1968), p. 62.

19   See also *The Lessons of the London Government Reforms,* op. cit., in regard to Hillingdon. 'At present only one officer can be released for training each year, and this is seen as a constraint of the size of the authority', although they add, 'our research in South-East England suggests that this constraint is easily overcome in high performance authorities of comparable size' (p. 63).

20   *Royal Commission on Local Government in England 1966–69.* See *Report*, Vol. 1 (Cmnd. 4040), para. 140, p. 39, and Vol. III, Appendix 12. It is important to note the nature of the Home Office evidence. This was based upon the confidential assessments of particular departments made by the Inspectors.

21   The staffing data used in this section are mainly based upon the Home Office returns.

22   Nationally the improvement was from 30 per cent to about 38 per cent.

CHAPTER 6

1   See Appendix, p. 212.

2   See Figure 3(a) on p. 150. Only one district sub-committee is shown because the other, Potters Bar, was not included in the GLC.

3   There is an unfortunate similarity of terminology here. The LCC's divisional organization was quite different from the counties' divisional administration. The former in no way involved the lower-tier bodies which was the whole point of the latter.

4   See *The Government of London*, pp. 33–6, 55–6.

5   Cmnd. 1164, para. 1002.

6   Ibid., paras. 517, 518.

7   Ibid., para. 517.

8   Ibid., paras. 263, 264.

9   Cmnd. 1164, para. 800.

10 Cmnd. 1562 (1961).

11 There was a case for making the boroughs fewer and larger than the Commission proposed even apart from the needs of education but there can be little doubt that education was the most important factor in the changes.

12 *The Government of London*, pp. 121–4, 186–9.

13 Except North Woolwich.

14 *The Government of London*, chapter 11.

15 It must be remembered that the main Middlesex county offices were in central London, not in the county itself.

16 The Association of Education Committees. For an account of the activities of this Association see J. A. G. Griffith, *Central Departments and Local Authorities*, London, 1966, p. 33, *et. seq.*

17 IMTA *Education Statistics* 1968–9.

18 *Children and Their Primary Schools: A Report of the Central Advisory Council for Education (England)*, HMSO 1967, Vol. I, paras. 151–8.

19 Para. 158.

20 DES Circular 11/67.

21 Admin. Mem. 6/68.

22 Source *HC Deb*, 28 November 1968, cols. *163–6.*

23 DES Circular 7/65.

24 Circular 1/65.

25 ILEA 959 – *Education of Immigrants in Primary Schools.*

26 DES Circular 1/70.

27 1965–6 AEC Yearbook.

28 See Table 6.1.

29 See Table 6.2.

30 See *Local Government in South East England*, Research Studies I of the Royal Commission on Local Government in England, HMSO 1969, pp 128–32.

31 Of course this does not apply to the ILEA with its remarkable continuity of staff.

32 For Merton and Harrow the boundaries are the same as those of the former divisional administrations.

33 Cmnd. 1164, para. 485. Of course the Herbert Commission were thinking of 52 new l.e.a.'s.

34 Deborah Ann Lewis: *Effects of London Government Reform on Primary Education in the London Borough of Haringey*, unpublished M.Sc. (Social Administration) thesis, 1969, p. 27.

35 Cmnd. 1164, paras. 471, 472.

36 See Table 1.

37 Cmnd. 1164, para. 512.

38 Only one primary school was lost to Newham.

39 S.31(8). In brief this makes residence in another l.e.a. not a valid ground for refusing admission of a pupil to a school or college. It also allows recoupment payments, to be sought for further education students from other l.e.a's, even where their home l.e.a's do not consent to their attending other authorities' colleges. In 1967–8, however, Harrow refused to meet all the fees of some of its residents attending institutions of further education outside the borough.

40 Most of the inter-authority secondary school movements of East Ham and West Ham were between the two authorities themselves, hence Newham is largely self-contained.

41 S. 31.

42 Not surprisingly the latter was not made explicit at the time but of the former the circular stated: 'It would not be realistic for authorities to plan on the basis that their individual (building) programmes will be increased solely to take account of the need to adopt or remodel existing buildings on a scale which would not have been necessary but for (comprehensive) reorganisation.'

43 At Kingston the Education Committee proposed a half-hearted comprehensive scheme although even this was rejected by the Council.

44 Richmond, of course, is composed partly of ex-Surrey and partly of ex-Middlesex territory.

45 1966–7 *Comprehensive Schools Survey*, published by Comprehensive Schools Committee.

46 HC Deb, 1966–7, Vol. 746, 12 May 1967, cols. 1954–6.

47 Hounslow appears to have invited outside bodies (of various kinds) to submit possible schemes rather than asking their c.e.o. to draw up proposals. Consequently his influence had to come in 'sideways' – but he still got the scheme he wanted.

48 Mr Ronald St John, Secretary of the Enfield Parents' Association was co-opted to the Education Committee after 1968.

49 Although three candidates stood against Conservatives because of dissatisfaction with the party's education policies.

50 For instance, Mr Christopher Chataway had to be recruited hurriedly to be its chairman.

51 See *ILEA Press Notice*, 29 April 1970.

52 Sutton, for instance, agreed to build a new comprehensive school as a result of negotiations with the DES about its building programme.

53 See DES circular 10/70.

54 Ss. 33, 34.

55 The term 'educationally subnormal' is going out of favour and, encouraged by the DES l.e.a.s are tending to refer to such children as 'slow learners'.

56 Cmnd. 3703, paras. 358–68.

57 AEC Yearbook 1969–70.

58 Education Act 1944, s. 48.

59 *Inner London Education Service: Proposals for an Education Welfare Service*, duplicated document issued with ILEA Press Notice, 7 January 1969, p. 1.

60 As recommended by the Seebohm Committee (Cmnd. 3703, Chapter VIII) and implemented in the Local Authorities Social Services Act 1970, s. 2.

61 Cmnd. 3703, para. 244.

62 ILEA *Education Committee Minutes* 5 February 1969, p. 13.

63 For example, one of their children being handicapped.

64 *The Social Welfare Services of the ILEA.*

65 Ibid., Vol. I, para. 118.

66 Cmnd. 3703, paras. 247, 248.

67 ILEA *Education Committee Minutes*, 5 February 1969, pp. 12–17.

68 For a full explanation of the new service including reasons for omitting certain functions see *Inner London Education Service: Proposals for an Education Welfare Service*, duplicated document issued with ILEA Press Notice 7 January 1969.

69 *Technical Education*, Cmnd. 9703.

70 *Report of the Committee on Higher Education*, Cmnd. 2154.

71 *A Plan for Polytechnics and Other Colleges*, Cmnd. 3006 of 1966.

72 See Table 6.1.

73 At the time it was the county's only general college of education; it had besides a PE one.

74 For a good short description of these bodies see R. D. Jamieson: *The Present and Future Role of Regional Advisory Councils in the Field of Further Education*, Association of Technical Institutions, 1968.

75 *Technical Education*, Cmnd. 9703.

76 See Ministry of Education Administrative Memorandum No. 545 of 1957.

77 See Figure 4, p. 196.

78 Together with the creation of Luton as a county borough this meant a jump from 18 to 35 l.e.a.s.

79 The Distribution of Courses Committee consists at present of the 35 l.e.a. representatives plus an HMI and the chairman of the Regional Academic Board *ex officio*. The latter body is part of the RAC structure and is meant to ensure close co-operation between the universities and the technical colleges in the provision of advanced courses.

80 Advanced courses are broadly those above ONC, OND or 'A' level standard, plus certain specialized City and Guilds Courses which use scarce resources.

81 In all its activities the RAC does not treat Greater London as a separate entity.

82 The Northern RAC is indeed now dealing with non-advanced courses in the same way as advanced.

83 The Engineering Board was operational from 1965.

84 The DES adjudicates on whether a course is a new one or a substitution for an old.

85 Cmnd. 2514 of 1965.

86 The no-area pool applies to the provision of primary, secondary and further education (other than courses of advanced further education) for pupils who do not belong to the area of any local authority, e.g. the children of servicemen serving overseas sometimes fall in this category.

87 Rate Support Grant (Pooling Arrangements) Regulations S.I. 467 (1967).

N.B. The AFE Pool applies to post-graduate and post-diploma courses, HNC and HND courses, diploma and final professional examination courses above the standard of ONC and GCE advanced level instruction. It also includes facilities for research and certain courses of a lower academic standing but which use scarce resources and equipment.

88 Two DES advisory committees are re-examining the problem.

89 Under regulations issued by the Minister recoupment should be paid within eighteen months of the end of the relevant session.

90 In the financial year 1967–8, Hillingdon paid out to other l.e.a.s £340,000 whilst Redbridge, whose college had not then opened, paid out £445,000.

91 There are now new government model arrangements for the governing bodies of colleges and these should help to ease problems of this nature.

92 For detailed accounts of the troubles at Hornsey College of Art see Students and Staff of HCA, *The Hornsey Affair*, Penguin, 1969 and Peter Laurie and Roger Law 'What Really Happened at Hornsey' in *Sunday Times Magazine*, 13 September 1970, pp. 36–47.

93 Laurie and Law *op. cit.*, p. 46.

94 A hotly-opposed proposal to divide all their secondary schoolchildren into bands of ability and to allocate a mix of all bands to each school.

95 See, for instance, Anne Corbett 'Is London Education Too Big' in *New Society*, 17 December 1970.

96 Cmnd. 4040, Vol. I, paras, 270 and 275. Of course the Redcliffe-Maud Commission's maximum population figure applied not only to units administering education but to those administering any of the 'personal' services.

97 Of course the popularity of an l.e.a. with its teachers is not only a function of its closeness or remoteness.

98 Cmnd. 4040, Vol. I, para. 249.

99 Although it is declining.

100 See 'Tories Plan to Break Up ILEA' by Muriel Bowen in *Sunday Times*, 31 January 1971.

101 For contradictory views see the respective evidence to the Redcliffe-Maud Commission of the Association of Education Committees and of NALGO.

CHAPTER 7

Herbert Commission, Cmnd. 1164, paras. 396, 785 (cf. paras. 588, 630).

2 Seebohm Committee, Cmnd. 3703, 1968, para. 388; Redcliffe-Maud Commission, Cmnd. 4040, 1969, paras. 250, 331–6.

3 As an example see Professor J. B. Cullingworth, *Report to the M.H.L.G. on Proposals for the Transfer of G.L.C. Housing to London Boroughs*, Vol. II (MHLG 1970).

4 Milner Holland Report, Cmnd. 2605, 1965, p. 228, para. 10.

5 Cmnd. 1164, para. 378.

6 Cmnd. 1164, para. 396.

7 Cmnd. 1164, para. 396.

8 Cmnd. 1164, paras. 794–5.

9 London Government Act, 1963, Section 21(4).

10 London Government Act, 1963, Section 61.

11 London Government Act, 1963, Section 22(1).

12 London Government Act, 1963, Section 22(5).

13 *Report to the M.H.L.G. on Proposals for the Transfer of G.L.C. Housing to London Boroughs*, Vol. II, MHLG evidence, p. 2.

14 London Government Act, 1963, section 21(5): MHLG circular 1/65, para. 9.

15 London Government Act, 1963, section 21(4): MHLG circular 1/65, para. 11.

16 London Government Act, 1963, section 23(4).

17 London Government Act, 1963, section 22(3).

18 See Report of Standing Committee F, 5th Sitting, 14 February 1963, cols. 246–51.

19 Report, col. 238.

20 Report, cols. 265–6: See 1963 Act, S.23(6): the subsidy was to be tapered so that although it would fall entirely on Inner London in 1965–6, the incidence would be gradually shifted, falling on the whole of Greater London by 1972–3 and in subsequent years.

21 Standing Committee F, 6th Sitting, 19 February 1963, cols. 269–75.

22 *Report of the Committee on Housing in Greater London* (Milner Holland Committee), Cmnd. 2605, HMSO 1965.

23 Report, Cmnd. 2605, Chapter 4.

24 The target figure is supposed to represent the number of contracts approved, a somewhat earlier stage than starts; the Ministry of Housing and Local Government's published statistics record only starts and completions.

25 Cullingworth Report, Vol. I.

26 MHLG circular 1/65, 20 January 1965, para. 3.

27 Figures refer to new tenants only: see 1968 Annual Abstract of Greater London Statistics (GLC 1970), Table 7.15.

28 'Inner London', i.e. the old LCC area, was the area of worst housing conditions generally although some outer boroughs (e.g. Newham) should also be included.

29 For example, early in 1967 Mr Mellish, no doubt with an eye on the forthcoming GLC election, was claiming that London's slums would be practically eradicated within 5 years (the *Guardian*, 24 January 1967).

30 Other consequences of the Government's economic policy, particularly on the rents situation, are discussed below (p. 244).

31 See report in the *Guardian*, 2 April 1969.

32 GLC Press Release No. 470, 30 July 1968.

33 'London under pressure' (*Town and Country Planning*, February 1970, p. 101).

34 The *Guardian*, 22 November 1966: negotiations were not concluded until the following February.

35 See Judy Hillman '£10-a-week flats leave needy out in the cold' (the *Observer*, 1 June 1968).

36 *The Times*, 1 July 1969.

37 See the *Evening Standard*, 30 October 1969; the *Guardian*, 9 October, 29 October, 7 November 1969.

38 HC Deb 1969–70, Vol. 797, 3 March 1970, cols 243–4.

39 London Government Act, 1963, s.23(3) and (4).

40 GLC Minutes, 28 January 1969: the figure of 500 represents the 35 per cent of lettings given up to the boroughs on the total of 71,000 transfers.

41 *Report to the Minister of Housing and Local Government on Proposals for the transfer of G.L.C. Housing to the London Boroughs*, Vol. 1.

42 Cullingworth Report, Vol. II, p. 102.

43 Report, p. 107.

44 *Report to the Minister of Housing and Local Government on Proposals for the Transfer of G.L.C. Housing to London Boroughs* (MHLG 1970), p. 129.

45 GLC Minutes, 13 May 1969, p. 318: in addition to the approximately 12,000 people to be accommodated in local authority housing, the agreement provided for a further 6,000 to be housed by private enterprise.

46 In addition to town expansion schemes the GLC has continued to build estates in areas outside but near Greater London (e.g. at Cheshunt and Waltham Holy Cross).

47 GLC Minutes: 11 March 1969.

48 GLC Minutes, 11 March 1969.

49 Published as *Housing in Camden* by Camden LB in 1969. A new rent scheme is being prepared following this. One of the main areas of inquiry was an assessment of the rent-paying capacity of tenants (see Vol. I of the report, paras. 6–9).

50 See Evidence of London Boroughs Association to Cullingworth the *Transfer of GLC Housing to London Boroughs*. Vol. II, p. 60, 'For the tenants ... [it would mean the end of] differing rent structures ...'

51 GLC Minutes: 4 July 1967, p. 442.

52 London Boroughs' Association, Minutes, 18 October 1967.

53 Minutes, 26 November 1968.

54 Section 22; in practice, this has led in many boroughs to the institution of two waiting-lists, one statutory and containing the names of everyone who has applied, the other an 'active' list of those who have been assessed as having some degree of priority.

55 This Joint Working Party was set up in 1964 to prepare the way for the introduction of the new system.

56 MHLG also suggested that a borough should consider for rehousing people who worked there but lived in another borough; this suggestion, fraught with difficulties both practical and political, was left by the London Borough Committee to be decided at the discretion of each borough (LBC Minutes, 24 March 1965).

57 LBC Minutes, 10 February 1965.

58 Minutes, 5 May, 25 June, 20 October 1965.

59 It is significant that the borough which has most conspicuously failed to adopt the Common Points Scheme (or indeed any points scheme at all) is Harrow, the one borough which did not have to re-examine its practices in 1965.

60 In the case of Harrow, this was simply a continuation of the pre-1965 arrangement.

61 *Report to the Minister of Housing and Local Government on Proposals for the Transfer of GLC Housing to London Boroughs* (MHLG 1970), Vol. I, paras. 55 and 58.

62 *Report to the Minister of Housing and Local Government on Proposals for the Transfer of G.L.C. Housing to London Boroughs* (MHLG 1970), pp. 78 and 88.

63 *Report to the Minister of Housing and Local Government on Proposals for the Transfer of G.L.C. Housing to London Boroughs* (MHLG 1970) Vol. II p. 55.

64 Jennifer Dale, unpublished paper on British Housing Policy 1919–70, pp. 78 and 88.

65 *Report to the Minister of Housing and Local Government on Proposals for the Transfer of G.L.C. Housing to London Boroughs* (MHLG 1970), Vol. II, p. 121.

66 *Report to the Minister of Housing and Local Government on Proposals for the Transfer of G.L.C. Housing to London Boroughs* (MHLG 1970), p. 110.

67 *Report to the Minister of Housing and Local Government on Proposals for the Transfer of G.L.C. Housing to London Boroughs* (MHLG 1970), p. 107.

CHAPTER 8

1 The Minister was advised by the London and Home Counties Traffic Advisory Committee (on which, see below, p. 279).

2 From 1947 to 1962 London Transport was part of the British Transport Commission.

3 The Greater London Group is currently engaged on a detailed study of the institutional arrangements for traffic and transport in Greater London which will cover much of the ground not explored here.

4 On this see *The Government of London*, p. 22.

5 Cmnd. 1164, para. 352(2).

6 On this, see below, p. 305.

7 Cmnd. 1164, para. 346.

8 Cmnd. 1164, para. 347.

9 Cmnd. 1164, para. 346.

10 Cmnd. 1164, paras. 438–40.

11 Cmnd. 1164, Ch. IX, esp. paras. 405, 410–1, 415, 427.

12 Cmnd. 1164, para. 442(1).

13 Cmnd. 1164, para. 779.

14 Trunk roads, which were of course the responsibility of the Minister, numbered 182 miles (out of a total mileage of just over 8,000) in the Commission's review area; there were no trunk roads in the County of London.

15 *London Government: Government Proposals for Reorganization* (Cmnd. 1562, 1961), paras. 36–7.

16 London Government Act, 1963, s.16 and Schedule 7.

17 Road Traffic Act, 1960, s.27: under the Road Traffic Act, 1962 (s.11) the Minister took concurrent powers to make orders on speed limits.

18 1963 Act, s.9(6) and 10, The GLC were required to consult the police and the boroughs before making orders.

19 1963 Act, s.10(8).

20 1963 Act, s.13(1).

21 Road Traffic Act, 1960, s.85.

22 1963 Act, s.25(3).

23 See below, p. 279.

24 It does not, of course, follow that had the Commission been examining the post-1960 situation they would have come to a different conclusion; merely that they would probably have put greater emphasis on other arguments e.g. the desirability of bringing traffic management within the control of local government.

25 Mr Hay, Parliamentary Secretary to the Ministry of Transport, speaking on the Committee stage of the London Government Bill (Standing Committee F, 12 February, 1963, col.145). Nevertheless, as a result of strong criticism, the Minister agreed to delete from the Bill many of these concurrent powers.

26 cf *The Government of London*, p. 181.

27 eg the unusual requirement on the GLC to consult the Minister of Transport before 1 April 1965 about their administrative arrangements (1963 Act, s9(3)).

28 Subsequently, under Regulations relating to the Greater London Development Plan, the GLC secured the power to include in the GLDP a statement of policy on the provision of public and private car parks.

S

29 M. K. D. Goldrick, *The Administration of Transportation in Greater London* (unpublished Ph.D. Thesis, University of London, 1967).

30 GLC Minutes, 8 December 1964.

31 Town and Country Planning (Development Plans for Greater London) Regulations, 1966, (SI 1966 No. 48), Reg. 11(e); there were also requirements to include statements on traffic interchange points, on policy regarding the relationship of road and traffic proposals to other forms of land-use and on policy on car parking (Reg. 11(f), (j) and (o)).

32 *Report of the Committee on London Roads* (Cmnd. 812, HMSO 1959); it dealt only with the County of London.

33 For a full discussion of the origin of the LTS, see Goldrick, Ch. 3.

34 In 1964, the London Labour Party accepted the motorway plan as part of its policy for the GLC election (Goldrick, p. 196).

35 GLC Minutes, 6 April 1965, pp. 347–8, 352.

36 In 1964 London Transport and British Railways, realizing the need for a more positive move to put the case for public transport, began to prepare a railway plan for London; this was not, however, published (Goldrick, pp. 205–22).

37 It should be pointed out that many of the proposals for new roads in London were far from new, going back in some cases to the Bressey Report of 1937 or even earlier. On this see W. A. Robson, *The Government and Misgovernment of London* (2nd ed 1948), especially pp. 139–55 and 421–26.

38 *London's Roads – The Council's Objectives* (GLC Minutes, 21 November 1967, p. 694; the motorway box (now known as 'Ringway 1') was to form the innermost ring. See also Figure 5, p. 274.

39 *Greater London Development Plan: Statement* (GLC 1969) para. 5. 11.

40 *A Secondary Roads Policy* (GLC 1969): the secondary network consists of approximately 1,000 miles of existing major traffic routes; it is to be distinguished from the local or borough roads.

41 cf the report *Parking Policy in Central London* which declares that such a policy must be looked at in relation to the question 'how Londoners could be better served by improved public transport services and interchange facilities, a matter to which we attach the very highest importance.' (GLC Minutes, 22 February 1966, p.

115). See also *Greater London Development Plan: Statement*, para. 5.2: *Report of Studies*, para. 6. 207.

42 *Tomorrow's London* (GLC 1969), p. 67. A handsomely-produced free booklet entitled *Transport in London: a Balanced Policy* which the GLC published in March 1970, tries hard to put over the view that they are aiming at a balance between different measures. 'The Council is greatly concerned that all forms of transport should be brought into balance as an integrated system' (p. 3).

43 GLDP, paras. 5.57, 5.59.

44 cf *Tomorrow's London*, p. 67, where the point is made that the bulk of London's road traffic is outside the central area and this is 'the core of a problem which public transport alone cannot solve'.

45 The five were urban motorways; decentralization of activities; traffic management; improvements in public transport; improvements in secondary and local roads (*Tomorrow's London*, p. 67).

46 See, for example, J. Michael Thomson, *Motorways in London* (London Amenity and Transport Association/Duckworth 1969), esp. pp. 161–4: Peter Self, 'What's wrong with the G.L.C.?' *Town and Country Planning*, Vol. 37, No. 5, May 1969, p. 196.

47 The London Transportation Study has not been published in its entirety but an extensive summary was published as *Movement in London* by the GLC in 1969.

48 GLC Minutes, 6 April 1965, p. 348.

49 Transport Act, 1962, s. 3(1), 7(1), 18.

50 See below, p. 287.

51 On this see Goldrick, pp. 282–6.

52 See below, p. 285.

53 Cmnd. 1164, para. 428.

54 Sir Gilmour Jenkins, *The Ministry of Transport and Civil Aviation* (Royal Institute of Public Administration/George Allen & Unwin, New Whitehall Series, 1959), pp. 135–7.

55 *Roads in England and Wales* (HC 346, HMSO 1965), para. 5.4.

56 Road Traffic and Roads Improvement Act, 1960, s.8; Road Traffic Act, 1962, s.11, 12, 27.

57 HC 346, 1965, paras. 5.4 and 6.11.

58 *Annual Abstract of Greater London Statistics, 1968* (GLC 1970), Table 4.14.

59 *Roads in England and Wales, 1960–1* (HC 321, HMSO 1961), p. 3: designation of a clearway involves waiting and loading restrictions (whether at peak hours only or for the whole day) along main routes.

60 GLC Minutes, 6 October 1970.

61 See LHCTAC Reports *London Traffic Congestion* (HMSO 1951) *Car Parking in the Inner Area of London* (HMSO 1953).

62 Under the Road Traffic Act, 1956.

63 *Parking – the Next Stage: A New Look by the Ministry of Transport at London's parking problem* (HMSO 1963).

64 See below, p. 290; the GLC, for example, had virtually no control over off-street parking under the 1963 Act.

65 *Report from the Select Committee on Nationalized Industries on London Transport*, Minutes of Evidence (HC 313–1, 1965) questions 1430–2.

66 *Greater London Development Plan: Statement* paras. 5.2, 5.10, 5.11.

67 *Parking Policy in Central London* (GLC Minutes, 22 February 1966, pp. 115–24). More detailed discussion of the Inner London Parking Area will be contained in the forthcoming study by the Greater London Group of the institutional arrangements for traffic and transport in Greater London.

68 J. M. Thomson, 'The Cost of London's Traffic Management', *The Times*, 22 January 1968.

69 See below, p. 290.

70 On the machinery, see GLC Minutes, 16 December 1969 and 7 July 1970; a green paper, on *The Future of London Transport*, was published by the GLC in October 1970.

71 1963 Act, s.9(2).

72 1963 Act, s.9(2).

73 This is one of the disadvantages of the division of planning powers in Greater London which is discussed further in the chapter on Planning.

74 London Boroughs Association, *Statement of Evidence to Royal Commission on Local Government in England* (October 1966), para. 34.

75 The Conservative election manifesto *Let's Get London Moving* contained a number of proposals for dealing with traffic and transport including the appointment of a Traffic Commissioner (below, p. 289).

76 See Chapter 11, p. 406.

77 Evidence to Royal Commission on Local Government in England, para. 13 (GLC Minutes, 15 November 1966).

78 See *The Government of London*, pp. 181–2.

79 The Government claimed that it would be necessary to get the GLC's view since transfer of trunk roads would involve the GLC in having to meet 25 per cent of expenditure instead of the whole cost being met by the Ministry (see London Government Bill Standing Committee Proceedings, 12 February 1963, cols. 165–6).

80 Cmnd. 3057.

81 For example, the Planning Advisory Group's report *The Future of Development Plans* (HMSO 1965), laid stress on the integration of land use and transport planning.

82 Cmnd. 3057, para. 139. For this purpose 'conurbation transport authorities' were to be set up.

83 Cmnd. 3057, paras. 64, 70.

84 GLC Minutes, 13 December 1966, 713.

85 GLC Minutes, 13 December 1966, 713.

86 For example, the Minister had to authorize pedestrian crossings on metropolitan or borough roads.

87 GLC Minutes, 14 March 1967, 214–20.

88 *Let's Get London Moving*, pp. 7–8.

89 The 'package deal' was announced by Mr Desmond Plummer, the Conservative Leader, on 30 March 1967 (Conservative Central Office Press Release 285/67).

90 Conservative Press Release 02/4, 18 April 1967.

91 Mr Stott, the Director of Highways and Transportation, was given

the new title of Traffic Commissioner and Director of Transportation, and an Executive Director was appointed under him to deal with day-to-day management (GLC Minutes, 21 November 1967).

92 See GLC Minutes, 19 December 1967, p. 751.

93 *Transport in London* (Cmnd. 3686, HMSO 1968).

94 The LTE retained responsibility for the whole of the underground system although part of it ran outside Greater London.

95 In 1967 the Ministry of Transport revised its classification of roads for grant purposes; principal roads were the important traffic routes for which the Ministry continued to pay specific grants; there are 872 miles of principal roads in Greater London including only 547 miles of metropolitan roads.

96 Above, p. 268.

97 A joint traffic executive between the GLC and the Metropolitan Police was proposed in the 1968 White Paper to co-ordinate traffic administration; this did not require legislation and has now been set up. The 1969 Act required consultations and liaison between the GLC, the new LTE and the British Railways Board on certain specific points but did not otherwise affect the formal powers of the British Railways Board.

98 An official committee in 1970 commented on 'the paucity of financial and statistical information and its lack of comparability and consistency' and on 'the large part played by subjective judgments in producing the present differences between authorities' (*Report of the Committee on Highway Maintenance*, HMSO 1970) paras. 5–7.

99 This will be even more evident when the 1969 Act provisions for GLC responsibility for principal roads are implemented.

100 It is relevant here to note that the Buchanan Report, *Traffic in Towns*, was published in 1963, that is, just at the time when the London reforms were being given statutory form.

101 See *The Government of London*, pp. 140–3.

102 *Transport Planning: The Men for the Job*, a report to the Minister of Transport by Lady Sharp: January 1970 (HMSO 1970), para. 44.

103 *Transport Planning*, paras. 38–9.

104 It is noteworthy, for example, how the Redcliffe-Maud Commission nearly ten years later than the Herbert Commission stressed that in

the three metropolitan areas which they identified, the metropolitan authority, corresponding to the GLC in Greater London, should have all statutory planning powers and should control all aspects of transportation since 'unified policy and execution are essential; and they must be in the hands of the authority responsible for planning'. (Royal Commission on Local Government in England, 1966–9, Report, Cmnd. 4040, 1969, paras. 326, 328).

105 See Transport (London) Act, 1969, sections 25–6.

106 *Transport in London* (Cmnd. 3686, 1968) paras. 55–6: Although the Conservative Government accepted these proposals, they provided a subsidy in late 1971 to limit 1972 fare increases.

107 Once again reference should be made to the forthcoming study of these questions to be published by the Greater London Group.

CHAPTER 9

1 A fuller version of this chapter has already been published (see Peter Self, *Metropolitan Planning*, Greater London Paper No. 14, LSE/Weidenfeld and Nicolson, 1971).

2 See J. Brian McLoughlin: *Urban and Regional Planning, A Systems Approach* (Faber 1970), Chs. 6 and 10.

3 ie the County of London and certain adjacent boroughs which were also congested.

4 Cmnd. 1164, paras 765–778.

5 Cmnd. 1164, para 769. Outline of evidence by Middlesex County Council to Royal Commission on Local Government in Greater London – *Memorandum of Evidence*, Vol. 2, p. 133.

6 London Government Act, 1963, Section 25.

7 Official Report, Standing Committee F, London Government Bill, 19 February 1963, col. 326.

8 *Greater London Development Plan: Statement* (GLC 1969), *Greater London Development Plan: Report of Studies* (GLC 1969).

9 See in particular *London Under Stress* (Town and Country Planning Association, 1970).

10 See Planning Advisory Group, *The Future of Development Plans* (HMSO 1965).

11 Official Report, Standing Committee F, London Government Bill, 19 Feb. 1963, col. 325.

12 Especially as the GLC has no development control powers in these areas, save for a right of consultation. See below, p. 320.

13 See GLDP Statement Section 3 for population, Section 4 for employment, Section 6 for amenity areas, Section 8 for town centres, Section 5 for transport.

14 For example Lady Pepler, who was Conservative spokesman on housing on the LCC from 1952–6 and 1958–60.

15 J. Michael Thomson, *Motorways in London* (London Amenity and Transport Association, Duckworth, 1969).

16 For fuller analysis see W. F. Luttrell 'Employment in Greater London' in *London Under Stress,* op. cit.

17 These arrangements rested upon Sections 24 (4) and 24 (8) of the London Government Act, which provided for the GLC to decide certain applications, and Section 24 (6) which authorized the procedure for notifications and directions. The types of case to be covered by these provisions were settled by the Town & Country Planning (Local Planning authorities in Greater London) Regulations 1965 (S.I. 679). The consultation procedure was settled by amendment to a General Development Order.

18 Camden, City, Islington, Kensington & Chelsea, Lambeth, Southwark and Westminster.

19 See McLoughlin, pp. 297–305.

20 On this see M. K. D. Goldrick *The Administration of Transportation in Greater London* (unpublished PhD Thesis, University of London, 1967), especially p. 281.

21 On this see the first director's (Dr Benjamin) letter to the *Guardian* (1 February 1971)

22 *New York Metropolitan Region Study:* 9 Vols. published by Harvard UP. See especially *Metropolis 1985* by Raymond Vernon (1960).

23 Judgments in this section are based upon the experience of the author as chairman (later vice-chairman) of the executive of the Town and Country Planning Association, and as a member of the South-East Regional Economic Planning Council, between 1964 and 1970. Information given to the Council is restricted by the

Official Secrets Act, and no use has been made of unpublished papers.

24 *The South East Study* 1961–1981, MHLG (HMSO 1964). *A Strategy for the South East:* A first report by the South East Economic Planning Council (HMSO 1967). *Strategic Plan for the South East*, Report of the S.E. Joint Planning Team (HMSO 1970).

25 South Hants (Solent City); Ashford; Ipswich; Milton Keynes – Northampton.

26 South Essex corridor; Reading–Basingstoke–Aldershot area; the Crawley–Burgess Hill area.

27 *Report of the Commission on the Third London Airport* (HMSO 1971).

28 *Town Planning in Greater London* (Greater London Paper No. 7, London School of Economics and Political Science, 1962), pp. 20–1.

CHAPTER 10

1 See Report of Royal Commission on London Government Cmd. 1830, 1923 also *The Government of London,* pp 5–7.

2 Cmnd. 1164, para., 280.

3 Cmnd. 1164. para. 290.

4 See *The Government of London*, pp. 75–6.

5 It is significant that the recent attempt by Cheshire County Council to assess the probable financial consequences of the Redcliffe-Maud proposals, bases a large proportion of its estimates on evidence from the London and West Midlands reorganizations.

6 Some opposition spokesmen did emphasize the Royal Commissions' claim that most Inner London boroughs would experience increased rate burdens; but this was generally combined with the broader arguments in favour of maintaining the LCC's equalising functions. See for example Lord Morrison of Lambeth: House of Lords debates vol. 227, 21 December 1960, cols. 1037/8.

7 Cmnd. 1562. para. 51.

8 See for example W. Bowdell, *Problems of Local Government Reorganization* IMTA 1970.

T

9 Cmnd. 1164, para. 953.

10 *Rate Equalization in London* (IMTA 1968) paragraph 47 (see below).

11 In 1964–5 for example, the highest and lowest rates of the metropolitan boroughs before equalization were 12/10d (64·2p) (Bethnal Green) and 6/1d (30·4p) (Westminster); after equalization, these became 9/4d (46·7p) and 7/2d (35·8p).

12 Section 66, 1963 London Government Act. See Standing Committee proceedings 14 March 1963.

13 See for example articles in 'Local Government Finance', December 1957, p. 287 by G. M. Gratten-Guiness, Deputy Chief Education Officer of Huddersfield; also July 1959 page 164 B. Quick, Treasurer of Ruislip-Northwood.

14 1963 London Government Act, Schedule 2. paras. 19–24.

15 The 1966 Local Government Act replaced the general grant with the needs element of the rate support grant. It also terminated the schools meals and milk grant which was thereafter to be included in the needs element of the rate support grant, a further complexity as far as inner London was concerned.

16 Among other things inner London boroughs would lose some advantages previously enjoyed by the metropolitan boroughs (e.g. the special housing subsidy from the LCC).

17 Report of TAC to London Boroughs' Committee 1964.

18 A. R. Ilersic, *Rate Equalization in London* (IMTA 1968) para. 28.

19 For a full account of the scheme and of the effects upon individual boroughs see A. R. Ilersic, *Rate Equalization in London,* chapter 2.

20 See L. E. Holmes, *Local Government Finance*, June 1967, p. 209.

21 A. R. Ilersic, op. cit., para 63.

22 A. R. Ilersic, *Rate Equalization in London.*

23 A. R. Ilersic, *Rate Equalization in London,* para. 56.

24 See Appendix to this chapter for a fuller account of the Scheme (p. 398).

25 Ilersic proposed that the scheme should be introduced by means of four annual transitional stages.

26 H. C. Deb. 1967–8, Vol. 758; written answer 6 February 1968, cols. 68–9.

27 See *The Government of London*, pp. 192–3.

28 London Government Act 1963, Section 70.

29 London Government Act 1963, Schedule 2, paras 25–9.

30 From 1 April 1971 a new loan sanction procedure has operated and similar conditions applied to all education authorities (Department of the Environment Circular 2/70, 1970).

31 Above p. 102.

32 See, for example, Royal Commission on Local Government in England, Research Study 3, *Economies of Scale in Local Government Services;* 4, *Performance and Size of Local Education Authorities;* 5, *Local Authority Services and the Characteristics of Administrative Areas* (HMSO 1968); also Bleddyn Davies, *Social Needs and Resources in Local Services* (Michael Joseph, 1968).

33 *Report on the Cost of Local Government Reform* (Cheshire County Council, 1970).

34 Below, p. 379.

35 Above, p. 101.

36 Below, pp. 387–8.

37 Gross capital expenditure on services (including ILEA), but excluding loans.

38 Now the Policy & Resources Committee.

39 The housing functions of the GLC are of course similar to those of the LCC only for a transitional period.

40 In 1963–4, 70% of the LCC's net capital expenditure on services was for housing and education; in 1968–9 the corresponding figure for the GLC was 74%.

41 Cmnd. 1164, para. 950.

42 Cmnd. 1164, para. 954.

43 The ten county boroughs are: Bradford, Coventry, Kingston-upon-Hull, Leicester, Newcastle upon Tyne, Nottingham, Portsmouth, Plymouth, Southampton, Stoke-on-Trent. (Populations between 200,000 and 350,000.)

44 See Figure 8, p. 368.

    Below, p. 383.

46 See for example the evidence of the LBA to the Royal Commission on Local Government in England. Report of Advisory Committee of Treasurers – Financial Effects of Reorganization in Inner London, London Boroughs Association December 1966.

47 Except for Newham.

48 See below, p. 380.

49 Newham also included small parts of Barking and Woolwich.

50 See above, pp. 93–94.

51 See page 250.

52 See, for example, reports in the *Guardian*, 'Borough fears housing crisis' (23 July 1969) and 'Lambeth crisis over housing' (25 February 1970).

53 See above, p. 4.

54 See below, p. 388.

55 Cmnd. 1164, para. 649.

56 Gross expenditure has been taken in order to minimize differences in levels of income obtained by authorities from sale of salvage, etc.

57 Birmingham, Bradford, Cardiff, Coventry, Kingston-upon-Hull, Leicester, Manchester, Newcastle, Nottingham, Stoke-on-Trent.

58 See for example *Greater London Development Plan – Report of Studies*, paras. 5:71, 5:74, page 134.

59 A feature emphasized by the recent *Report of the Committee on Highway Maintenance*, H.M.S.O. 1970.

60 The Study was commissioned by the Cheshire County Council and forms part of the examination of reorganization costs published as *The Cost of Local Government Reform*, 1970.

61 It was assumed that 'operational costs' (eg teachers' salaries) would not change as a result of reorganization but only as a result of changes in the standard of provision of services.

62 The County of London was excluded from the investigation. Outer London boroughs were regarded as being broadly comparable in terms of responsibilities with Redcliffe-Maud metropolitan districts. They were also formed from parts of existing counties and county boroughs as were Redcliffe-Maud's proposed metropolitan districts.

63 Cheshire County Council, *The Cost of Local Government Reform*, Pt. II, Section 4, Table 3.

64 *ibid.*, Table 7; housing was not analysed at this stage.

65 *ibid.*, Table 8.

66 It would not, of course, alter the fact that *in total* the reforms had increased costs.

67 Above, p. 165.

68 Section 70.

69 The area would have been 614 sq. miles as compared with the 620 sq. miles of Greater London.

70 Cheshire Study, para. 5: 10.

71 Fire Service Circular 43/1958.

72 National Board for Prices and Incomes, *Report No. 32 Fire Service Pay*, Cmnd. 3287, May 1967.

73 Eg the high proportion of areas of high fire risk in London as defined in Home Office Circular 43/1958.

74 *Departmental Committee on the Fire Service*, chairman Sir Ronald Holroyd, Cmnd. 4371, 1970.

75 Cheshire, Derbyshire, Durham, Essex, Herts, Kent, Lancashire, Nottinghamshire, Staffordshire, Surrey, Warwickshire, Worcestershire, Yorkshire West Riding, Glamorgan.

76 Birmingham, Leeds, Liverpool, Manchester, Sheffield.

77 GLC Minutes 30/11/65.

78 Department of Education and Science, Circular 4/65 1965; under the Public Library and Museums Act 1964, the Secretary of State had assumed overall responsibility for the Service.

79 See for example S. K. Ruck, *Municipal Entertainment and the Arts*, (Allen & Unwin, 1965), page 59. It was suggested that where the ratio of full time, non-manual staff to population was less than 1: 2,000, and where expenditure on new books (in 1960–1) was less than 20d per head of population the service was due for improvement. On this basis 20 metropolitan boroughs and all of the 46 Outer London library authorities fell short on staff, but on book expenditure only 3 metropolitan boroughs and 27 Outer London authorities fell short.

80   London Boroughs Association – Statement of evidence submitted to Royal Commission on Local Government in England.

81   See for example the 'Budget Statement' made by Mr Horace Cutler to the Greater London Council 23 February 1971. Land costs in London were estimated to be about four times the average for the rest of the country.

82   Mr B. T. Jennings in *Local Government Finance*, August 1970.

83   Report, paras. 7.1 – 7.6.

84   In the Cheshire study calculations of these capital costs were based not on previous reorganizations but on estimates of the number of staff and the amount of accommodation which they would require; it could of course be argued that these costs may be reflected in increased administrative costs in London at the present time, eg through the need to rent additional accommodation, but the two are by no means equivalent.

85   Report of Royal Commission on Local Government in England, 1966–9 (Cmnd 4040, HMSO 1969), para. 523.

86   Above, p. 94.

87   Above, p. 329.

88   Information from the Clerk of Essex CC.

89   On this see the evidence of Dartford MB to the Royal Commission on Local Government in England (Written Evidence of Non-County Borough Councils, HMSO, 1968, p. 94). A further example is the prolonged dispute in 1970 between Essex CC and its firemen over the proposal to pay London weighting to men serving on the borders of Greater London in an effort to improve recruitment there.

CHAPTER 11

1   Cmnd. 1164, para. 853.

2   The LCC had had a Director of Establishments (within the Clerk's Department) between 1951 and 1957, when the post was discontinued.

3   He had left Middlesex in 1962 to become secretary of the Commission for New Towns. The LCC had not had a separate planning department, as was indicated above.

4 There was one elected member from each of the other standing committees plus 21 other members: in addition there was a member elected by the ILEA.

5 GLC Minutes, 6 July 1965, p. 470.

6 See LCC Minutes, 6 February, 1962, pp. 58–61.

7 The GLC had, however, decided at the beginning to examine the internal structure of the departments once they had been functioning for some time.

8 GLC Minutes, 17 May 1966, pp. 259–60.

9 Again, a device based on an LCC precedent.

10 GLC Minutes, 21 November 1967, pp. 684–7.

11 *Committee on Management,* Report, paras. 170–80: the Committee's view of the Clerk's role was, however, linked to their idea of a management board to which the Clerk would be responsible.

12 GLC Minutes, 21 November 1967, p. 686.

13 GLC Minutes, 13–14 February 1968, pp. 93–6.

14 On this see below, p. 409.

15 GLC Minutes, 23 July 1968, p. 428.

16 Above, p. 341.

17 There had previously been a Scrutiny sub-committee of the Finance and Supplies Committee; the promising innovation of a separate committee was, however, short-lived; following the 1970 election, with the transfer of responsibility for the annual budget from the Finance and Supplies Committee to the Policy and Resources Committee (see below), the Scrutiny Committee disappeared and its functions were merged in those of the renamed Finance and Scrutiny Committee.

18 A year earlier the Leader's Co-ordinating Committee had been set up as an unofficial committee of the Council; its antecedents were in Leader's Conferences used by both the GLC and the LCC; but these conferences were ad hoc meetings and not part of the regular machinery (see GLC Minutes, 18 July 1967). (For information on similar committees in county boroughs, see Greenwood, Norton, and Stewart, *Recent Reforms in the Management Arrangements of County Boroughs in England and Wales* (Institute of Local Government Studies, University of Birmingham, 1969), p. 9.

19 Greenwood, Norton and Stewart, *Recent Reforms in the Management Arrangements of County Boroughs,* pp. 5–12.

20 This provoked reaction from the opposition, Sir Reginald Goodwin claiming that it put 'too much power in the leadership of the Council as well as strengthening the director general's department at the expense of the Treasurer's' (*The Times,* 23 April 1970).

21 This was the name given to the Planning and Transportation Committee following the 1970 election.

22 GLC Minutes, 13–14 February, 1968, p. 95.

23 See below.

24 Architect, Joint Directors of Planning and Transportation, Valuer and Estates Surveyor, but not, interestingly, the Director of Housing.

25 Now known as Traffic Commissioner and Director of Development; he is responsible under the joint directors for the day-to-day running of the department.

26 It was thus logical to transfer responsibility for the budget in 1970 from the Finance to the Policy and Resources Committee.

27 Strategic Town Planning, Transportation, Housing, Health and Safety, Leisure and Amenity, General Services.

28 GLC Press Release 373, 9 July 1969.

29 GLC Minutes 13–14 February 1968, p. 94.

30 See below, p. 414.

31 It is interesting that several months after the merger the GLC were seeking to appoint a Co-ordinator of Transport Policy, an Engineer Planner or Economist on the staff of the Joint Directors to act as a kind of go-between who would draw attention to the policy issues raised by the preparation of transport plans in relation to the GLCs strategic planning.

32 London Government Act, 1963, s.30.

33 Especially on informal relations between the Leader of the GLC and the Leader of the ILEA; cf also the fact that the GLC Treasurer was appointed Treasurer of the ILEA and that a member of the ILEA serves on the GLC General Purposes Committee.

34 GLC Minutes, 21 May 1968, p. 287.

35 *The Times,* 22 September 1969.

36 *The Times,* 14 January 1970.

37 And, as already noted, this is particularly true of Inner London members because of the ILEA.

38 See GLC Minutes, 7 October, 1969, p. 513.

39 *The Government of London,* pp. 71–2.

40 Cmnd. 1164, para. 763.

41 Cf L. J. Sharpe, *Research in Local Government* (Greater London Paper, No. 10, LSE. 1965), p. 8–9.

42 Cmnd. 1164, paras, 761–2.

43 London Government Bill, Standing Committee F, 19 March 1963, col. 874.

44 London Government Act, 1963, section 71.

45 *Research in Local Government,* p. 3.

46 Cmnd. 1164, para. 760

47 Cmnd. 1164, para. 758.

48 GLC General Purposes Committee Paper GP 326, 24 May 1965, para 3.

49 GP 326, 24 May 1965, para. 9.

50 GLC Minutes, 14 December 1965, p. 843 and 22 March, 1966 p. 176.

51 His salary was the equivalent of that of second-rank officers in the departments of planning and highways and transportation.

52 This concession was recognized in the re-designation of Dr Benjamin as 'Director of Research and Intelligence' in October 1966.

53 On this see Eric J. Thompson, *The Implications of Recent Population Trends for Greater London* (Quarterly Bulletin of GLC Research and Intelligence Unit, No. 2, June 1968), pp. 15–17: also, the further discussion by P. H. Levin and E. J. Thompson in No, 5, December 1968, pp. 49–54.

54 *Greater London Development Plan: Statement,* paras. 3.10–3.17 speaks of planned overspill of about 20,000 per year, ie a much reduced scale.

55 After leaving the GLC Dr Benjamin referred to some of these

fundamental disagreements and criticized the GLC for not taking sufficient notice of the unit's findings (letter in the *Guardian*, 1 February 1971).

56 Cf GLC Minutes, 22 July 1969, p. 450.

57 L. J. Sharpe, *Research in Local Government,* pp. 12, 18.

58 Excluding teachers and the police, Royal Commission on Local Government in England, Research Study 7, Appendix 1 (2).

59 The department is responsible generally for staff management, recruitment, training, etc., but the Director General retains certain responsibilities, e.g. relating to chief officers.

60 Mr D. S. Mitchell, formerly Chief Personnel Officer of the UK Atomic Energy Authority.

61 Mr A. E. Conventon; on his retirement in 1968 he was succeeded by another outsider, Mr J. Charles, personnel officer of the UK Atomic Energy Authority.

62 GLC Minutes, 12 May 1965, p. 41.

63 The head of the programme office is, however, a former assistant clerk of the LCC.

64 Manual workers come within the ordinary national negotiating machinery and are not considered here.

65 The national machinery consists of a number of joint negotiating bodies with representatives of employers and staff associations; in particular, there is the National Joint Council for administrative and professional staff.

66 See *The Government of London,* pp. 219–20.

67 Report No. 45, *Pay of Chief and Senior Officers in Local Government Service and in the Greater London Council* (Cmnd. 3473, 1967), para. 50.

68 Report No. 45 Cmnd. 3473 (1967), para. 50.

69 GLC Minutes 22 March 1966, p. 193.

70 Above, p. 213.

71 Above, p. 343.

CHAPTER 12

1   Cmnd. 1164, para. 743.

2   Cmnd. 1562, para. 8.

3   The reference is, of course, to the present system; from 1974 a new system of county and district authorities will operate.

4   There are 32 London boroughs; the City of London, which is not separately treated here, also has the powers of a London borough in addition to the powers which it possessed before 1965 (eg it controls its own police force); it thus has an anomalous position in the new system.

5   Croydon 327,000; Kingston upon Thames 143,000 (Registrar General's Estimates of Population at 30 June 1970).

6   Tower Hamlets, Newham, Barking, Hackney, Islington, Southwark.

7   Kensington and Chelsea, Barnet, Bromley, Harrow, Westminster, Kingston upon Thames, Richmond upon Thames, Sutton.

8   Sample Census, 1966, Greater London, Table 14 (HMSO 1967). The figures refer to occupied males only.

9   See *Greater London Development Plan, Report of Studies* (GLC 1969), paras 2.38–2.42; also Frances Kelly, *Classification of Urban Areas* (GLC Research and Intelligence Unit Qly. Bull. 9, p.13).

10 Cmnd. 1562, paras, 19–20.

11 Cf Greater London Group, *Lessons of the London Government Reforms* (Royal Commission on Local Government in England, Research Study 2, HMSO, 1968), pp. 43–5.

12 Information from *Management of Local Government*, Vol. 5, Local Government Administration in England and Wales (HMSO 1967), Appendix A, Tables II, XXXVIII, IIa and XXXVIIIa.

13 R. Greenwood, A. L. Norton, J. D. Stewart, *Recent Reforms in the Management Structure of Local Authorities: The London Boroughs* (Institute of Local Government Studies Occasional Paper 2, University of Birmingham, 1969), pp. 11–12.

14 Institute of Local Government Studies, Occasional Paper No. 2, pp. 8–9.

15 R. Greenwood, A. L. Norton, J. D. Stewart, *Recent Reforms in the*

Management Arrangements of County Boroughs in England and Wales (Institute of Local Government Studies Occasional Paper No. 1, University of Birmingham, 1969), pp. 8–9.

16 Institute of Local Government Studies, Occasional Paper No. 2, pp. 15–16.

17 See *Local Government Chronicle* 8 November 1969, p. 2113: 15 November 1969, p. 2159, 6 December 1969, p. 2348; also an article on the reorganization by the Leader of Haringey Council (*Local Government Chronicle*, 17 January 1970, pp. 112, 127).

18 Institute of Local Government Studies, Occasional Paper No. 2, p. 41.

18 *Local Government Chronicle*, 27 June, 1970, p. 1271: 4 July 1970, p. 1333.

20 Cmnd. 1164, para. 694.

21 Cmnd. 1164, para. 688.

22 Cmnd. 1164, paras. 234–5.

23 J. M. Lee, *Social Leaders and Public Persons* (Oxford, 1963); A. M. Rees and T. Smith, *Town Councillors* (Acton Society Trust, 1964).

24 Cf G. W. Jones, *Borough Politics* (Macmillan, 1969), esp. chapter 7.

25 *Local Government Administration in England and Wales* (Committee on the Management of Local Government, Vol. 5, HMSO 1967), p. 48.

26 Cmnd. 1164, para. 236.

27 R. V. Clements, *Local Notables and the City Council* (Macmillan, 1969).

28 *Local Government Administration in England and Wales*, p. 48.

29 Cmnd. 1164, para. 240.

30 They thought that the Council for Greater London should have about 100 councillors but they stressed the importance and wide range of its functions (Cmnd. 1164, para. 853).

31 *The Government of London* p. 178.

32 See, for example, Research Study No. 9 Royal Commission on Local Government in England, *Community Attitudes Survey: England* (HMSO 1969), Section F. pp. 137–42.

33 From 1973 onwards: until then GLC councillors have been elected for borough areas.

34 One may compare these figures with those proposed by the Red-cliffe-Maud Commission; they suggested a maximum size of 75 councillors for their authorities, which would result in populations of over 10,000 per councillor in the largest metropolitan districts.

35 Mary Horton, *The Local Government Elector* (Maud Committee Vol 3, HMSO 1967) p. 72.

36 Eg Community Attitudes Survey: England (Section C).

37 Louis Moss and Stanley H. Parker, *The Local Government Councillor* (Maud Committee, Vol. 2, HMSO 1967), p. 234.

38 Dilys M. Hill, *Participating in Local Affairs* (Penguin Books, 1970), especially pp. 187, 194–5.

39 But see Enid Wistrich, *Local Government Reorganisation: The First Years of Camden* (Camden L. B. 1972).

40 Above, p. 75.

41 Above, p. 388.

42 Bleddyn Davies, *Social Needs and Resources in Local Services,* (Michael Joseph, 1968).

43 Above, p. 236.

44 The Minister's decision was announced in March 1971 and the transfers operated from 1 April 1971.

45 Cmnd. 1164, para. 752/3.

46 The London boroughs are 'treated as county boroughs' for the purpose of the AMC's internal organization (see J. A. G. Griffith, *Central Departments and Local Authorities* (Allen & Unwin/RIPA, 1966), p. 37.

47 See GLC (General Powers) Act, 1970. A rate of $\frac{1}{10}$ penny would yield about £250,000. ($\frac{1}{10}$d $= \cdot$24p)

48 The detailed task of preparing proposals fell on the advisory body of London borough treasurers, one of the numerous chief officer groups which service the LBA; they have devoted a considerable amount of time to this question.

49 Above, p. 219.

50　*Report of Royal Commission on Local Government in England, 1966–69,* Volume 2 (Cmnd. 4040–I, HMSO 1969), para 365.

51　'(Social Services)' was not added to its title until September 1967.

52　Haringey had previously been a member but withdrew in 1970.

53　The number of courses provided has risen from 65 in 1965–6 to 117 in 1969–70; the number attending courses from 1,680 to 2,646 in the same period.

54　Inner London members – Camden, Greenwich, Hackney, Hammersmith, Islington, Lewisham, Southwark, Tower Hamlets, Wandsworth: Outer London – Barnet, Brent, Haringey, Kingston, Newham, Redbridge, Richmond, Sutton, Waltham Forest.

55　150 staff in 1965–6, 310 in 1967–8, according to the Annual Reports of the Unit: no reports have been issued since 1967–8 but by 1970 the staff of the Unit numbered 344.

56　By 1967–8, 22 authorities outside London were using the unit; these activities were entirely paid for by the authorities concerned.

57　N.E. London (Hackney, Tower Hamlets, Haringey); S. E. London (Bexley, Greenwich, Southwark).

58　Above, p. 211.

59　Above, p. 165.

CHAPTER 13

1　Cmnd. 1164, para. 969.

2　*London Government: Government Proposals for Reorganization* (Cmnd. 1562, 1961), para. 8.

3　In Inner London education was of course not subject to a transfer of this kind.

4　Below, p. 489.

5　Above, p. 211.

6　Third Report of Standing Working Party on London Housing: J. B. Cullingworth, *Report to the Minister of Housing and Local Government on Proposals for the Transfer of GLC Housing to the London Boroughs:* John Greve, Dilys Page, Stella Greve, *Homelessness in London* (Scottish Academic Press, 1971).

7　The Greater London Group is proposing to carry out a further research project into democracy and politics in Greater London.

8 Cf *The Government of London,* p. 132.

9 i.e. excluding the transitional year 1964–5.

10 See *The Government of London,* pp. 121–4 and 225–6.

11 Cmnd. 1164, para. 281.

12 Above, p. 348.

13 Above, p. 262. See also John Greve, *Homelessness in London.*

14 Above, p. 323.

15 See below, p. 487 for proposals made by the Government in 1970 with regard to planning.

16 For further discussion of this issue see Greater London Group, *Local Government in South East England* (Royal Commission on Local Government in England, Research Study I, HMSO, 1968), pp. 428–32.

17 *Report of Royal Commission on Local Government in England, 1966–9* (Cmnd. 4040, HMSO, 1969) para. 243.

18 See Figure 7, p. 330.

19 Above, p. 338.

20 See preface, p. xiii.

21 See also: Sir Malcolm Trustram Eve, *The Future of Local Government*, a lecture delivered before the University of London on 6 February 1951.

22 See *The Government of London,* pp. 97–9.

23 Cmnd. 1164, para. 793.

24 Eg *London's Housing Needs up to 1974* (Standing Working Party on London Housing, Report No. 3, MHLG, 1970) para 5.

25 With some exceptions (London Government Act, 1963, s.21(4)).

26 *London's Housing Needs up to 1974*, para. 6: J. B. Cullingworth, op. cit. p. 11. In 1971 the LBA set up a London Housing Office; it will be interesting to see how effective this proves to be.

27 See preface, p. xiii.

28 Report of Royal Commission on Local Government in England 1966–69 (Cmnd. 4040, HMSO, 1969); Reform of Local Government

in England (Cmnd. 4276, HMSO, February 1970); Local Government in England: Government Proposals for Reorganization (Cmnd. 4584, HMSO, February 1971).

29   Cmnd. 4040, paras. 257, 271.

30   Cmnd. 4584, para. 12.

31   Above, p. 343.

32   *The Government of London,* p. 243.

Index